# ESPRIT Basic Research Series

Edited in cooperation with
the Commission of the European Communities, DG XIII

# Foreword

Geographical Information Systems are becoming increasingly important to geologists, geographers, and urban and environmental planners, amongst others. Although there is an increasing number of such systems available on the market, many questions of a theoretical nature need to be addressed, because of the inherently large amounts of data and the lack of appropriate models and languages. Also, there is a need for flexible and efficient usage of GIS, thus permitting their application in a broad range of new areas, such as navigation systems for transport and the automotive industry.

ESPRIT Working Groups provide a structure that enables researchers to meet, interact and share research results, thus providing a platform for addressing fertile research areas with significant industrial promise. This particular volume is the result of a workshop organised by the "Basic GOODS" Working Group. The challenge has been to define richer and more powerful data definition and manipulation languages, and also to develop new and efficient algorithms. The challenge has been met, as presented in this volume, and the Working Group has succeeded in laying a good foundation for future research and eventual applications. We look forward to seeing the results of the recently launched "AMUSING" project, which will build on this foundation.

June 1992 G. Metakides

G. Gambosi M. Scholl H.-W. Six (Eds.)

# Geographic Database Management Systems

Workshop Proceedings
Capri, Italy, May 1991

Springer-Verlag
Berlin Heidelberg New York
London Paris Tokyo
Hong Kong Barcelona
Budapest

Volume Editors

Giorgio Gambosi
Department of Mathematics, University of L'Aquila
Vetoio di Coppito, I-67100 L'Aquila, Italy

Michel Scholl
CNAM, 292 rue St. Martin
F-75141 Paris Cedex 03, France
and
INRIA, B. P. 105
F-78153 Le Chesnay Cedex, France

Hans-Werner Six
Praktische Informatik III, Fernuniversität Hagen
Postfach 940, W-5800 Hagen, FRG

CR Subject Classification (1991): H.2, H.4, E.1, I.3.5

ISBN-13:978-3-642-77607-6 e-ISBN-13:978-3-642-77605-2
DOI: 10.1007/978-3-642-77605-2

Publication No. EUR 14477 EN of the Commission of the European Communities, Scientific and Technical Communication Unit, Directorate-General Telecommunications, Information Industries and Innovation, Luxembourg
Neither the Commission of the European Communities nor any person acting on behalf of the Commission is responsible for the use which might be made of the following information.


Softcover reprint of the hardcover 1st edition 1992



Typesetting: Camera ready by authors
45/3140 - 5 4 3 2 1 0 - Printed on acid-free paper

# Preface

The International Workshop on "Geographic Database Management Systems" has been organized in the framework of the activities of the Basic GOODS (Geographic Object–Oriented Database Systems) ESPRIT Basic Research Working Group 3191 with the aim of gathering research experiences, both from members of the Working Group and from other research groups active in the field of spatial data management.

Thirteen papers were selected for presentation at the Workshop (5 from members of the Working Group and 8 from external groups). Moreover, two invited talks by leading experts in the field were given.

The contributed papers cover most fundamental aspects of geographical database management, ranging from data structures, and physical organization of data, to data models, user interaction issues, and query resolution.

The Workshop succeeded thanks to the contribution of a number of people. Giorgio Ausiello first established this ESPRIT Basic Research Working Group on Spatial Data Management and is responsible for the coordination of the Basic GOODS activities; he has directed the Workshop organization. The Program Committee would also like to thank Carlo Gaibisso, Enrico Nardelli and Maurizio Talamo for their planning support, and Paolo Franciosa and Fabrizio d'Amore for their assistance in all the organization activities.

We also wish to thank the TWT s.r.l. for their behind-the-scenes organization and the Capri Tourist Board (Azienda Autonoma di Turismo e Soggiorno) for their help in making the time spent in Capri so pleasant.

Last but not least, we would like to thank Springer–Verlag and the Division A7 of DG XIII, ESPRIT Basic Research, for the excellent cooperation and support in the publication of this volume.

May 1992

Giorgio Gambosi
Michel Scholl
Hans–Werner Six

# Geographic Information Management – The ESPRIT "Basic Goods" Action

Giorgio Ausiello

Geographic information systems are becoming increasingly important in a wide variety of applications, such as those related to land use monitoring, environmental assessment, urban planning, resource management, etc.

Despite a growing role in the software market and the existence of commercial products that are suitable for particular types of applications, geographic information systems are still far from having achieved a fully satisfactory technological development.

Among the weaknesses of most currently available systems we may list:

- the lack of models and languages which allow a management of geometric entities and their relationships together with the conventional alphanumeric type of information,
- the lack of physical data organization structures which allow the user to store, retrieve and manipulate spatial data with heterogeneous formats efficiently and to integrate geometric and conventional information,
- integration and interoperability problems which have made it very difficult, until now, to speak in terms of "general purpose" environments for geographic applications.

Future generations of geographic information systems will have to overcome such difficulties in order to satisfy the demanding requirements arising from the application domains that have been previously indicated. For this reason research in the field of geographic systems deserves greater attention inside the general area of information system research.

When geographic information systems are kept in consideration, many advanced research subjects in information processing may get particular emphasis and better focusing: user interaction and navigation techniques, integration with knowledge bases, statistical databases, mathematical models, new database models such as object oriented models, query languages, spatial data structures, efficient geometrical algorithms. For all these research issues and for various technological developments such as multimedia workstations, open system integration, and three-dimensional graphics, geographic applications provide a paradigmatic and challenging application testbed and a source for new concepts and new solutions.

Within the ESPRIT programme, geographic information management appears as an application domain in various projects. Nevertheless the creation of a specific Working Group on this subject, in the framework of ESPRIT Basic Research, allowed us to address some of the most paradigmatic fundamental research topics related to geographic information systems directly for the first time.

Objectives of the Basic Goods Working Group are study and information dissemination concerning the following issues.

**Data models for geographic information systems.** Particular attention is devoted to the study of relational models (suitably extended to deal with geometrical entities and relationships), of object oriented database models (capable of representing complex objects characterized by geometrical features and by alphanumeric attributes) and of statistical database models (needed to deal with aggregated data associated with geographical regions, such as in thematic mapping).

**Data definition languages and data manipulation languages.** Here the main focus is on functionalities and implementation of object oriented languages that are widely considered the most versatile and powerful languages suitable for dealing with highly structured data, for supporting user interaction, and for allowing the integration of a large variety of software modules for complex applications on a unique platform. Besides this, considerable interest is devoted to the analysis and definition of styles of interaction and new metaphors to support effective human system interaction in geographic applications.

**Efficient algorithms and data structures.** In order to support the implementation of both definition and manipulation languages for geographic information systems and to guarantee the efficiency required by most applications (especially those related with mathematical modelling and simulation) it is necessary to provide new general purpose data structures on secondary storage that allow to deal with geometrical data in different formats (both vector and raster) or at least to perform fast format transformations. Also, such data structures should be tested for efficiency with fundamental geometrical primitives needed for the implementation of more sophisticated algorithms.

Various research groups participate in the Action, working on the various aspects according to their specific scientific background:

- INRIA, (Paris, France), data models and languages,
- University of Freiburg (Germany), physical organization and data structures,
- Fernuniversität Hagen (Germany), physical organization and data structures,
- University of Rome "La Sapienza" (Italy), data models and data structures,
- IASI–CNR (Rome, Italy), data models and human computer interaction,
- ISRDS–CNR (Rome, Italy), statistical database models and languages,
- Algotech s.r.l. (Rome Italy), geographic database systems and system integration aspects.

The activity of the Working Group started in July 1989 and will finish in July 1992. During the first two years the partners at the various sites carried on the investigation on the topics of the Action. Besides some small informal meetings, two main workshops were organized for the presentation of research results achieved by the partners within the Working Group, for the discussion of specific issues and for the exchange of ideas with some of the main experts in the field.

The present volume collects papers presented at the Capri Workshop on Database Management Systems for Geographical Applications covering basically all research areas considered in the working group, as listed above. Its content provides a comprehensive description of the current state of the art in the field of spatial data management, as resulting from activities both performed within the Basic GOODS working group and in other relevant worldwide research institutions.

# Beyond Query Languages for Geographic Databases: Data Cubes and Maps*

**Andrew U. Frank**

## Abstract

After several years of efforts to extend query languages for Geographic Information Systems (GIS), the limits of SQL-like languages for the selection of geographic data to be displayed in map form became evident. While SQL provides a means for selecting data based on well-specified criteria, it does not support the exploratory access geo-scientists need. Icons and direct manipulation are appropriate user interface techniques. The traditional work methods of cartographers provided the concepts for the layer-based GIS, where overlay is the central operation (similar to the physical overlay of maps drawn on transparent material). The inspiration for a visualization of these operations for the user interface can be borowed from the same source. Two different concepts are identified, one is 'data sets' the other 'maps', and a representation as icons is proposed. Operations can be succinctly expressed by dragging data cubes onto maps and the icons show by 'repelling forces' if the user is about to do an illegal or meaningless operation. The interface and its operations are described and we conclude with open detail questions and describe a formal method that will be used to further develop the interface design to achieve consistency.

## 1. Introduction

Ease of access to the data in a Geographic Information System (GIS) is of utmost importance and determines the usability of the system. Attempts to extend standard query languages like SQL (ANSI/X3/H2 1985) to include sufficient facilities for selection and presentation of spatial data were not entirely satisfactory (Egenhofer 1991a). They require that users understand the structure of the data stored and how they relate to their information needs, and are able to write a boolean expression to qualify the data to be selected. The presentation of the data retrieved is crucial for the effective comprehension of the information and graphical, map-like formats are indispensable for a GIS (Frank 1982). Most of the spatial extensions to query languages do not include tools for the user to direct the presentation of the data retrieved (Egenhofer 1991b, Egenhofer 1991c). A proposal for a query language with rich tools to select the rendering of the data as a map showed that such an extended language is very complex and the query expression a user had to type unacceptably long (Egenhofer and Frank 1988).

The current query languages, which are essentially retrieval languages, are applicable only for the few cases in which the user completely understands the structure of the data stored and needs only a very few data elements displayed in a default format. They may also be useful as a

* Funding from NSF for the NCGIA under grant SES 88-10917, from Intergraph Corporation, and Digital Equipment Corporation is gratefully acknowledged.

standardized interface between programs that exploit spatial data and the database management system; however, they are inappropriate for *ad hoc* access to GIS data in situations when problems to be solved are poorly defined, the available data is not well understood, and highly trained users — geographers, planers, oceanographers or geologists — explore the data guided by their intuition and expertise. For these very important applications, a tool supporting an *exploratory behavior* is necessary. Such tools exist for commercial GISs and they allow the manipulation of thematic data sets and their visualization in the form of computer maps (Buttenfield and Ganter 1990), but their user interfaces are difficult to learn and cumbersome to use. The user interfaces are typically based on typed commands and systems provide users with very large sets of commands — more than 1000 in a commercial system (Antenucci and Brown 1989) — each with its own set of parameters and modifiers. Their use is not intuitive, requiring substantive planning from the users to translate the exploratory steps into system commands, thus breaking the direct link between the user and the data. Often a GIS specialist is necessary as a mediator between the scientist and the data to guide the data processing.

There is an urgent need for a visual, direct manipulation interface to support exploratory spatial data analysis and presentation by experts. Most notable are the widely publicized research efforts to detect Global Change and the enormous databases that will be necessary (Mounsey and Tomlinson 1988). The user community requests such exploratory interfaces which support an unstructured experimentation with the data (Smith and Frank 1990). Scientists ask for 'browsing' capabilities which current generation GIS do not provide.

We present a user interface, based on a visual metaphor and direct manipulation (Shneiderman 1983, 1987). Observing the traditional, manual work methods of planners and cartographers provided the concepts for many GIS programs, which organize the data in thematic layers. These are then overlaid over each other (Tomlin 1983, 1989). We identified three kinds of objects: (1) the data sets the user is interested in, (2) the visualization parameters, and (3) the actual rendering in map form as the result of applying the visualization parameters to the data set.

The traditional manual method can also be used as a source of inspiration for the visualization of the objects for the user interface. The design proposed is similar to the familiar desktop metaphor (Smith, et al. 1982), widely publicized by the Macintosh and now also the Windows user interface. The interface described here is substantially richer than the 'files and folders' on the Macintosh desktop. We use two different icons to differentiate the two different object types, namely *data cubes* for data sets and *maps* for the visualization parameters. Several specific operations are expressed by direct manipulation, i.e., a data set is rendered by dragging the data cube icon on top of the map icon. We also introduce reactive icons, which use visual "force feedback" to indicate to the user that she is about to perform a meaningless or illegal operation.

New GIS interfaces often use "windows, icons, menus, and pointing device" (WIMP) (Nielson 1990), but their structure reveals that they are mostly a translation of command interfaces into menus, which, surprising to many, do not make learning or using the system substantially easier. Menus and icons help users select data and provide the necessary parameters for operations, but do not use a visually attractive and intuitive metaphor like the desktop with files and folders, which makes previous experience with the physical world applicable to the computer environment and therefore reduces the learning time needed (Kuhn and Frank 1991). Just translating commands into menu choices cannot reduce the effort of the user to translate from his concepts to the system commands.

The approach presented here is currently restricted to GISs that organize their data in thematic layers and use the special concept of "space with properties" as opposed to "objects with attributes embedded in space" (Mark and Frank 1989, Frank 1990).

In the next section we briefly characterize exploratory behavior and exploration of geoscientific data. Next the workspace and manual methods of cartographers, planners, and geographers are sketched. Section 4 details the concept from the user's point of view and Section 5 we describes the visualization of these concepts. In the conclusion we indicate the next steps necessary to develop a consistent user interface.

## 2. Exploratory Behavior

A scientist using a GIS to explore some data is looking for the "unusual" and searches for "interesting" facts. Visualization (Mandle 1990, McCormick, et al. 1987, McLaren 1989) is an eminent tool, because it brings to bear the human's outstanding ability to detect subtle visual patterns. Visual exploration benefits from a direct control of the user over the display. Scientific visualization has developed tools to render a data set and to allow the scientist to control the display parameters easily. The GIS environment can use some of the same tools, but must extend them to include the manipulation and combination of multiple data sets.

As a source of inspiration we can ask, "How do humans cope with situations when they look for something 'unusual' or 'interesting' without further clues?" and "Where is one exposed to such situations?" A basic assumption in our interface design work is that the design of a human interface should be modelled after "standard" human behavior, as it is found in non-computer problem solving and interaction. Human beings are surprisingly flexible and willing to adapt to artificial environments, but at a high cost. A better alternative is the metaphorical use of familiar situations to structure new situations (Lakoff and Johnson 1980, Lakoff 1987). The Xerox Star and Apple Macintosh desktop metaphor is a prime example.

An example of exploratory behavior is observed if a human is confronted with a physical object (or collection of objects) of unknown purpose. How does one explore the object? One typically looks at it from different angles and perspectives, from far and near, one may even get a magnifying glass to see details, may change the illumination, or move the parts of the object to see how they could connect to each other and what their interaction could be. Detection of patterns leads to hypotheses, which are then tested by comparing them with observations, using the same tools. This behavioral paradigm is also found when geoscientists work with their data, most typically seen on a cartographer's or planner's desk.

## 3. The Geo-Scientist's Workspace

A planner or cartographer, like any geo-scientist, assembles data from different sources describing different aspects of an area. Some of this data is in tabular form, some is already mapped and only graphically available. This source material is combined to reveal "interesting" patterns, which are then further examined. The major technique for combination is graphical overlay, where data sets are drafted on transparent material and then placed over each other (Chan and White 1987, White, et al. 1989). It is then possible to see boolean combinations of data values. In Figure 1c one can see that most of the forest is publicly owned, while most of the residential land is in private ownership. Often a new map is drawn as a result of the analysis of an overlay, and this map can then be used in the overlay process again (Tomlin 1990).

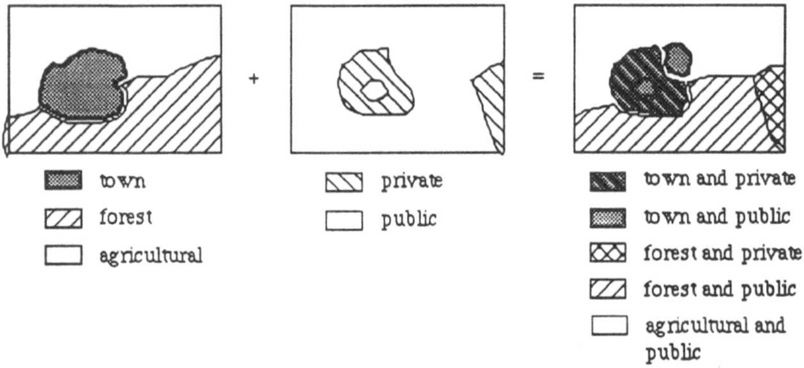


Figure 1: a. Land use b. Land ownership c. Overlay

The method relies on a concept of space (Mark and Frank 1989, Frank 1990) where each point is characterized by some data values and data values describing the same properties for an area are arranged in a layer or coverage. This concept stresses the space as the organizational structure and the data that characterize space. This is different from a concept of space, where objects and their relations are located in space.

Computers have been used to formalize and improve this process. Dana Tomlin described it in his seminal thesis (Tomlin 1983) as a MAP algebra. The basic concept is the combination of spatially corresponding values in data sets by boolean or other operations.

The concept of map overlay is very powerful and has been successfully moved to computer implementations. There are widely distributed, but limited public domain systems (Sen and Radhakrishna 1988) used for teaching and powerful, feature laden systems for professional use (Aronson and Morehouse 1983, Dangermond 1988, Morehouse 1989). None of the implementations we know of uses a visualization of the objects and operations. A straightforward implementation displaying a desktop with maps that can be overlaid is obvious. It would be an improvement over the current command driven interfaces, but does not capture the appropriate operations well. A careful analysis of the overlay process reveals that the geo-scientist deals with two different kinds of objects, namely the data sets and their renderings as maps and the applicable operations are different.

## 4. User Interface Concept Level

The design of the user interface must start with an analysis of the concepts used to structure the tasks from the user's view point. Unnecessarily complex, unclear structures at the concept level cannot be corrected with impressive icons or colorful menus. Clean concepts will lead to clean interfaces that are easy to learn. There is some evidence that learning a system depends, among other things, on the number of concepts a user has to master.

In this section we detail the three object concepts we identify to structure the task and explain the most important operations on them.

### 4.1 Data Sets

The geo-scientist collects data sets that describe some aspects of the world in which he is interested. These data sets contain a data value for each point in an area and are called thematic layers or coverages. Such data sets are produced by remote sensing instruments, but other data collection methods often result in coverages as well - from statistics that give the population data per census block to maps that show the soil types of an area. This concept of spatially registered thematic layers is so prevalent that it is often identified with the concept of a GIS and is visualized as a stack of layers as a symbol for a GIS (figure 2).

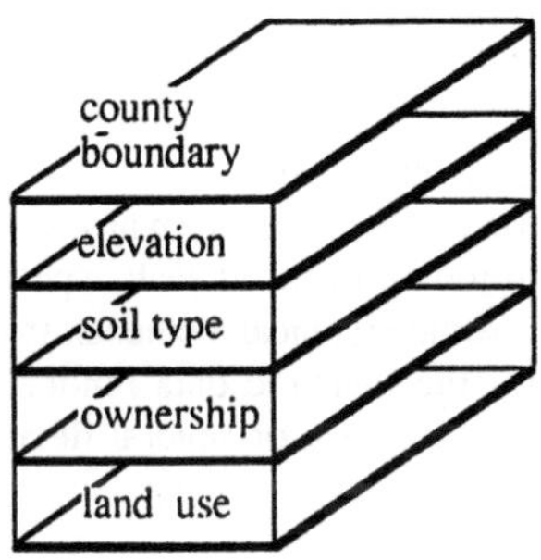


Figure 2: A stack of layers

Each data set has a title and some description of the data it contains. It covers a certain, typically rectangular area, where for each point in space a specific value is recorded. Conceptually these values are available for an infinite number of points but practically both space and attribute values are discretized (Goodchild 1990). Such values may be uniform for regular or irregular shaped areas. Additional descriptors are the scale on which the values are measured, the source of the data and its spatial resolution, the quality or lineage of the date, and so on.

### 4.2 Maps

Maps are renderings of data values in a graphical format. On a geo-scientist's desk, maps are often used as the primary means of data collection and the difference between data values and graphic encoding is not always considered (Peuquet 1988). Operations on data values that yield new data values must be separated from mappings of data values onto graphical variables (Bertin 1983) for visualization.

Maps are a stable graphical product, that can be reproduced, stored and retrieved for later usage. Physical maps can be copied and some minor changes in their appearances can be made - small variations in scale, in color intensity, and so on. Major changes require a reconstruction of the map, either using the original data or using the existing map as source of data for construction of a new map.

### 4.3 Visualization Parameters

Maps can be seen from two different perspectives:
• The graphical presentation of some data in order to make them visually comprehensible (see above),
• The map, separated from the data it represents, as an instruction (or procedure) of how to render a data set. It is a mapping (!) from data values to graphical representations.

The second point of view - even if abstract - is quite natural; in the U.S. one speaks about 'a 7 1/2 minute quad' meaning the general appearance of such a map sheet without reference to a particular map sheet rendering a specific data set. A cartographer is often given a data set and an existing map and asked to draw the data in the style of the example - the map example stands for the abstract rendering instructions, which are more easily specified by example than by detailed instructions.

An example map determines how a data set is rendered. It controls the following aspects:
- The area and the scale used for graphical representation.
- The translation of data values to graphical variables (i.e., the map legend).
- The placement of map title and other auxiliary elements on a map (scale, north arrow, etc.).
Operations on the rendering instructions include detailed tools to control each aspect of the rendering process. The rendering instructions may include some graphical elements, from map neat lines, grid to backdrops, which provide context for the thematic data rendered. The rendering instructions, as a complex object, can be applied to one or several data sets to produce a graphical map.

## 5. User Interface Visualization Level

Once we have identified the concepts the user deals with, we have to decide on the visualization for them. We strongly prefer to visualize conceptual objects with icons and the operations as 'direct manipulation' (Shneiderman 1983) of the objects or menu items. Therefore we use icons for data and for maps.

### 5.1 Data Cubes

One can visualize a single data set as a surface over an area, often shown as block diagrams (figure 3). The concrete example of a real valued function f (x,y) -> z can be generalized to data sets that tabulate values in dependence of more variables, while the domain can be any kind of value - nominal, ordinal, etc. (Stevens 1946).

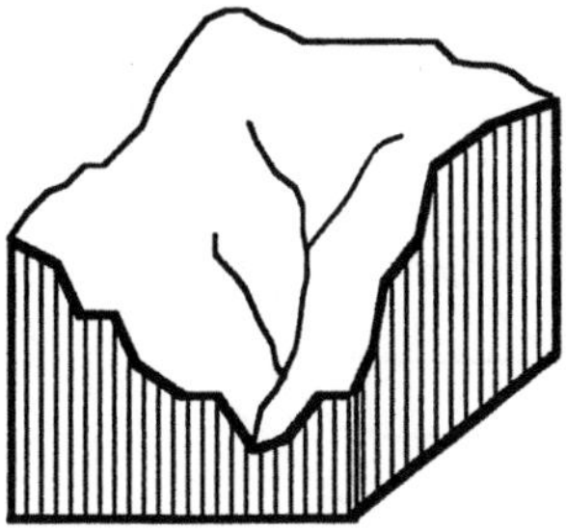

Figure 3: Block diagram

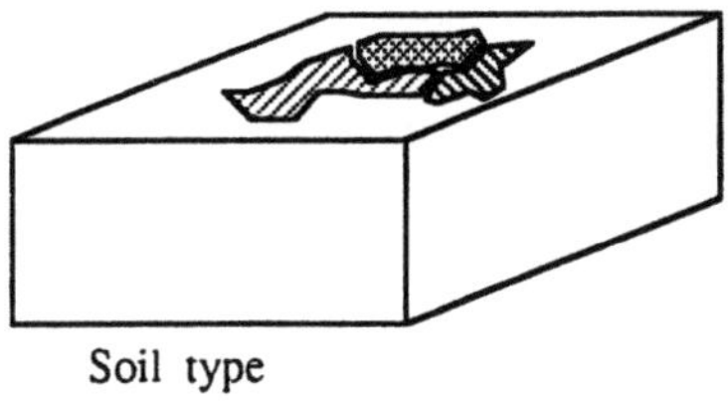


Figure 4: Data cube

Variables are associated with a default rendering, for example height data as a color range from green to brown, temperature data as a color range from blue to red, and so on, and these associations are global for all data sets with the same data type on the desk. Each data set is represented on the 'map desk' as a data cube icon (figure 4) with a title that describe the data set (similar to a file name on the desktop). It contains some description of the data. On the top surface of the cube, a much reduced rendering of the data values is given. This exploits the ability of humans to recognize maps from color and rough shape. This idea has been proposed for text files (Nielson 1990) but it was found that a reduced picture of a title page of a text is not a good representation; this is not a problem in the map domain.

Data cubes can be opened by double clicking, which allows access to a spreadsheet like editor for data values. Shift-clicking on the edge of a data cube turns it one quarter around this edge, which affects the surface on which the data are projected for rendering (figure 5).Data cubes can be stacked one atop of another to indicate that the data should be combined (figure 6).

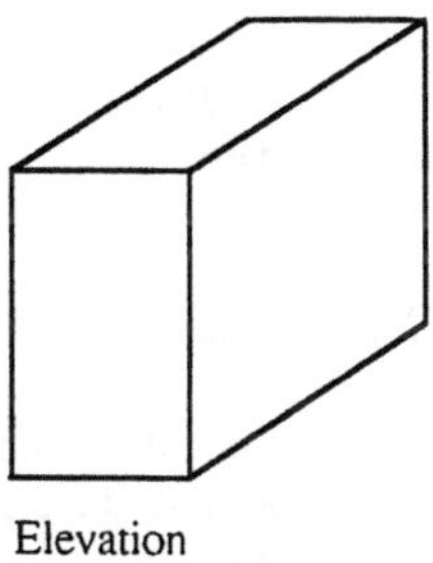


Figure 5: Data cube projected on y-elevation plane

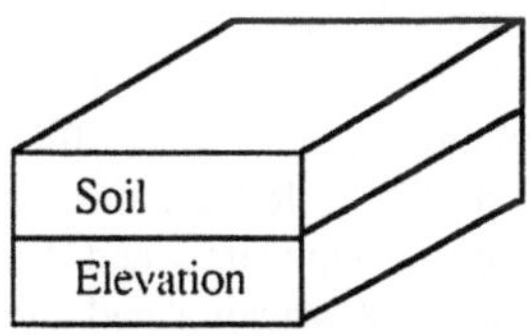


Figure 6: Two data cubes stacked

## 5.2 Maps

The rendering instructions are mapped as an icon and called maps. The visualization of a map is simply a map sheet (figure 7) as a flat surface in perspective. A data set (or a stack of data sets) is rendered by stacking it on a map that describes how it should be rendered(figure 8).

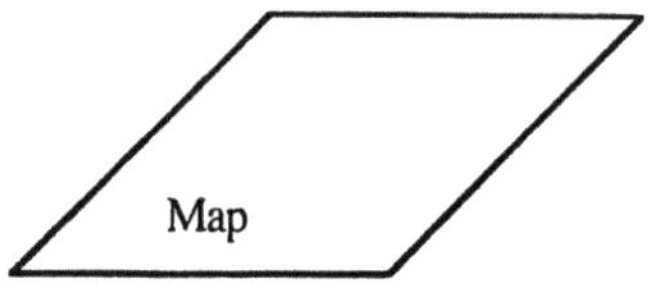


Figure 7: A map icon

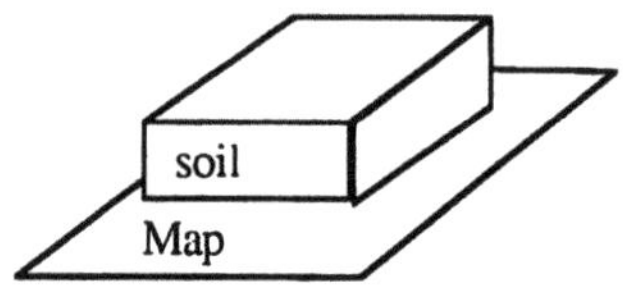


Figure 8: A data cube stacked on a map

The combination of map and data sets stacked on it can be opened (by double-clicking) to see the rendition. Once a map is open for viewing, the editor used to influence the graphical rendering parameters is available and its effects can be directly seen (immediate feedback). For example we can use pan and zoom (Jackson 1990) to visualize the selection of area and the scale. Map legends can be 'active' such that the user can change them and thereby changes the underlying mapping from data values to graphical variables.The results of the editing are stored in the map (but do not affect the data in the data cube).

The link between the data set and the map is 'hot', meaning that changes in the data are immediately reflected in the map. So a user can access, in two different windows at the same time, the data set editor to change data values and the map editor to change the graphical representation.

It is also possible to construct a graphical map, which contains only the graphical rendering and not the original data. This is the equivalent of a traditional paper map (a map instance). The data are 'squished' into the map and irreversibly converted into graphics according to the transformation rules. Only very limited changes to the rendition of the map (comparable to the operations applicable to a paper map) are then possible.

Maps can be copied and the copies contain all the properties of the original map (but without the data) and can then be combined with other data sets. Maps can also be stacked and represent the sum of all the rendition rules and graphics of the individual parts.

## 5.3 The Mapper's Desk

The data cubes and the maps sit on the familiar desktop. It is an implementation question, if this is the standard desktop that is used for other operations, like editing files, or if this is a separate 'mapper's desk'. While the icons respond individually to the standard operations on the desktop, e.g., 'open', 'duplicate', and can be arranged in folders if desired, there are also the visual manipulations of 'stacking' and 'turning' that evoke specific operations:

Data cubes can be **stacked** to indicate that they should be combined along the common range (figure 6). Most obvious is the stacking using the x-y plane as the common base, but they can also be first **turned** and then stacked to construct a diagram which shows two different variables (for example temperature and rainfall) against the y axis - effectively constructing two overlaid profiles. Data cubes can be stacked on maps (figure 9), or maps can be stacked on each other - all evoking consistently the same general operation of 'overlay', but modified depending on the specific object type (polymorphism). The visualization of the metaphor serves both for visualization of objects as icons and for visualization of operations by actions (not by icons).

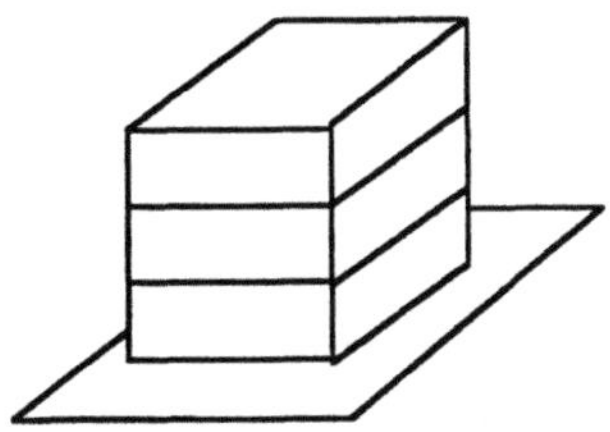

Figure 9: A set of data cubes stacked on a map

The pasting of a reduced sketch of the data in a cube or map on its surface has already be mentioned as an mnemonic aid. It is a 'passive' information about the content of the object. There is also an opportunity for active feedback to the user when attempting illegal or meaningless operations. Each data cube's data cover a certain geographic area; if one tries to stack cubes which do not substantially overlap in the area, the result of the combination operation is meaningless. Cubes can warn the user of this problem, by sliding partially away from each other if one stacks cubes which are not covering a common area (figure 10). A similar behavior applies to the interaction of maps and data cubes. If the map is fixed in its geographic coverage - which it may or may not be - then only cubes that cover substantially the same area sit squarely on the map when put there. Lack of substantial overlap between areas is again indicated by the cube 'sliding' away (figure 11).

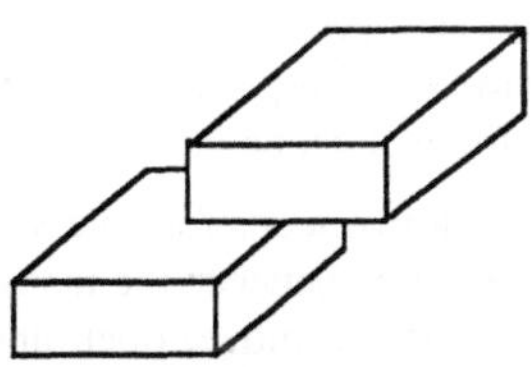

Figure 10: Cubes not covering same area

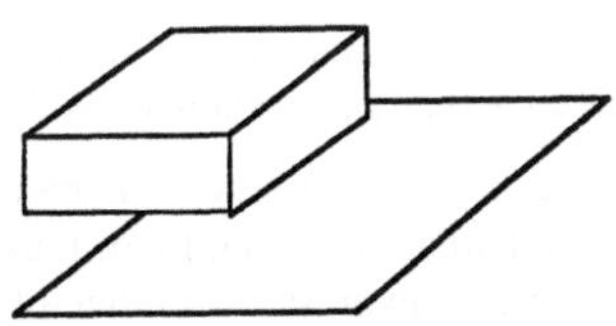

Figure 11: Cubes not covering area of map

The 'squishing' of a data set into a map is an important feature to fix a map and detach it from potential changes in the data values. This may be important to assure that one can keep an unaltered copy of a map once produced. It is of even more importance to construct backdrop maps which are used to provide context to the map reader. A backdrop map is produced by stacking the relevant topics from the topographic data cubes (for example the county boundaries, and major city cubes) on a map. Next the rendering parameters are set to produce a black and white line map, in gray (say 60% screened). Then the data are 'squished' into the map to produce a map instance that is independent of the data cubes and can be used as a backdrop with other data cubes (figure 12). Such maps can be duplicated and overlaid (stacked) with other maps to combine rendition methods and a backdrop.

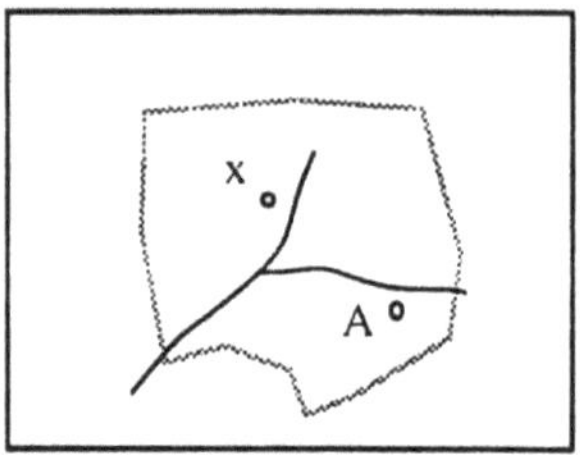


Figure 12: Backdrop map with county boundary

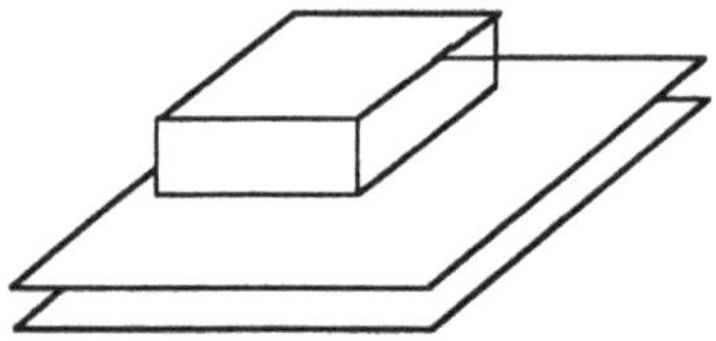

Figure 13: Two maps and one cube stacked for rendering

## 6. Conclusions and Future Work

Exploration of the data collected in a GIS is not sufficiently supported by the current extensions of query languages. New tools are necessary to give the user control over the selection of data when the result is a large data collection and to allow him to direct the graphical presentation of the data in map form. Scientific visualization has shown how the enormous potential of the human eye to detect patterns can be exploited to support exploratory data analysis. A direct and intuitive user interface is required to have minimal distraction of the expert.

Starting with the traditional methods of data analysis and map making, as used by planners, geologists, geographers and cartographers, we identify a source domain for a metaphor, which is the rendering of data sets in graphical form and the graphical overlay of such maps to construct boolean combinations. These methods have been implemented and generalized in computer programs, but the interfaces to the available systems are mostly command driven and require memorizing large numbers of commands with parameters. They are difficult to learn and lack the direct and intuitive character which is important for exploration.

We presented a visualization of this metaphor, introducing two icons for data sets (data cubes) and maps. Direct manipulation of these icons and interaction between the icons on the desktop allow one to express a number of operations beyond the move, open and copy operations, familiar from the Macintosh desktop. In addition, the icons can react and warn the user of inappropriate actions.

In order to confirm the suitability of the proposed interface we will write algebraic specifications for the source domain describing the concrete behavior of the traditional work of a cartographer. The target domain - the MAP algebra - is already formalized (Dorenbeck 1991). Then specifications for the visual interface will be drawn up. There should exist good mappings between these domains (specifically morphism) to make it easy for the user to manipulate and understand the abstract objects in terms of the concrete visualization (Kuhn and Frank 1991). These morphisms show how to design the details in the semantics of the operations and the visualization such that a consistent interface results.

The proposed interface has a major limitation in its reliance on the organization of data in layers or coverages, either in raster or in a vector data structure. It is not clear how a similar intuitive method can be provided for data that is structured in spatial objects with attributes. It is an open question if a single interface concept can cover both view points or if two related but separate methods are the best we can do.

## References


ANSI/X3/H2. 1985. *American National Standard Database Language SQL.* American National Standards Database Committee.

J. Antenucci and K. Brown. 1989. An interview with Jack Dangermond. *URISA* 1 (1) : 50-59.

P. Aronson and S. Morehouse. 1983. "The ARC/INFO Map Library: A Design for a Digital Data Base". In *Sixth International Symposium on Automated Cartography.* Edited by B. S. Wellar. 372-382. Ottawa, Ontario, Canada: The Steering Committee for the Sixth International Symposium on Automated Cartography.

J. Bertin. 1983. *Semiology of Graphics.* Madison, WI: The University of Wisconsin Press.

B. P. Buttenfield and J. H. Ganter. 1990. "Visualization and GIS: What Should We See? What Might We Miss?". In *4th International Symposium on Spatial Data Handling.* Edited by K. Brassel. 307-316. Zurich, Switzerland: International Geographical Union IGU, Commission on Geographic Information Systems.

K. K. L. Chan and D. White. 1987. "Map Algebra: An Object Oriented Implementation". In *International Geographic Information Systems (IGIS) Symposium (IGIS'87).* Edited by R. T. Aangeenbrug and Y. M. Schiffman. 127-150. Arlington, Virginia:

J. Dangermond. 1988. "A Technical Architecture for GIS". In *GIS/LIS'88, Third Annual International Conference.* 561-570. San Antonio, Texas: ACSM, ASPRS, AAG, URISA.

C. Dorenbeck. 1991. "Spatial Algebras in Geographic Information Systems." Dipl. inf. thesis (unpublished), University of Bremen.

M. J. Egenhofer. 1991a. "Deficiencies of SQL as a GIS Query Language". In *Cognitive and Linguistic Aspects of Geographic Space.* Edited by D. M. Mark and A. U. Frank. (in press): Kluwer Academic Publishers.

M. J. Egenhofer. 1991b. Extending SQL for Graphical Display. *Cartography and Geographic Information Systems* (in press)

M. J. Egenhofer. 1991c. Interacting with Geographic Information Systems via Spatial Queries. *Journal of Visual Languages and Computing* 1 (4) : 389-413.

M. J. Egenhofer and A. U. Frank. 1988. "Towards a Spatial Query Language: User Interface Considerations". In *14th International Conference on Very Large Data Bases.* Edited by D. DeWitt and F. Bancilhon. 124-133. Los Angeles, CA.

A. U. Frank. 1982. MAPQUERY: Database Query Language for Retrieval of Geometric Data and its Graphical Representation. *ACM SIG Computer Graphics* 16 (3) : 199 - 207.

A. U. Frank. 1990. "Spatial Concepts, Geometric Data Models and Data Structures". In *GIS Design Models and Functionality*. Edited by D. Maguire. Leicester, UK: Midlands Regional Research Laboratory, University of Leicester.

M. F. Goodchild. 1990. "A geographical perspective on spatial data models". In *GIS Design Models and Functionality*. Edited by D. Maguire. Leicester: Midlands Regional Research Laboratory.

J. Jackson. 1990. "Developing an Effective Human Interface for Geographic Information Systems Using Metaphors, ACSM-ASPRS Annual Convention 1990, Denver, CO, pp. 117-125.

W. Kuhn and A. U. Frank. 1991. "A Formalization of Metaphors and Image-Schemas in User Interfaces". In *Cognitive and Linguistic Aspects of Geographic Space*. Edited by D. M. Mark and A. U. Frank. in press. Dordrecht: Kluver Academic Press.

G. Lakoff. 1987. *Women, Fire, and Dangerous Things: What Categories Reveal About the Mind*. Chicago, IL: University of Chicago Press.

G. Lakoff and M. Johnson. 1980. *Metaphors We Live By*. Chicago, IL: University of Chicago Press.

P. Mandle. 1990. "A Shell for a Geographical Data Visualization System (GDVS) on PC Using Some Commercial Available Programs". In *4th International Symposium on Spatial Data Handling*. Edited by K. Brassel. 443-449. Zurich, Switzerland: International Geographical Union IGU, Commission on Geographic Information Systems.

D. M. Mark and A. U. Frank. 1989. "Concepts of space and spatial language". In *Auto-Carto 9*. Edited by E. Anderson. 538-556. Baltimore, MD: ASPRS & ACSM.

B. H. McCormick, T. A. DeFanti and M. D. Brown. 1987. Visualization in Scientific Computing - Report of the Panel on Graphics, Image Processing and Workstation. *ACM SIGGRAPH Newsletter* 26 (6)

R. A. McLaren. 1989. "Visualization techniques and applications within GIS". In *Auto-Carto 9*. Edited by E. Anderson. 5 - 15. Baltimore, MA: ASPRS & ACSM.

S. Morehouse. 1989. "Architecture of Arc/Info". In *Auto-Carto 9*. Edited by E. Anderson. 266-277. Baltimore, MD: ASPRS & ACSM.

H. Mounsey and R. Tomlinson ed. 1988. *Building Databases for Global Science - Proceedings of the IGU Global Database Planning Project. Tylney Hall, Hampshire, UK*. London: Taylor & Francis.

J. Nielson. 1990. Traditional Dialogue Design Applied to Modern User Interfaces. *Communications of the ACM* 33 (10) : 109 - 118.

D. J. Peuquet. 1988. Representations of geographic space: toward a conceptual synthesis, Annals of the Association of American Geographers 78 : 375-394.

L. Sen and A. V. Radhakrishna. 1988. "Expansion of the analytical capabilities of GIS: Integrating an optimization programming package. Package with OSU MAP-for-the-PC". In *Third International Symposium on Spatial Data Handling*. Edited by D. F. Marble. 269-292. Sydney, Australia: International Geographical Union, Commission on Geographical Data Sensing and Processing.

B. Shneiderman. 1983. Direct Manipulation: A Step Beyond Programming Languages. Computer 16 (8) : 57-69.

B. Shneiderman. 1987. *Designing the User Interface: Strategies for Effective Human-Computer Interaction*. Reading, MA: Addison-Wesley.

D. C. Smith, E. Harslem, C. Irby, R. Kimball and B. Verplank. 1982. Designing the Star User Interface. 7 (4 (April))

T. R. Smith and A. U. Frank. 1990. Very Large Spatial Databases - Report from the Specialist Meeting. *Journal of Visual Languages and Computing* 1 (3) : 291 - 309.

S. S. Stevens. 1946. On the theory of scales of measurement. *Science Magazine* 103

C. D. Tomlin. 1983. "A Map Algebra". In *Harvard Computer Graphics Conference*. Cambridge, MA:

C. D. Tomlin. 1989. *Geographic Information System and Cartographic Modeling*. New York: Prentice Hall.

D. White, K. Chan, M. Maizel and J. Corson-Rikert. 1989. "Polygon overlay to support point sample mapping: The national resources inventory". In *Auto-Carto 9*. Edited by E. Anderson. 384-390. Baltimore, MA: ASPRS & ACSM.

# Implementation of a User Environment for a Spatial DBMS

Jonathan F. Raper
Michael S. Bundock

Development of a standard user environment for use with Geographic Information Systems (GIS) has been prompted by the recognition that many of todays commercially available GIS products are firstly, difficult to learn and use, secondly, time-consuming to customise, and thirdly the knowledge gained in using one product is not readily transferable to another. The aims of the research within the Universal Geographic Information eXecutive (UGIX) project are to produce a prototype environment which is independent of the underlying GIS and DBMS, which provides high level analysis, design and customisation facilities, and presents the user with an adaptable, extensible, easy to learn and easy to use interface. The research has the following specific objectives:

- to identify the common functional components that must be supported within a generic spatial language for GIS operations
- to define the form of a generic spatial language processor to support the functions while permitting input from a variety of sources such as voice, WIMP and command line interfaces
- to build a prototype generic user interface environment with interfaces to a number of commercially available GIS products and their underlying DBMS's
- to test user response to the user environment.

This paper outlines the work performed to date and discusses in more detail the conceptual model proposed within the graphical user interface, the identification of a "standard" set of common GIS functions and the problem of interfacing the generic environment to any underlying spatial DBMS.

## 1. Key Problems for GIS User Environments

### 1.1 Goals, Tasks and Functions

The use environment is a vital element of any GIS. Long ignored as an esoteric aspect of GIS design whilst GIS development was driven by the need to extend functionality, the user environment is now beginning to attract its due attention. The development of the Universal Geographic Information eXecutive (UGIX) (Rhind, Raper and Green 1989) is a response to the widely expressed need to improve the usability of GIS, especially through the improvement and extension of the user interface. However, the improvement of the human-computer interaction (HCI) process. Since GIS are conceptually complex and involve diverse operations ranging from data modelling to geometric transformations, improving the HCI cannot be a complete solution to the improvement of GIS use.

A number of general problems afflict many commercially available GIS which can be characterised as failings of the user environment. One of the most pervasive is the blurring of the distinction between goals, tasks and system functions in the language and process of interaction with the system. This means that the user cannot easily appreciate the structure of the user environment. In the design of UGIX the following definitions are used, and the concepts implemented in the interface:

| | |
|---|---|
| GOAL | a user target for spatial data manipulation expressed in terms of application-specific outcomes<br>e.g. finding which stands of trees in a forest will come to maturity in each of the next 5 years |
| TASK | a spatial data manipulation procedure expressed in terms of system implementable steps<br>e.g. searching for certain spatially referenced items in a database and displaying the results in a map |
| FUNCTION | a low-level system operation to manipulate spatial data<br>e.g. plotting a symbol at specified X,Y coordinates on an output device. |

In this scheme tasks and functions refer to system operations, while goals apply to conceptual operations which are conceived of without reference to a computing environment.

Using this terminology, most GIS offer a command language composed of functions which are spatial tools and algorithms of various kinds. These commands are often modifiable with arguments and the complete expressions used are complex and often obscure. As part of a general movement to improve this situation a number of commercial systems have begun to offer graphical user interfaces (GUI's) built using window-icon-mouse-pointer (WIMP) techniques. However, these developments have illustrated the difficulties inherent in assigning icons or menu items to functions i.e. while the range of options available is now stated, the system structure is still no easier to understand. In particular, it is difficult to convert a goal into a series of tasks, each made up of the appropriate functions. Added to these implicit difficulties are the problems of overfilling the screen with icons or creating very long menus, the use of inappropriate screen metaphors and the lack of activity indicators to indicate the status of an operation to the user.

## 1.2 Non-transferability of knowledge

A further problem is that there are many different user languages for space in use, such as those defined by professionals working with spatial data (see examples in table 1). The table (with reference to two contrasting applications) shows how the relationship between terminology used for entities in the application domain, user descriptions of space and the geometry supported by GIS products can be complex and difficult to understand.

The process by which the user links the concept and implementation can lead to confusion and to users making errors in the way they specify operations. A well designed user environment requires an interface which permits the customiser or expert user to link the appropriate user language for space to the system architecture in a way which is transparent to the end users. In other words the interface should allow the user to manipulate objects that are meaningful in terms of the application, like sub-divide a parcel rather than split a polygon.

**Table 1** Some examples of user language for space for land and mining surveying

| **Application terminology** | **Geometric Data** | **Comment** |
|---|---|---|
| Land surveying | | |
| Monument | Point | Fixed point located by physical mark |
| Centroid | Point | Centre point to which reference code is linked |
| Metes | Line | Boundaries of parcel defined by distance & direction |
| Bounds | Line | Boundaries defining position of adjoining land parcels |
| Strip | Line | Corridor of fixed width either side of a centreline |
| Abuttal | Line | Boundary of a parcel on an adjoining parcel |
| Easement | Line/polygon | Corridor or area of land parcel set aside (e.g.for access) |
| Parcel | Polygon | Unit of land ownership |
| Aliquot | Polygon | Subdivision of parcel |
| Tract | Polygon | Land segregated by resurvey |
| Mining Surveying | | |
| Face | Line | Section of mine boundary used for excavation |
| Entry | Line/polygon* | Centreline/section of tunnel forming access to face |
| Cross-cut | Line/polygon* | Centreline/section of tunnel at right angle to entries |
| Pillar | Polygon | Area of unexcavated material within mine |
| Room | Polygon* | Section of tunnel ending in face |

* Polygon defined by closure of an open polygon e.g across the end of a tunnel

User environments of all forms have long lacked good visualisation tools for the spatial database data model. Ideally the user should be able to see the entities and their interrelationships graphically expressed so that they can formulate queries more easily. Finally, the poor quality of help systems for many GIS has also become a major drawback for many use environments. Frequently the help is simply a formal statement of the command syntax and arguments, and not an explanation of its wider usage.

Each GIS product on the market today incorporates its own distinctive environment, being substantially different from virtually all other available products. Each system tends to have its own unique command language, icon set, menu organisation and form layouts. The methods of interaction with the system vary considerably, even for such simple actions as selecting an object, obtaining help information or indicating confirmation of an action. Each vendor tends to use their own set of jargon, often in a manner which is inconsistent with other GIS vendors.

Even worse, the underlying system architectures show through and must be understood by the user before effective system usage is possible. In the absence of any other strong conceptual model for the system (as might be presented in a fully customised environment), the underlying architecture (files, layers, coverages, points, lines and polygons) becomes the mental model understood by the user. The application problem (e.g. forest resource management) becomes mapped to the problem of manipulating the components of the GIS architecture (i.e., the coverages, polygons etc.). Consequently the skill set acquired by a user is specific to the jargon and architecture of a particular product. Since each GIS uses different jargon and different architectures, the user's knowledge of one system is not readily transferable to another.

### 1.3 Customisability

Commercial GIS software packages are normally designed to be fairly general purpose in nature - they are not designed for a specific well-defined application within a particular organisation. Consequently, they need to be adapted to fit the specific application and user requirements of the organisation within which they are implemented. This adaptation of the as-supplied system is termed *customisation*. The term *customisability* is used to describe the ease and extent to which a system may be customised.

The objectives of the customisation process are to provide a system for the user that supports both the datamodel and functionality demanded by the application requirements; that presents to the end-user an interface that is specific to the user's application, language and experience; is uncluttered by non-required functions; icons and menus, is easy to learn and easy to use.

At present the base products delivered by GIS vendors are little more than a box of low-level spatial tools. These general purpose tools do not directly satisfy the user's functional requirements which are determined by organisational and application specific objectives. Furthermore, the tools will often have little meaning or applicability to the end user who must be educated in the language, interface and conceptual model supported by the product. The customisability of existing GIS products is poor, especially in the areas of database design and implementation, task definition and user interface design and development. The result is that effective customisation of a GIS product to satisfy corporate GIS requirements involves enormous expense and effort. This problem is so acute that it is not unusual to see organisations struggling to use a system that is uncustomised and uncustomisable with the available resources. Effectively, the application requirements are largely discarded so that the functions the GIS supports *becomes* the application. (Bundock 1991a)

## 2. Objectives of the UGIX project

### 2.1 Identification of Common Functional Components

The first phase of the research involved identification of a set of common (generic) functions that should be supported within the UGIX interface. These functions must be independent of any underlying GIS implementation strategy (e.g. object, relational or sheet based) and dependent rather on the goals of the user. These functions will be accessed via a set of icons, menus and forms within the GUI, and as a set of commands, operators and procedures within the command language. A "standard" set of icons and command names will be provided for the generic functions which may be modified (when required) within the customised environment. It should be noted that the generic functionality will not be implemented within UGIX, but rather, it will be accessible through UGIX. This distinction is important, since it restates the concept of separating the application from the user interface.

By identifying the functions required to satisfy generic goals, and providing access to these via icons and commands, the interface should become more consistent and less dependant on the underlying GIS data structures and architecture. It is believed that this will make the system easier to learn and use, and once learnt will provide the user with knowledge that should be more readily transferable to other systems.

## 2.2 Definition of a Generic Spatial Language Processor

A command language for interaction with the GIS database that supports the generic functions is being developed. It is intended that the spatial language will be embedded within both 3GL and 4GL languages which will provide the program logic and control structures. A spatially extended form of SQL (SQL-SX) has been designed to provide a standard, transportable language suitable for database definition, query, insertion, deletion and update (Bundock 1991b). SQL-SX is to be supplemented by the set of generic GIS functions identified above. The overall architecture for the spatial query processor has now been sketched out. It includes layers for SQL-SX, an equivalent iconic query language, an inter-process communications interface and a customisation environment.

## 2.3 Development of a Prototype Generic Use Environment

To test the feasibility of the concepts described here and the usability of such an interface, a prototype use environment must be developed. Further, it must be interfaced to a number of commercially available GIS products to prove that the user interface can be detached from the underlying GIS product architectures. The more difficult aspects of the development are likely to be:

- creating an efficient system to map between the functions supported within the generic spatial language and the matching functions supported by the underlying GIS,
- hiding the user interface supplied with the underlying GIS.

## 2.4 Testing User Response

The resulting prototype must be evaluated to determine that it does indeed provide a superior user interface environment to the standard interfaces provided by each of the underlying GIS products. To test the acceptability of the UGIX environment, a number of trials are proposed. Methods for evaluation of user interfaces include:

- formal analytical methods where the interface is evaluated in isolation from users (Grudin 1989)
- empirical methods where users are requested to perform the same tasks using different interfaces, and the performances measured and compared
- ethnographic methods where actual users are observed and the context and users observations are elicited (Crellin, Horn and Preece 1990).

During design of the graphical user interface we propose to explicitly adopt accepted guidelines (e.g. proposed ISO guidelines Williams 1989, Strijland 1990) as an aid to user interface specification. Analytical methods, operating on data gathered by automated monitoring of user interaction, may be used subsequently to determine interface effectiveness, identify frequent command usages, common errors and the relationships between use patterns and error occurrences. Chen (1990) describes the use of monitoring facilities built into the Xt toolkit to automatically gather appropriate information. Empirical studies are also proposed to compare the effectiveness of the new environment in direct comparison to the interface provided by the underlying GIS. Finally, user acceptability will be evaluated based on user interviews. It is intended that this evaluation will lead to suggestions for improvement for subsequent developments.

# 3. The UGIX Development

## 3.1 Background

The UGIX system design as described by Rhind, Raper and Green (1989) contains 3 main modules, viz. (A) containing the screen interfaces, dialogues and command processor; (B) containing a help and information system for a GIS; and (C) an expert system shell or high level system access module. The structure of UGIX is illustrated in Figure 1. This section describes the approach taken in the UGIX project, through first and second generation implementations. The major distinction between the generations is that the first aims to improve the usability of a specific GIS implementation, whilst the second aims to provide a generic user environment supporting transfer of skills between GIS and allowing easier customisation.

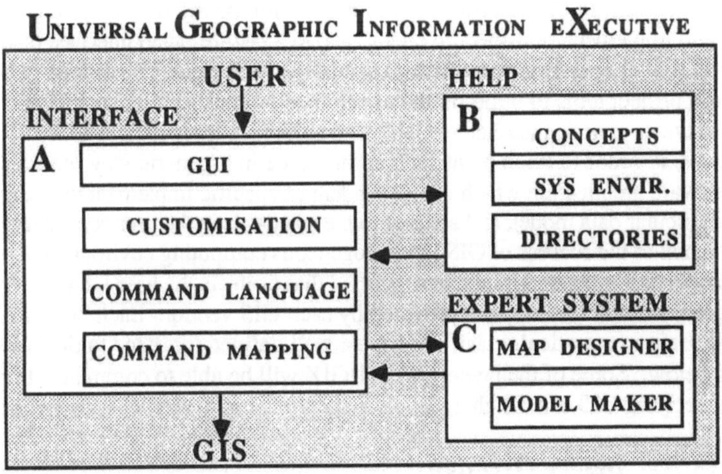


Fig. 1. The three primary modules of the UGIX architecture

The first generation approach to interface design within the UGIX project has been to prototype using Hypercard for the Apple Macintosh, where the Hypercard application (complete with in-built communications software) acts as a client to a host processor running the GIS application software (Raper, Linsey and Connolly 1990). This approach is similar to the one used by Cowen and Love (1988) to create an interface to the South Carolina Historic Preservation Office GIS database. Hypercard with its standard set of buttons, scrolling boxes and cards makes use of the GIS less daunting for the less technically-aware user. In addition, with the rich graphics environment available in Hypercard it is possible to show a graphic to illustrate the effect of various options available at any one point. It is also desirable to display all the commands available to the user in one place, with a pop-up explanation for each option.

Screen design has involved the standardisation of button and text field formats as well as card and background layout for different areas of activity such as:-

- Introduction and explanation (using a map guide);
- Map and file selection (using standard Macintosh file selection dialogue);
- Session screens for command processing;
- Help environment (UGIX (B) based on GISTutor version 2);
- A Gallery for maps and images generated in the GIS (along with buttons to redraw them).

Screen metaphors have been developed for each of these areas to make location in the system a graphical attribute. The interface also displays an activity index to give continuous feedback to the user on the status of the session. Currently this system interface 'shell' is being implemented for the GIS ARC/INFO, and is known as 'HyperArc': it is currently under test with users at 'novice' and 'competent' levels of expertise. In addition to feedback on the use of the system, the aim of the evaluation phase is to define a core area of functionality in common use to help optimise the UGIX system structure.

Another UGIX design principle is that improving access to existing GIS can be achieved by converting the current function-orientation of the native system interface (primitive and implementation-specific operations) to a task-oriented interface (sequences of high level spatial operations) usable by a spatially aware user. The second generation of the UGIX project aims to build on the experience of constructing such task-oriented interfaces to create generic interfaces capable of communication with any GIS.

However, in order to implement such an interface in a generic way requires a new form of software architecture which is independent of specific implementations, does not enforce a particular data model, and adheres to the standards in the user community which are most crucial to the success of GIS in heterogeneous computing environments.

To achieve this a layered model is suggested that protects the user interface from the actual implementation mechanisms provided by each GIS vendor. Each layer within the model will perform a particular task and have a well defined interface to the layers both above and below. Some of the layers within UGIX will be able to communicate directly with the underlying GIS at a matching level.

## 3.2 UGIX(A) Design Overview

Separation of the user interface from the application is not a new concept. Early work resulted in what is often known as the "Seeheim model" (Green 1985) developed during a 1983 workshop on architectures for interactive systems at Seeheim. Subsequently, the identification of components of the overall system corresponding to semantic, syntactic and lexical aspects, and the relationships between them has lead to various alternative architectures.

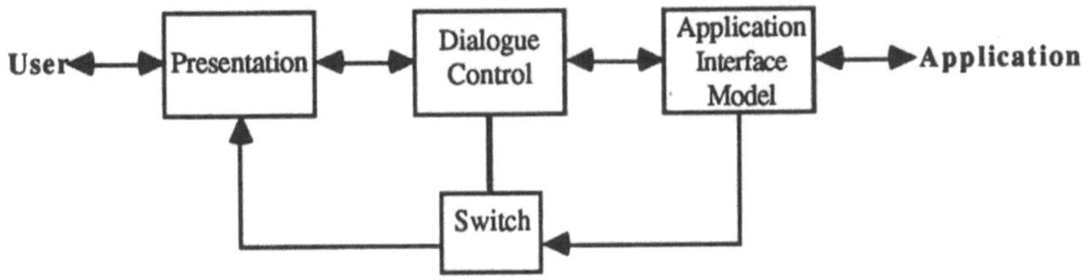


Fig. 2. The Seeheim model

The overall architecture for UGIX(A) is similar to the Seeheim model in many aspects. The requirement for a high bandwidth communications channel from the GIS application is supported to allow efficient graphics display and manipulation. Figure 3 illustrates the overall structure proposed for the UGIX environment.

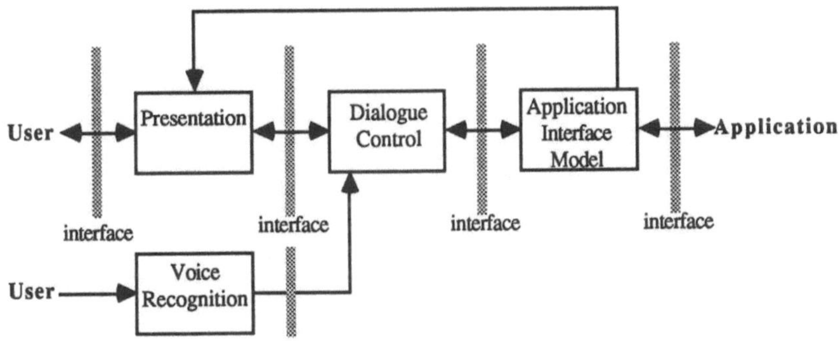


Fig. 3. Overview of UGIX(A) architecture

## 3.3 Description of Components

The presentation layer incorporates a standard user interface toolkit (e.g. Motif, OpenLook etc.) a widget design facility, a screen design facility and a screen execution facility. The widget and screen design facilities operate within the constraints of the toolkit, and will be implemented as a set of executable screens. The customisation environment itself and the actual resulting end-user application, will also simply be a set of screens with which the user may interact. The screens will be designed in terms of a set of windows, a set of widgets within each window and a set of forms. The behaviour of the windows, widgets and forms will be described in terms of the 4GL command language.

The application interface module accepts generic spatial language commands and maps them onto the command language of the underlying GIS. It then issues these commands to the GIS via an inter-process communications mechanism. The GIS responses may include alphanumeric information (which may be used to fill a form), status information (error and function status) and/or graphics.

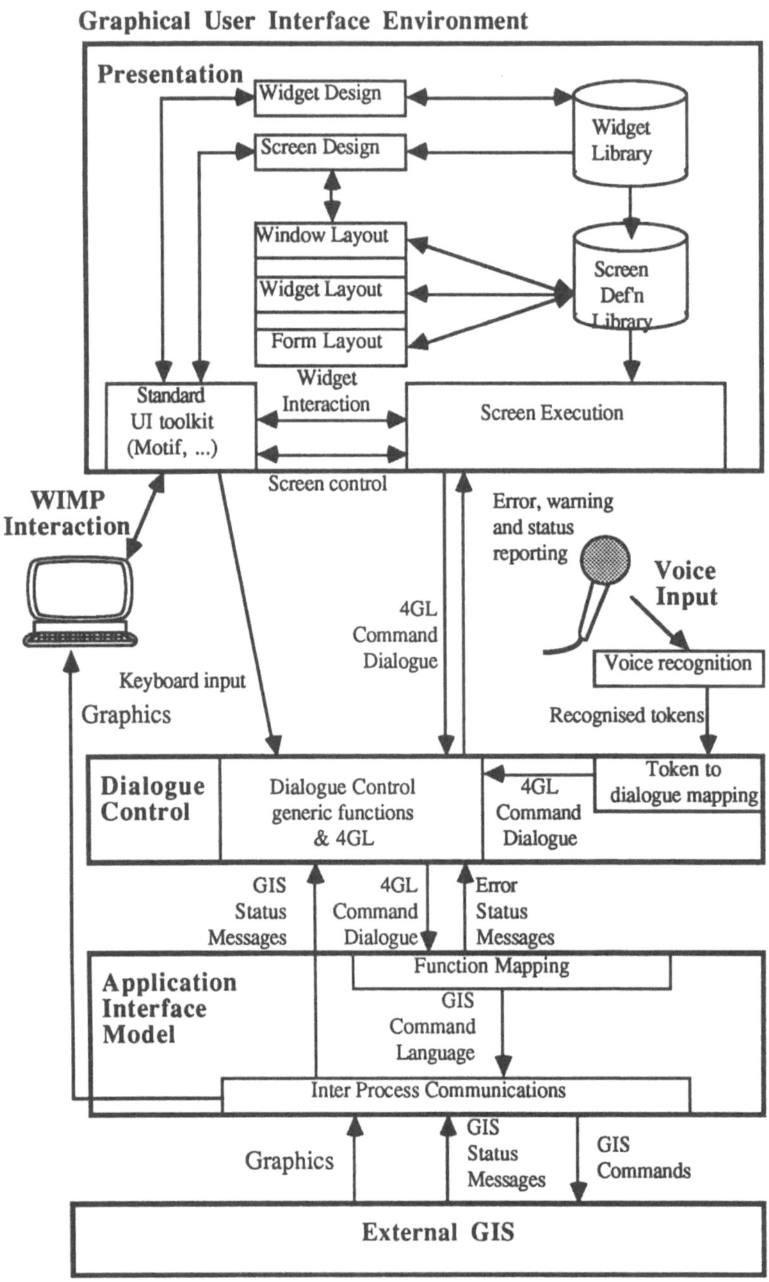


**Fig. 4.** More detailed breakdown of UGIX architecture

To support a highly interactive graphics environment requires that a high speed channel be provided to display the graphical data. However, alphanumeric and status data may be routed through the dialogue control module for further processing and display. Figure 4 illustrates the proposed architecture of UGIX(A) in more detail.

## 3.4 The Graphical User Interface

Wilson (1990) reviews the use of graphical user interfaces within GIS and the applicability of the desktop metaphor. He suggests guidelines for building suitable user interfaces. The GUI is described as having three components:

- an underlying conceptual model,
- a command structure comprising codes, function keys, buttons etc. with which to create a syntax, and
- the visible screen graphics, such as command lines, menus and icons.

To date, within the GIS world, most emphasis has been placed on the development of a command syntax and the design of menus, icons and screens.

Initial attempts at improving usability concentrated on reducing the number of interactions, menus, forms and icons that the user had to deal with. This approach either reduced the available functionality or produced menu hierarchies that were difficult to use. An alternative approach is to use existing knowledge of a related field that may be applied to the new problem domain.

*The Underlying Conceptual Model* A major contributing factor towards the non-standardisation of the GIS user interface is the lack of an underlying conceptual model for the interface. It has been suggested (Gould & McGranaghan 1990) that the primary mechanism by which a user learns to use GIS is by metaphoric learning. Existing graphical user interfaces for non-GIS applications have often been developed using the desktop metaphor as the underlying conceptual model. The desktop metaphor is suitable for many business related applications since the activities performed by the computer based application have direct equivalents with the manual methods. However, it may not be readily applicable to many GIS applications due to the lack of spatial and mapping related activities that normally occur on and around a desk.

Current GIS technology imposes on the user a conceptual model of geographic space that is a function of the internal structures supported by the GIS (e.g. layers, points, lines, polygons). What we should be aiming for is a user interface that permits the system customiser to present a conceptual model to the user that is relevant and applicable to the both the user's background and the application in hand.

The strength of the desktop metaphor as used within the Macintosh and other PC environments for the underlying conceptual model, is that it provides an organising framework within which other operations and metaphors may exist. Gould and McGranaghan (1990) have extended this idea to suggest the need for an organising metaphor within which there may be other nested metaphors (which may themselves be organising metaphors). This approach has promise since it provides a structure within which applicable and relevant metaphors may be applied, rather than trying to apply a single metaphor to all situations.

*The Organising Metaphor within UGIX* The current thinking for the UGIX GUI is to develop an environment supporting nested metaphors. The proposed overall organising metaphor is a building, within which there are a set of rooms, each accessible via a door. It should be noted that the idea of using the room/building metaphor has been independently conceived by a number of different groups including experimenters at the University of Waterloo (Chan and Malcolm, 1984) and even built into a number of existing products (e.g. Rooms from ENVOS and even X11 rev 4 attempts to provide a Rooms-like system).

Within UGIX, each room may possess its own organising metaphor. Most rooms will be directly accessible from the entrance hall although some special-purpose rooms may require access from within another room. On entering the system the user is located in the entrance hall, a neutral, public space through which the user moves to a particular environment. Doors provide access to the environments the user has access to. The door metaphor is a strong metaphor for access into and out of different environments (Cátedra 1990) and provides features such as locks, opening and closing. These features may be used directly for access control, entering and leaving. Within each room, a single type of activity occurs, and a single lower-level organising metaphor is employed.

*The Iconic Query Language* Access to the functionality provided within each environment (room) will be predominantly via icons, menus and forms. The icons should fit the organising metaphor for that environment so that they have relevance and preferably direct applicability. Consequently, the drafting office will support map cabinets from which maps may be extracted, drafting boards on which map updates and viewing may be performed and light tables on which map overlay operations may be carried out.

## 3.5 The Command Level Interface

The command level interface incorporates a 4GL command language, a function mapping facility and an inter-process communications facility. It accepts commands from the GUI and the voice recognition system in terms of 4GL command sequences. It can also be accessed directly to perform ad hoc functions and applications development. Magik, an object-oriented (OO) development facility from Smallworld Systems has been selected for the development of the prototype.

*Magik: An object-oriented development environment.* The objective is to support a single command line and development environment in which application development, database definition, ad hoc queries, menu, form and icon commands, database access and graphics are all available. It is also desirable that the full power of the underlying GIS, DBMS and UIMS are available. The Magik language (Chance, Newell and Theriault 1990) is a hybrid of the procedural and OO approaches and supports its own interactive development environment. It is fairly readable (certainly more so than 3GLs such as C and C++), comes with a comprehensive set of standard object classes, methods and procedures, and provides the ability to transfer applications between hardware and operating system platforms with a minimum of effort. It utilises the X-Windows standard for all interaction with the workstation. For the UGIX development, we are extending Magik by adding a new set of language constructs to support a spatially extended version of SQL.

*SQL-SX: A spatially Extended version of SQL* The relational language SQL (Date 1989). forms a suitable base on which to develop spatial extensions since the use of relational database management system (RDBMS) technology within the existing GIS user community is virtually universal. Further, it is likely that RDBMS will be the major data management technology for at least the 1990s, and that SQL will be the major language for interaction with those databases. The investment by user organisations in training staff in the use of SQL is significant so there is consequently a sizeable body of expertise available both within GIS user organisations and in the general computing industry. The use of a non-standard query language for GIS implementations does not appear commercially viable, nor practical in the near future.

A number of GIS vendors are already developing spatially extended versions of SQL and have reported their work in the research literature including Kork (Ingram & Phillips, 1987), Intergraph (Herring, Larsen & Shivakumar, 1988), Wild Heerbrugg (now Prime) (Charlwood, Moon & Tulip, 1987) and GeoVision (Westwood, 1989). Each has attempted to provide facilities supporting spatial predicates and spatial data manipulation facilities within SQL (or SQL-like query languages). Unfortunately, the basic query language in each case has been an incomplete implementation of the ANSI/ISO SQL standard (ISO 1987, ISO 1989), and the spatial extensions were fairly minimal and elementary. To make matters worse, the extensions in general do not maintain consistent syntactic and semantic constructs with the rest of SQL. For example, spatial predicates are not in general supported within the WHERE clause, but rather within a separate clause.

Other researchers including Pong-Chai Goh (1989), Sacks-Davis, McDonell and Ooi (1987) and Egenhofer (1987) have also provided useful contributions towards the development of spatial extensions to SQL. However, until recently there has been no proposal for a standard set of extensions put forward for discussion. In a recent paper, Bundock (1991b) proposed a set of spatial extensions for SQL that could form the basis for an agreed standard between GIS vendors and the GIS user community. The spatial extensions are based on the existing proposals for SQL3 being studied by the combined ISO-ANSI SQL standards working group (ISO-ANSI 1990). These proposals include a number of object-oriented concepts, including support for entity hierarchies, user defined data types, operators and functions. In particular, the detailed proposal in support of user defined data types (Kerridge 1989) if implemented, would provide the framework in which to develop spatial data types, spatial operators and spatial functions, whilst remaining completely within the SQL3 standard.

The extensions necessary to make SQL usable within GIS applications for query and update of both spatial and aspatial data include:

- spatial data type(s) e.g. point, node, line, polygon
- spatial operators (predicates) e.g. at, inside, encloses, connected_to
- spatial functions e.g. area_of, length_of
- long transaction management statements
- report specification facilities - both textual and graphical

It should be noted that current SQL3 proposals (ISO-ANSI 1990) do not include support for long transaction management, management of design alternatives and report specification facilities.

In addition, functional requirements demand that:

- in an interactive SQL session, the spatial data types should be displayed graphically as a result of being SELECTed
- the data dictionary and DDL support the extensions
- spatial access control (protection) may be provided by inclusion of spatial data types, predicates and functions within VIEW definitions
- spatial integrity maintenance may be provided by support of spatial data types, predicates and functions within the CONSTRAINT clause.

*The Function Mapping Facility.* This facility accepts UGIX commands in the form of function specifications and data selection specifications, and transforms these to the command language of the underlying GIS. For functions supported by both UGIX and the underlying GIS, the mapping should normally be moderately straightforward. Using OO techniques, the function mappings become methods for the function objects. It is hoped that this technique will provide a straightforward approach to allow the support of multiple underlying GIS products.

However, not all functions known to UGIX may be available in the underlying GIS. Consequently, there will be holes in the UGIX interface for those unsupported functions. Icons representing unsupported functionality will be displayed in grey, and the matching command level functions will possess a method that reports on the function unavailablity.

Where the underlying GIS supports functions unknown to UGIX, a general purpose facility will be provided that allows UGIX to send the system dependant command string directly to the underlying GIS. The command string will be explicitly declared and may be associated with icons, menus and forms in the normal way. In this way, special purpose functionality may be readily included in the UGIX GUI, even if it is not considered generic enough to warrant inclusion in the generic function list.

# 4. Common Functional GIS Operations

Analysis of the functions to be supported directly within the interface is based on goal analysis and hierarchical decomposition (top-down) techniques rather than task analysis techniques since the interface, as delivered, must retain its generality. Task analysis might provide a more detailed definition of requirements for a particular (well defined) application but is considered too specific when attempting to define general requirements. Abstraction of the general functionality from a task based analysis might be possible given sufficient access to a wide range of actual users in a wide range of applications. However, operating within a tight set of time and resource constraints, requires the analysis to be performed quickly with minimal contact with real users. To overcome this limitation, the study initially involved :

- a review of existing research literature,
- a review of a number of GIS tenders,
- a review of the functionality supported by a number of commercially available GIS products.

A number of authors have reported attempts at classifying GIS functionality, including Dangermond (1983) and Maguire and Raper (1990). However, Dangermond based his analysis firmly in terms of the map concept which partitions the world into discrete tiles. Consequently, much of the functionality described is concerned with managing these discrete units, to allow queries across multiple map sheets, to join multiple maps to form a new composite map, and to perform edge matching. The functionality necessary to support a map-sheet based GIS implementation is quite distinct from the goals of the user which are implementation independent. Maguire and Raper describe the functional components of a GIS in a more generic manner. Below each identified high level function they separate the functionality that applies to the spatial data, from that which applies to the attribute (aspatial) data. In the current study we attempt to retain the user's concept of objects within their application domain which may possess spatial and/or aspatial components. This suggested set of generic functions will be presented with the first UGIX prototype.

## 5. Conclusions

The development of a GIS independent user interface environment capable of interfacing with a number of commercially available products whilst still providing an adaptable, consistent, easily learnt and easy to use interface, appears at first sight a difficult task. A structured approach to this problem is however beginning to indicate the feasibility of the project.

The usability of the prototype system will be evaluated and compared to the interfaces offered by the underlying GIS products. If successful, we would like to promote the concepts incorporated within UGIX to initiate discussion between users and vendors on the development of standards and guidelines for GIS user interface design and construction.

## 6. Acknowledgements

The authors of this paper acknowledge the help of ESRI in the funding of work leading to the prototyping of HyperArc, the first generation of UGIX. Apple Computer (UK) have also assisted in this work by establishing a laboratory of Macintosh computers which form the Apple Mapping Centre at Birkbeck College. Receipt of an ESRC/NERC grant from the collaborative programme in GIS is acknowledged by MSB. Smallworld Systems have assisted the project with hardware and software tools.

## References

Apple Computer (1987) *Human interface guidelines.* Addison Wesley, Amsterdam.

Bundock M.S. (1991a). GIS Customisation: From Tools to Efficient Working Systems, *Proceedings Mapping Awareness '91*, London, UK.

Bundock M.S. (1991b). SQL-SX: Spatially Extended SQL - Becoming a Reality, Paper presented at EGIS '91, Brussels, Belgium.

Cátedra M. (1990). "Through the Door": A view of space from an anthropological perspective. *Proceedings of NATO ASI Cognitive and Linguistic Aspects of Space*, Las Navas, Spain

Chan P.P. and Malcolm M.A. (1984). Learning considerations in the Waterloo port user interface. *Proceedings IEEE First International Conference on Office Automation*, IEEE, New York

Chance A., Newell R.G. and Theriault D.G. (1990). An Overview of Magik, Technical Paper 9/1 from Smallworld, Smallworld Systems Ltd, Cambridge, UK.
Charlwood, G. Moon, G. and Tulip, J. (1987). Developing a DBMS for Geographic Information: A Review, *Proceedings Auto-Carto 8*, Baltimore, Maryland.
Chen J. (1990). Providing Intrinsic Support for User Interface Monitoring, *Proceedings Human-Computer Interaction - INTERACT '90*, Cambridge, UK.
Cowen D.J. and Love S.R. (1988). A Hypercard based workstation for a distributed GIS network. *Proceedings GIS/LIS '88*, San Antonio, TX, USA.
Crellin J., Horn T. and Preece J. (1990). Evaluating evaluation: A case study of the use of novel and conventional evaluation techniques in a small company. *Proceedings Human-Computer Interaction - INTERACT '90*, Cambridge, UK.
Dangermond J. (1983). A Classification of Software Components Commonly used in Geographic Information Systems. *Proceedings United States/Australia Workshop on Design and Implementation of Computer-Based Geographic Information Systems*, Amherst, New York
Date C.J. (1989). *A Guide to the SQL Standard*, Second Edition, Addison-Wesley, Reading, Massachusetts
Egenhofer M.J. (1987). An extended SQL syntax to treat spatial objects, *Proceedings 2nd International Seminar on Trends and Concerns of Spatial Sciences*, New Brunswick.
Green M. (1985). Report on Dialogue Specification Tools. In Pfaff, G.E (ed) *User Interface Management Systems*. Springer Verlag. pp 9-20.
Gould M.D. and McGranaghan M. (1990) Metaphor in Geographic Information Systems, *Proceedings 4th International Symposium on Spatial Data Handling*, Zurich, Switzerland.
Grudin J. (1989). The Case Against User Interface Consistency, *Communications of the ACM*, 32, 10.
Herring J.R. Larsen R.C. and Shivakumar J. (1988). Extensions to the SQL Query Language to Support Spatial Analysis in a Topological Data Base, *Proceedings GIS/LIS '88*.
Ingram K.J. and Phillips W.W. (1987). Geographic Information Processing Using a SQL-Based Query Language, *Proceedings Auto-Carto 8*, Baltimore Maryland.
ISO (1987). Report IS: 9075 Information Processing Systems-Database Language SQL
ISO (1989). Report IS: 9075 Information Processing Systems-Database Language SQL
ISO-ANSI (1990). (ISO/ANSI Working Draft) Database Language SQL2 X3H2-90-398 and ISO DBL LON-3, J Melton (editor), October 1990
Jacobson R. (1990). Virtual Worlds, Inside and Out. *Proceedings of NATO ASI Cognitive and Linguistic Aspects of Space*, Las Navas, Spain
Kerridge J.M. (1989). A Proposal to add User Defined Datatypes to SQL, ISO report ISO/TC97/SC21/WG3-DBL-FIR-3.
Maguire D.J. and Raper J.F. (1990). Design Models and Functionality in GIS, *Preoceedings GIS Design Models*, Leicester, UK.
Pong-Chai Goh (1989). A GQL for Cartographic and Land Information Systems, *International Journal of Geographic Information Systems*, vol. 3, no. 3, 245-255.
Raper J.F. and Bundock M.S. (1990) UGIX: A Layer Based Model for a GIS User Interface. *Proceedings of NATO ASI Cognitive and Linguistic Aspects of Space*, Las Navas, Spain
Raper J.F. and Green N.P.A. (1989). Development of a Hypertext Based Tutor for Geographical Information Systems. *British Journal of Educational Technology*.
Raper J.F., Linsey T. and Connolly T. (1990). UGIX: A Spatial Language Interface for GIS : Concept and Reality. *Proceedings of EGIS '90*, Amsterdam, The Netherlands
Rhind D.W., Raper J.F. and Green N.P.A, (1989). First UNIX then UGIX, *Proceedings of AutoCarto 9*, Baltimore, MD, USA.
Sacks-Davis R., McDonell K.J. and Ooi B.C. (1987). GEOQL - A Query Language for Geographic Information Systems, Internal Report No. 87/2, Dept of Computer Science, Royal Melbourne Institute of Technology.
Strijland P. (1990). *Icons for use on screens Part 1: Specifications*. ISO/IEC JTC 1/SC18/WG9 Working Document (4 July 1990).
Talmy L. (1990). How Language Structures Space. *Proceedings of NATO ASI Cognitive and Linguistic Aspects of Space*, Las Navas, Spain
Tauber M.J. (1990) ETAG: Extended task action grammar. A language for the description of the user's task language. *Proceedings Human-Computer Interaction - INTERACT '90*, Cambridge, UK.

Westwood K. (1989). Toward the Successful Integration of Relational and Quadtree Structures in a Geographic Information System. *Proceedings National Conference - GIS - Challenge for the 1990's.*, Ottawa, Canada

Williams J.R. (1989). *Menu Design Guidelines.* ISO TC159/SC4/WG5 WD 9241-14 (17 March 1989)

Wilson P.M. (1990) Get your Desktop Metaphor off my Drafting Table: User Interface Design for Spatial Data Handling. *Proceedings 4th International Symposium on Spatial Data Handling,* Zurich, Switzerland.

# Extending a Spatial Database System by Graphs and Object Class Hierarchies

Ralf Hartmut Güting

**Abstract:** Two important open problems in spatial database systems are the modeling and implementation of *spatially embedded networks* and the handling of *heterogeneous collections of spatial objects*. We sketch a data model that together with relations provides *graphs* as explicit structures. Graphs allow to model spatial networks; they consist of nodes, edges, and *explicit paths*. Heterogeneous collections of objects can be represented in object type hierarchies which are in turn realized by tuple hierarchies. The resulting model contains the relational model; hence it is still possible to ask standard relational queries. At the implementation level, we describe a simple general purpose data structure for the representation of graphs on external storage. It is shown how such a graph representation module can be embedded into the architecture of an extensible relational database system, the Gral system, and also how it can be implemented with relatively little effort in such an environment. Hence we obtain a strategy to migrate from a given relational system with spatial data types to a system with graphs and object class hierarchies.

## 1. Introduction

So far, a large part of the research in spatial database systems has been concerned with extending data model and query language by geometric data types and operations, supporting spatial search by new types of index structures, and realizing complex geometric operations through efficient algorithms from computational geometry. Most often, data model extensions were based on the relational data model and its various approaches to query languages, and the implementation either tried to augment existing relational systems by adding front-ends or other extensions, or attempted a more thorough integration of new index structures and query processing methods with a relational architecture.

For example, extended data models and query languages have been proposed in [Fr82, LiN87, JoC88, RoFS88, ScV89, Eg89]. Among these [Fr82, RoFS88, Eg89] are based on SQL whereas [JoC88] use a QBE-like approach. In [LiN87] a query language is defined using an extended ER calculus. Scholl and Voisard [ScV89] develop an algebra for maps (partitions of the plane). User interface issues are discussed in [EgF88]. Some example prototype systems are described in [Ab89, OoSM89, RoFS88,

ShSN90] among which [Ab89, OoSM89, RoFS88] extend existing relational systems. Spatial query processing has been studied in some more detail in [Or86, OrM88, Or90].

These issues have also been addressed in the *Gral system* [Gü89] implementing the model called *geo-relational algebra* [Gü88]. The data model is based on the formal concept of a *many-sorted* algebra which provides a clean framework for the integration of geometric data types and operations. In Gral, there are types to represent points, lines, and regions in the plane. Operations include geometric predicates such as “intersects”, the computation of new geometric objects, for example, “intersection” or “overlay” (the latter combines two partitions of the plane), and numeric calculation on geometric objects, such as “distance”. The data model is extensible; it is allowed to add any atomic data type and any operation that takes relations and atomic values as operands and returns a relation or a value. The Gral system realizes this extensibility: it allows to add through well-defined interfaces (i) the representation of data types, (ii) procedures implementing operations on data types, (iii) new types of index structures (so the LSD-Tree [HeSW89] has been added as a spatial index structure), (iv) the implementation of new algorithms for query processing (for example, plane-sweep algorithms), (v) rules to describe the mapping of new operations in the query language to possibly new access methods or other executable operations (realized through a rule-based optimizer [BeG90]).

We would now like to attack two important issues arising in spatial databases that have not been addressed in Gral and to our knowledge also not been satisfactorily solved in other proposals. The first is the treatment of *spatially embedded networks*. Examples are:

- Streets etc.: Highways, roads, city streets, pedestrian ways, bicycle routes, ...
- Waters: Rivers, canals, lakes, ...
- Means of transport: Train, bus, underground, air plane, ...
- Electricity, telephone, gas, water, sewage, ...

Current systems, including Gral, are capable of modeling adequately the *geometry* of such networks, but not the *connectivity*, that is, the underlying *graph structure*. For example, in modeling a river network, we can determine which rivers intersect roads, but it is very hard to find out which waters are downstream from a certain site at one river and would be affected by pollution. Some other examples of queries involving spatial networks are the following:

- Plan a container ship trip from Rotterdam to Duisburg. How long will it take?
- Which fraction (of the total length) of city streets in Germany is more than 10 kms away from a highway exit?
- A fog area covers some part of the highway network. Which alternative routes can be recommended?
- Plan a public transport trip (train, bus, underground, taxi) from an address in Dortmund to an address in Munich.
- What is the total population of the area affected by the failure of a certain power plant?

Graphs as such have certainly been considered in database research. A classical problem is the computation of the transitive closure. Here graphs are usually modeled by the standard facilities of the given data model, for example by relations with special restrictions [Ag87] or on the basis of the functional data model [Rose86]. There are some interesting proposals to integrate searching on graphs into query languages [CrMW87, CrMW89, BiRS90]. However, in that research area the spatial embedding of graphs is not considered. The examples above illustrate that there are many interesting queries that have geometric as well as connectivity aspects. We feel this is a very promising area of research in spatial databases.

Our approach to deal with such spatial networks is the following:

- Extend the given data model by explicit modeling concepts for graphs. Furthermore, graphs will offer besides nodes and edges also *explicit paths*, that can have associated information. For example, rivers or highways are paths over the underlying graph, whose nodes have associated POINT objects and whose edges have associated LINE objects.
- Graph operations, such as finding a shortest path or determining connected components, are integrated into the query language as algebra operations.
- There are special external data structures available in the system to represent graphs. Such a structure represents the connections between the components of a graph in such a way that fast traversal is possible[1]. This structure is the basis for the implementation of graph algorithms that realize the query language operations efficiently.

The second issue concerns "generalization queries". Typical queries on a spatial database have the form "Find all spatial objects of kinds A, B, C, D, E, ... within region R" or "Which line objects of classes ... intersect line L?". Specific examples are "Find all cities, villages, rivers, and railway lines in Bavaria", or "Which streets, sewage canals, power, and telephone lines are affected by a planned highway section?".

It is well-known that this is an important class of queries (see e.g. [Fr82, Ab89, EgF89]). But most systems, including Gral, so far offer no or only a superficial solution to this problem. The reason is that the system is usually based on relational concepts and represents each class of objects by its own relation. A query that refers to k object classes needs then to be decomposed into k queries on k relations. If the system offers any support for these queries, then support is restricted to

- allowing to overlay the k results in a single graphic display,
- syntactic facilities in the query language so that one needs not to write k queries.

Even when syntactic support exists, the system will still decompose the user query into k queries, with the following disadvantages:

- Each of the k queries is optimized separately, with the corresponding overhead.
- Each class of objects has its own representation on secondary storage. In the best case, each of these representations is clustered by geometry. Spatial search for the

[1] It is far too costly to perform a join in order to follow an edge in a graph.

same query area needs to be done separately on each representation, which leads to a lot of superfluous page accesses.

- Since the k results can be combined only graphically but not "semantically", it is not possible to perform further calculations on the result.

We suggest to solve this problem in a clean and elegant way by introducing a concept of generalization among object types with an associated type hierarchy. This is probably not too surprising, as these concepts are quite popular in functional and object-oriented modeling. There is an interesting twist in modeling here arising from a connection between the object type hierarchy and the data type hierarchy (GEO > EXT > LINE) that was already present in Gral. But our main point is to show that such an object type hierarchy can be integrated relatively easily into a relational model and system, in our case, the existing Gral system.

Within this overall strategy, the purpose of this paper is to show how graphs and object class hierarchies can be embedded in a given relational system architecture. Clearly this needs to go along with a corresponding extension of the data model; this can only be sketched in this paper and will be described in more detail elsewhere [Gü92]. More specifically, the topic of this paper is a basic data structure for the representation of graphs in external memory; we show how this structure can be implemented within the Gral architecture. The structure has certain provisions to accomodate object class hierarchies. Of course, it is not necessary to start from a relational environment: the graph structure itself could be used as well in any other context, for example, a functional or object-oriented system.

The paper is structured as follows: In Section 2 we sketch some aspects of the intended data model with graphs and object class hierarchies. In Section 3 we first examine what it means to query a graph and then briefly introduce the Gral system architecture and in particular the parts needed here, that is, the modules underlying the graph representation module. The graph representation component is then designed bottom-up in several steps. A short discussion follows in Section 4.

## 2. A Sketch of the Data Model

Let us call the data model, which is at the time of writing still somewhat preliminary, the *Gral II data model.* It is built around the concepts of *data types*, *object types*, and *structures*. As in Gral, the query language has the structure of a *many-sorted algebra.* Essentially this means that we have a collection of "carrier" sets and functions between these sets; a query is simply a nested application of functions, or a well-formed term of this algebra (see [Gü88, GüZC89] for a precise definition).

***Data Types***. Let there be given a collection B of *base data types*, such as BOOL, INT, STR, etc. Each such type denotes a domain of possible values. The set of base data types is extensible; further types can be added to the system by implementing a representation for the type. In addition to B, the set of *data types T* contains types that can be viewed as generalizations of types in B. Let "≤" be a partial order on T that organizes the types in T into hierarchies. A specific design for Gral II looks as follows:

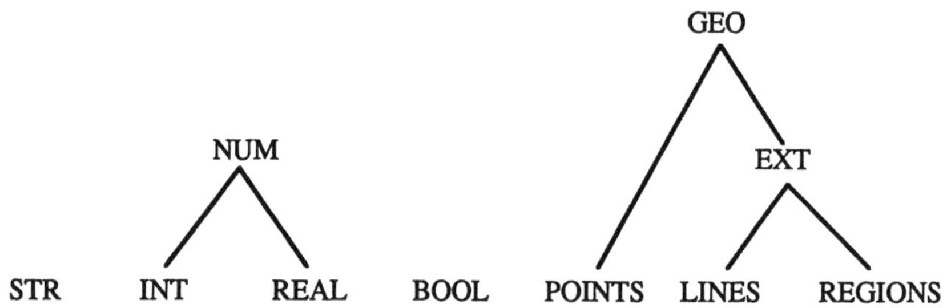


Figure 1

For example, we have LINES ≤ EXT ≤ GEO. For data types A, B we say, A is a *subtype* of B iff A ≤ B. Types A and B are *related*, iff there exists a type C such that A ≤ C and B ≤ C. In the example, POINTS and LINES are related, whereas INT and STR are not related. If two types A and B are related, then their *smallest common supertype*, denoted *lub* (A,B), exists and is uniquely defined.

Data types of course have associated operations and the purpose of the hierarchies is to describe the applicability of operations. If an operation is defined with functionality $T_1\ T_2 \dots T_n \rightarrow T'$, then it can be applied to any tuple of operands $U_1\ U_2 \dots U_n$ where all $U_i \leq T_i$. For example, an operation defined with functionality GEO × GEO → REAL can be applied to two LINES operands.

Readers familiar with Gral [Gü88, Gü89] should note that the geometric data types and their operations have been redesigned: A value of type regions may consist of several disjoint polygons, possibly with holes. Similarly a LINES value may be a collection of polygonal lines, a POINTS value a set of points. Basically, these types have been closed under union, intersection, and difference of the underlying point sets.[2] The **intersection** operator has now the functionality

| | | | | |
|---|---|---|---|---|
| LINES × LINES | → | POINTS | **intersection** | # ( _ , _ ) |
| REGIONS × REGIONS | → | REGIONS | | |
| EXT × EXT | → | LINES | | |

Type inference conventions have been slightly extended so that the three possible functionalities are checked in the order in which they are written. Hence the last functionality will be used precisely in the case that one operand is of type LINES, the

[2] This design was discussed in joint work with Scholl and Voisard [GüSV90]; a similar design is described in [ScV91].

other of type REGIONS. The notation to the right specifies prefix syntax ("**intersection** (a, b)") for this operator.

The data type hierarchies have already been present in Gral. But now we use them as a basis to define object type hierarchies.

***Object Types***. Basically, as in the relational model an object is described by a tuple of values of the base data types. Hence, an *object* is a list of pairs

$$o = < (v_1, T_1), \ldots (v_n, T_n) > \qquad \text{for } n \geq 1$$

where the $T_i$ are base data types and each $v_i$ is a value of type $T_i$. An *object type* is simply a list of (not necessarily base) data types

$$O = < T_1', \ldots T_m' > \qquad \text{for } m \geq 0.$$

(For m = 0 we have the *totally undefined* object type, denoted by ◊). Hence for the object o above, its type is $T_1 \ldots T_n$. The data type hierarchy induces a hierarchy on object types as follows. Let $O_1 = T_1 \ldots T_n$ and $O_2 = U_1 \ldots U_m$ be two object types. Then we say, $O_1$ is a *subtype* of $O_2$,

$$O_1 \leq O_2 \quad :\Leftrightarrow \quad n \geq m \text{ and } T_i \leq U_i \text{ for } 1 \leq i \leq m.$$

Note that the more special object type $O_1$ has *more* components ("attributes") than its supertype $O_2$. The idea is that an object of type $O_1$ can be used like an object of type $O_2$ since the types of its first m attributes are subtypes of those required by $O_2$. This is a simple way to formalize attribute inheritance and at the same time to couple it with the data type hierarchy. We say object o *conforms to* a type O' iff the type of o is a subtype of O'.

For example, <STR, INT, STR> ≤ <STR, NUM, STR> ≤ <STR> (the symbol "≤" should be read as "is a more special type than").

Let us briefly illustrate how this can be used. We assume a user can define *named object types*[3], for example, in a syntax as follows:

| | | |
|---|---|---|
| **define object type** city | = | name: STR,<br>center: POINTS,<br>pop: INT; |
| **define object type** geoobject | = | name: STR,<br>geometry: GEO; |

[3] We have here for brevity only shown the definition of unnamed object types. In a named object type there are also names for the type as a whole and for the attributes.

Clearly *city* is a subtype of *geoobject* since <STR, POINTS, INT> ≤ <STR,GEO>. Type *city* could alternatively directly be defined as a subtype of *geoobject* (assuming *geoobject* was defined already):

**define** geoobject **type** city = center: POINTS **refines** geometry: GEO,<br>
pop: INT;

This is equivalent with the first definition, since the *name* attribute is inherited. Let us define a further type *state:*

**define** geoobject **type** state = region: REGIONS **refines** geometry: GEO,<br>
pop: INT;

Now for any two object types $O_1 = T_1 \ldots T_n$ and $O_2 = U_1 \ldots U$ their *smallest common supertype* is defined as

$$lub\,(O_1, O_2) := < lub\,(T_1, U_1), \ldots, lub\,(T_k, U_k) >$$

where $k \in \{0, \ldots \min(n, m)\}$ such that for $1 \leq i \leq k$, $T_i$ and $U_i$ are related, and either $k \in \{m, n\}$, or $T_{k+1}$ and $U_{k+1}$ are not related. In other words, we take the longest common prefix of related types and therein for each pair of types their smallest common super data type. For example, we have *lub* (city, state) = <STR, GEO, INT>. If two object types have really "nothing common" then their smallest common supertype is ◊.

***Structures.*** With the definition of object types we have just described object structures, but not yet created any "containers" to keep objects. This distinction corresponds to the declaration of types and variables in programming languages. To keep objects, there are two types of *structures* available, namely *relations* and *graphs.*

In principle, a *relation*[4] is simply a collection of objects. A relation R has a schema, which is an object type O, and all objects in R conform to O. Any object that conforms to O can be inserted into R. Hence a relation can now hold a quite inhomogeneous collection of object tuples; these are considered, however, under the unifying view of the relation schema. For example, we might create relations:

**create relation** cities **of type** city;<br>
**create relation** countries **of type** state;<br>
**create relation** fed_states **of type** state;

[4] Perhaps another term would be more appropriate. However, we would like to emphasize that these collections can still be manipulated in the same simple way as relations in a query language.

For any two relations their "generalizing union" can be formed as a query language operation; the result relation's schema will be inferred to be the smallest common supertype of the two operand schemas. The well-defined schema of the result relation can be used as the basis for further operations in the same query. This solves the issue of "generalization queries", the second major problem mentioned in the introduction. For example, consider the query "Which streets, sewage canals, power, and telephone lines are affected by a planned highway section?" Suppose there are relations *streets, sewage_canals, power_lines,* and *phone_lines* all declared to be of types that are subtypes of <STR, LINES> and the planned highway section exists as an atomic data object called *track* of type LINES. Then we can formulate the query as:

<streets, sewage_canals, power_lines, phone_lines> **union**
**select** [route[5] **intersects** track]

In a textual representation of the result relation all attributes of all objects will be shown that can be textually represented. In a graphical representation all lines intersecting the track will be shown. The operations used here have the following functionality:

| | | | |
|---|---|---|---|
| $\text{REL}^+$ | $\rightarrow$ | REL | **union, select** |
| LINES × LINES | $\rightarrow$ | BOOL | **intersects** |
| GEO × REGIONS | $\rightarrow$ | BOOL | **inside** |

The notation $X^+$ denotes a list of one or more operands of type X. Such a list is written in brackets if it has more than one element, otherwise brackets may be omitted. The **select** operator performs a selection on the cartesian product of all its operands; hence it can be used much like the select-from-where in SQL. The **inside** operator is used in the next query "Show cities and phone lines in Bavaria!" Let us assume Bavaria has already been determined as one particular state object.

<cities, phone_lines> **union select** [_2 **inside** Bavaria.region]

Here the relation resulting from the union has schema <STR, GEO> since GEO is the smallest supertype of POINTS and LINES. *Bavaria.region* denotes a value of type REGIONS, hence the **inside** operator is applicable.

Note that whether cities, streets, phone lines etc. are represented physically separately or in a single relation representation structure (that might by clustered by location) is a matter of physical design; it is of no concern at the modeling level.

[5] The second attribute may have different names in the operand relations to the union. It is assumed that in at least one of them it is named "route". One may also refer to the n-th attribute by writing "_n", for example "_2".

However, at the modeling level one may introduce views such as

**define view** supply_lines
**as** <gas_lines, power_lines, phone_lines, sewage_canals> **union**

and then simply refer to *supply_lines* in a query.

The second type of structure available to keep objects is a *graph*. A graph has three kinds of components, namely *nodes*, *edges*, and *paths*. Here a path is not just any path induced by the edges that are present, but rather an *explicit path*, a specific entity within the graph that we wish to keep information about. A graph may in general consist of any number of *component types*, that is, it may have several *node types*, *edge types*, and *path types*. Each edge type involves two node types of the graph and may be directed or undirected. A path type has associated a regular expression over edge types of the graph. Each component type in a graph has at most one associated object type (as defined above). If component type C has an associated object type O, then each C component (for example, a C node) has precisely one O object associated.

To illustrate this let us consider the modeling of a highway network. Junctions and exits will be nodes of the highway graph, edges correspond to route sections. Named highways, such as "A1" or "Interstate 90" are paths over this network. We first define object types and then create a graph over these types.

| | | |
|---|---|---|
| **define object type** location | = | pos: POINTS; |
| **define object type** arc | = | route: LINES; |

The first two are general types for the modeling of any spatial network. These are now refined for highways. It is assumed junctions are identified by name, exits by number.

| | | |
|---|---|---|
| **define** location **type** junction | = | name: STR; |
| **define** location **type** exit | = | no: INT; |
| **define** arc **type** hw_section | = | no_lanes, top_speed: INT; |
| **define object type** highway | = | name: STR; |

**create graph** highways **as**
  **nodes** location,
  **edges** hw_section **between** location **and** location,
  **paths** highway **is** hw_section*;

Suppose we wish to find a highway connection from our home to some target location. We assume that *home* and *target* are values of type POINTS already known to the system. We proceed (simplifying a little) by finding the nearest highway exits to the home and target locations and then searching for a shortest path between these two exits.

Since finding paths is a frequent application, there is a special **shortest_path** operator in the query language algebra. It needs two parameter functions to be supplied. The first is needed to determine the cost of the path; this function must be applicable to an edge of the operand graph and yield a REAL value. Since **shortest_path** will be implemented by the A* algorithm (see [Ni80]) which performs a heuristic search, a second parameter function specifying the heuristic must be given. This function takes as operands two nodes of the graph and must return a REAL value. The idea is that this function estimates the "distance" from the given node to the goal node. Fortunately, in a *spatially embedded* network a quite good heuristic function is available, namely the Euclidean distance. The **shortest_path** operator has functionality

| | | | | |
|---|---|---|---|---|
| GRAPH × NODE × NODE | → | PATH | **shortest_path** | _ _ _# |

and parameters

[pf: PATHFORM, cost: EDGE → REAL, distance: NODE × NODE → REAL]

A PATHFORM is a regular expression over edge types; in a graph with multiple edge types it can be used to restrict the paths allowed as a result of the search. Parameters are named and can be omitted if they are not needed. In this case the PATHFORM is not needed since the *highway* graph has only a single edge type. It is also possible to set default values for parameters to be used in subsequent queries, for example:

**default** shortest_path **is** cost(e) = **length** (e.route),
distance (p, target) = **dist** (p.pos, target.pos);

Here the operators

| | | | | |
|---|---|---|---|---|
| LINES | → | REAL | **length** | #( _ ) |
| POINTS × POINTS | → | REAL | **dist** | #( _ ) |

have been used. The query can then be written as follows:

highways
(highways.exit home **closest** [pos])
(highways.exit target **closest** [pos])
**shortest_path**

The notation "<graph name>.<object type>" allows one to restrict attention to one particular class of objects within a graph, hence *highways.exit* denotes the class of components in the graph that have type *exit*. A class of objects so identified can be used like a relation. The **closest** operator takes as a first operand a collection of objects that have a POINTS attribute and as a second operand a POINTS value and returns the object that is closest to the value.

The parameters of the **shortest_path** operator can now all be omitted[6]. The result of the query is a PATH object in memory which, for example, can be displayed against the background of the highway network. There are also further operations available on PATH objects.

Finally we introduce the notion of a *complex graph:* It is sometimes useful to establish connections between objects in a graph that are themselves edges or paths rather than nodes. We allow to define such edges, for example, an edge between a path and a node. If this occurs, we call the resulting graph *complex.* For example, in a river network one might want to represent the fact that a river flows into another river explicitly by an edge, and define:

**create graph** rivers **as**
  **nodes** location,
  **edges** arc,
  **paths** river **is** arc*,
  **edges** flows_into **from** river **to** river;

It is assumed that object type *river* has been defined previously. In this graph, river objects are associated with paths over a network consisting of river segments as edges and river junctions as nodes. Hence "flows_into" is a directed edge connecting two path objects.

Note that the applicability of graphs is not limited to the modeling of spatial networks. We have now in fact a hybrid model with graphs and relations and the user has a choice of whether links between objects are established *by value*, in relational style, or through edges. For example, look at the following modeling:

| | | |
|---|---|---|
| **define object type** book | = | title, publisher: STR,<br>year: INT; |
| **define object type** writer | = | name, initials: STR, country: STR; |

**create graph** literature **as**
  **nodes** book, writer,
  **edges** wrote **from** writer **to** book;

The data model intended for Gral II has only been sketched here; it will be described in more detail in [Gü92]. A functional data model that also offers explicit graphs and integrates graph operations has been developed in [ErG91].

Some questions about the model and its implications are still open. We felt that to have a more solid basis for further modeling decisions we should have a better understanding of implementation options, and have started by studying graph representations. This is the subject of the following section.

---

[6] If there were no defaults specified, one would have to write
"... **shortest_path** [cost(e) = **length** (e.route), distance (p, target) = **dist** (p.pos, target.pos)]"

# 3. A Basic Graph Representation

In this section we develop a simple multi-purpose graph representation and show how it can be implemented within the given extensible architecture of the Gral system. But before we start, it is important to understand what it means at the physical level to query a graph.

## 3.1 Simple Graphs, Layered Graphs, and the Query Concept

The graph representation will be developed in two steps. In a first step we describe a structure to represent graphs that are not *complex graphs* in the sense defined at the end of Section 2; they have no edges between higher level objects. We call this a *simple graph (representation).* In a second step we introduce a further representation called a *layered graph (representation)*, which is a collection of simple graph representations that are connected by external references. Layered graphs have a two-fold motivation:

(i) They are used to represent complex graphs.

(ii) During evaluation of a query, temporary graphs may be constructed that are subgraphs of, and connected with, other graphs in the database. Layered graphs allow to isolate these subgraphs as somewhat independent components.

The first aspect is illustrated by the complex graph modeling a river network from Section 2. This would be represented in a layered graph by one simple graph representation with "meeting" nodes, "arc" edges, and "river" paths, and a second simple graph representation whose nodes are references to rivers in the first graph and whose edges represent the "flows_into" relation.

The second aspect deserves further study. The question arising is what it really means to query a graph. The simple initial assumption is that a query is a mapping that takes one or more graphs $G_1, ..., G_m$ as input and produces a new graph G' as a result. Difficulties arise when G' is not a completely independent graph but has links with graphs in $\{G_1, ..., G_m\}$.

Consider a query $q = q_o q_r$ which consists of two operations $q_o$ and $q_r$ that are applied to a given database D. $q_o$ maps D to D' and $q_r$ maps D' to D". The following requirements need to be fulfilled.

(1) A query does not change the database.

(2) Suppose $q_o$ constructs a temporary addition to D called T so that $D' = D \cup T$. The remainder of the query, $q_r$, must be allowed to operate on the whole database D'.

It is difficult to reconcile both requirements when D is a graph. (1) says that D should not change by the evaluation of a query. More precisely, if we view query q as a transaction, then after q has finished, D should appear unchanged, and while q is executed, it should appear unchanged to all transactions other than q. Requirement (2) says that during evaluation of q the database must change. Namely, if $q_o$ computes an addition T there must not only be pointers from T into D but also from D into T to let it all appear as a single graph to $q_r$.

This problem does not arise in relational databases. There, D as well as T are collections of relations. Relations in T may contain *values* that may be interpreted as pointers into D, but at the same time, values in D may also be used as pointers into T. Hence the relational model maintains implicitly pointers between the graphs D and T (both represented by sets of relations). In contrast, we need to maintain these pointers explicitly in our graph representations.

Our solution to these problems are *layered graphs* and *time stamps* in external references. Consider the example of a join[7] operation that establishes edges between graphs $G_1$ and $G_2$, say edges (b, 1) and (d, 4), as in Figure 2.

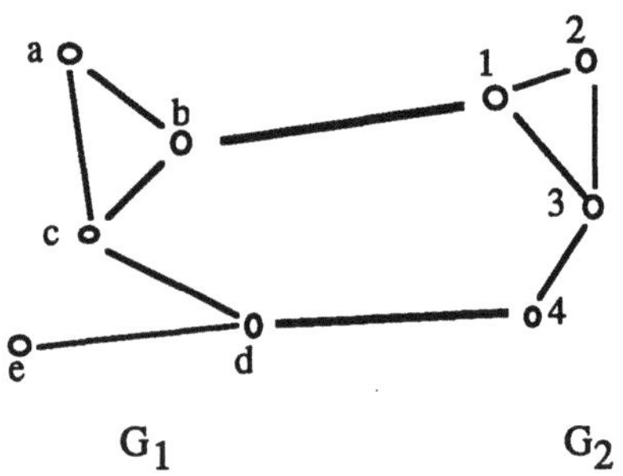


Figure 2

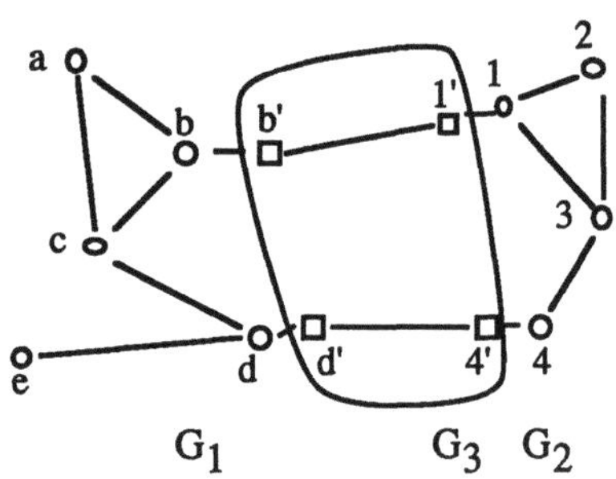


Figure 3

This is implemented by creating a new simple graph $G_3$ that has a special type of nodes (called *noderef*) as representatives for the objects in $G_1$ and $G_2$ that it connects, as shown in Figure 3 (where noderefs are shown as squares).

Now let us assume that the join operation constructing $G_3$ corresponds to the first part $q_o$ of query q. Then $G_3$ is the addition T constructed by $q_o$. We assume that query q receives a transaction number $n_q$ before execution. It uses this number as a time stamp within the pointers that it creates in b to b', d to d', etc. All subsequent or even simultaneous queries will have different transaction numbers and consider these pointers as invalid. Hence this scheme achieves all the stated goals:

- The remainder of the query $q_r$ can view $G_1$-$G_3$-$G_2$ as a single graph.
- The database appears unchanged after termination of q.
- It appears unchanged during execution of q to all other transactions.

On the other hand, when layered graphs are used to implement complex graphs of the data model, time stamps indicate that external pointers are valid permanently and to all queries.

---

[7] What is shown in this example will be the natural meaning of a join operation in our graph data model.

### 3.2 System Environment

It is our goal to describe not only in an "abstract" way a storage structure for graphs but also to show how it can be embedded into the architecture of an extensible database system and how that environment offers tools that make the implementation relatively easy. Let us briefly review Gral's extensible architecture that has been explained in some detail in [Gü89]. There it was represented as shown in Figure 4.

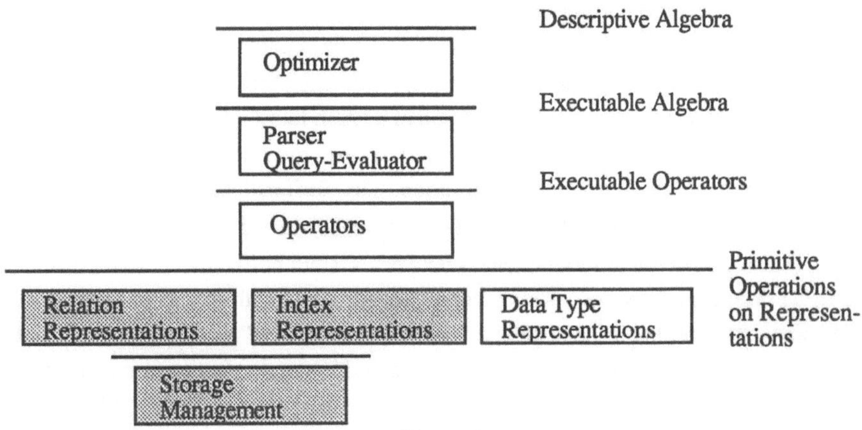


Figure 4

The second layer from the bottom deals with representations and it is this layer that we shall extend by a graph representation module. Note that the next layer above implements so-called *executable operations*, such as a sort-merge join, or a polygon intersection test. That layer will later be the proper place for the implementations of graph algorithms such as A*. For now, let us look at the structure of the three shaded boxes in more detail.

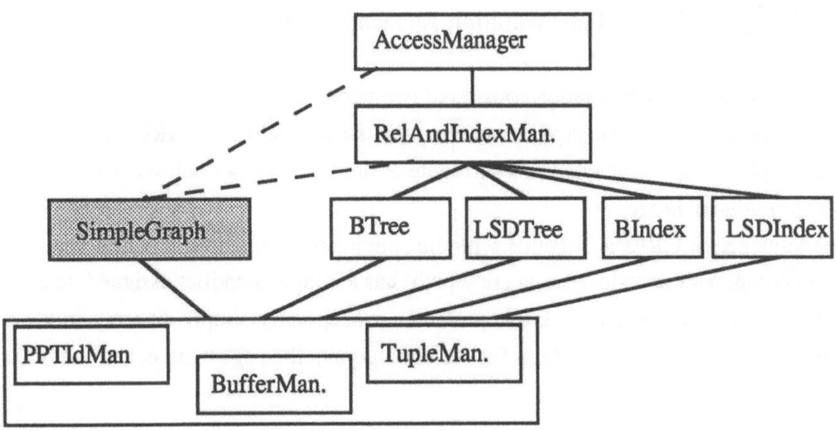


**Figure 5**

So far, primary and secondary index structures for relations are implemented on top of a group of modules comprising the BufferManager, the TupleManager, and the PPTIdManager, whose services are described below. The graph representation will be realized in a module "SimpleGraph" that also uses the services of these three modules. The RelAndIndexManager offers a uniform interface to the extensible collection of index structures. For example, an executable *scan* operator is implemented once on top of this interface and works on all relation representations. The AccessManager offers a "complete" view of a relation. For example, it offers an operation to insert a tuple and makes sure that corresponding entries into all indexes for the relation are made.The dashed lines in Figure 5 are meant to indicate that later we intend to connect the graph representation module to the RelAndIndexManager and AccessManager so that it is possible to access a collection of objects stored in a graph like a relation.

The services offered by the three underlying modules are crucial for the implementation of the SimpleGraph structure. The BufferManager, as usual, offers storage space in the form of *segments*, a segment can be viewed as an array of pages. Of course, it also takes care of moving pages between system buffer and external memory. The PPTIdManager provides various techniques for tuple addressing. In particular, it implements the "probable page pointer" or "hybrid pointer" concept (see [Chan82, LoS87]) that we shall use here to realize pointers within graphs. But most interesting for our purposes here is the TupleManager (TM). It performs two main tasks: (i) It hides the internal structure of a tuple, and (ii) it constructs a tuple representation as a bytestring that is guaranteed to be smaller than half a page. In Gral, a tuple represents a list of attribute values that are data objects of arbitrary size, hence a tuple representation could in principle be very large. To make sure a tuple representation is smaller than half a page, the TM constructs a *reduced* tuple by moving large representations of attribute values out[8] of the tuple string into their own page sequences. The BufferManager allows to access these values through special buffers.

### 3.3 Managing Graph Representation Elements (GREs)

The simple graph representation consists of two parts:

- a *sequence of graph representation elements* (*GREs*, a GRE represents, for example, a node or an edge or a surrogate of one of these, see below), and
- a *component index.*

The sequence of GREs is, together with the component index, stored in a segment. Each graph component (a node, an edge, or a path) has a unique identifier assigned on creation of the component. The *component index* is conceptually simply an array indexed by component identifier that contains for each component the page number and the index on

[8] More precisely, when an attribute representation is moved out, an *info* part is left back. For example, for a polygon representation the info part contains the bounding box and the area.

the page[9] where the component is stored. We first treat *graph representation elements* as black boxes, assuming a GRE is represented within a bytestring small enough to fit into one page. It contains in general pointers to other GREs. To implement a sequence of GREs, the following layers of abstraction (Figure 6) are built.

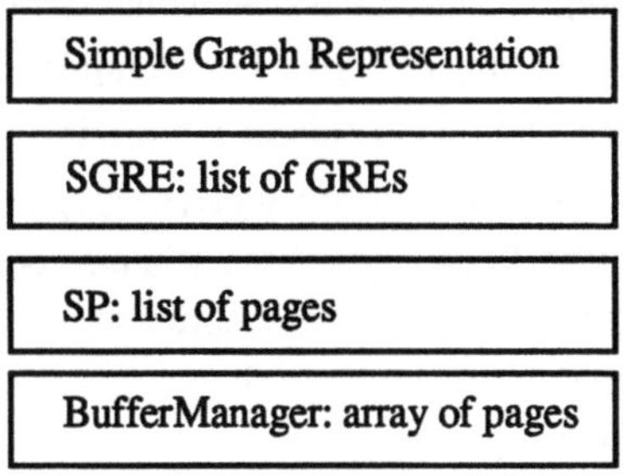


Figure 6

The SP layer offers a doubly linked list of pages on top of the BufferManager with operations to traverse the list and to insert and delete pages within the list. The SGRE layer embeds a list of GREs into a list of pages. It allows to insert or delete GREs at any place within the list of GREs and maintains a page filling > 50 % by inserting additional pages on overflow into the list of pages and by merging adjacent pages on underflow. We now turn to the top layer.

### 3.4 Structure of GREs

Let us consider the internal structure of GREs. There are six different types of GREs:

| | |
|---|---|
| - nodeobj | - noderef |
| - edgeobj | - edgeref |
| - pathobj | - pathref |

Since they are closely related, we shall also consider three further representations, namely those of NODE, EDGE, and PATH objects that may be present in internal memory (NODE, EDGE and PATH are sorts of the algebra, see for example the definition of the **shortest_path** operator in Section 2). GRE representations contain more elementary representation units called "ref", "eref", and "schema" that need to be explained first.

***References (ref).*** A ref object represents a pointer *within* a given graph. It combines through the PPP or "hybrid pointer" concept a logical and a physical pointer to some GRE in the graph. Hence it consists of an object ID, which can be used as an index to the component index table, and of a TID.

***External References (eref).*** An external reference is a pointer to a component of some arbitrary graph in the given database. It has the structure:

<eref> ::= <graph-id> <time-stamp> <ref>

[9] Those two together form what is known as a "tuple identifier" (TID).

We use regular expressions and a grammar-like notation to describe the structure of sequences, in particular of byte strings. Here "<ref>" denotes the byte string representing a "ref" object, and the structure of an "eref" object is defined to be the concatenation of the representations of a graph identifier, a time-stamp, and a reference.

***Schema Codes (schema).*** It is useful to keep within an object representation (which is a tuple) a description of the object's type. For example, one might want to scan a relation or a graph restricting attention to objects conforming to a certain type. A *schema code* is a compact representation of a relation, tuple, or graph component schema. For an object type $<T_1, \ldots, T_n>$ (the $T_i$ data types) the schema code is a byte string $b_1 \ldots b_n$. Each data type including generalization types is represented by some number in the range 1..255. It is not difficult to devise a coding scheme that represents the generalization hierarchy in such a way that the operations

| | | | |
|---|---|---|---|
| schema × schema | → | schema | object_supertype |
| datatype × datatype | → | bool | related |
| datatype × datatype | → | datatype | data_supertype |

can be implemented very efficiently. The numbers assigned to data types are hidden in a module SchemaCode that exports these operations. Extensibility by data types is maintained because changes in the coding scheme are not visible outside the SchemaCode module.

## GRE Structure

The structure of the six kinds of GREs and of NODE, EDGE, and PATH objects is shown in Table 1. All GREs have <length> and <kind> fields where <length> is the total number of bytes of the GRE, needed for GRE management, and <kind> encodes one of the six GRE types. Each of the first three kinds of GREs represents a component of a graph and the object associated with it (if this component type has an associated object type). Such GREs then have the unique object identifier of this node, edge, or path, then some references, then the schema code of this object and the list of attributes.

<nodeobj>::= <length><kind><obj-id><eref> <schema><$a_1$>...<$a_n$>

<edgeobj>::= <length><kind><obj-id><eref><ref><ref><schema><$a_1$>...<$a_n$>

<pathobj>::= <length><kind><obj-id><eref><ref>$^+$ <schema><$a_1$>...<$a_n$>

<noderef>::= <length><kind> <eref><eref><eref>

<edgeref>::= <length><kind> <ref> <ref> <ref>

<pathref>::= <length><kind> <ref>

<NODE> ::= <eref>

<EDGE> ::= <eref><ref> <ref>

<PATH> ::= <eref><ref>$^+$

Table 1

Considering the references within Xobj GREs let us for a moment ignore the external references; they are explained below. An edgeobj GRE has pointers to its source and target node GREs. A pathobj GRE contains a list of pointers <ref>$^+$ that point to the Xobj GREs of nodes and edges of the path $v_0e_1v_1 \dots v_{n-1}e_nv_n$.

Each Xref GRE is essentially a representative of some corresponding Xobj GRE. An edgeref GRE represents an edge object GRE within this graph. Its first <ref> pointer points to that edge object GRE, the other two to the source and target node GREs. A pathref GRE represents some path object GRE; its single pointer points to that object.

Noderef GREs play a special role, since they represent Xobj GREs *in other graphs*. They are used to implement layered graphs as described in Section 3.1.

The three <eref> pointers in noderef GREs and the single <eref> pointer in nodeobj, edgeobj, and pathobj GREs are provided to establish these external connections. For an Xobj component c in a graph G there may be any number of other simple graphs having noderef representatives of c. Two of the eref pointers in a noderef GRE are used to form a doubly linked list of all such representatives of c. The third eref pointer points to the object c itself to allow direct access to it from any graph. On the other hand, each Xobj GRE c has an eref pointer. If there are any noderef representatives of c, then this eref pointer points to the first of them.

Finally, a NODE value in internal memory is just a pointer to some node object GRE of some graph; note that it specifies which graph the node belongs to. Similarly an EDGE value points to an edge object GRE in some graph and additionally to its source and target nodes in the same graph, a PATH value to some path object GRE and additionally to all node and edge objects within the path. There are various purposes underlying the details of this design. For example, given two PATH values in memory, one can check whether they have common nodes without accessing the underlying graph. One can easily materialize a node, edge, or path from a NODE, EDGE, or PATH value by accessing the graph. Xref GREs have a fixed length and the structure of references in node, edge, and path object GREs matches the structure of NODE, EDGE, and PATH values. This allows the following simple implementation.

### Implementation of GRE Structure in the Gral Environment

NODE, EDGE, PATH, and SCHEMA (to represent schema codes) will be introduced as data types through the standard data type extension facility. For Xobj GREs the SimpleGraphManager (SGM) will itself construct the prefix <length><kind>. For the remainder it builds a list of attributes and lets the TupleManager construct a representation for it. For example, for a nodeobj GRE the data types of these attributes are as shown below:

nodeobj: <length> <kind> <obj-id> <eref> <schema> <a1> ... <an>

INT NODE SCHEMA $T_1$ $T_n$

bytestring by SGM

list of attributes for TupleManager (TM) bytestring constructed by TM

For edgeobj and pathobj GREs their sequence of references fits with types EDGE and PATH. Note how the TupleManager solves once and for all the problem of representing very large objects, including the handling of very long paths within pathobj GREs. So the SGM can easily guarantee that Xobj GREs are smaller than a page, since the tuple bytestring constructed by the TupleManager is even smaller than half a page. Note also that for example a PATH object in memory can be obtained very efficiently from its pathobj GRE by copying the PATH attribute value (from the point of view of the Tuple Manager) to memory or by fixing it in the buffer. It is only necessary to change the <eref> component within the PATH representation, since it has another meaning there (in the PATH value it points to the pathobj GRE whereas in the pathobj GRE it possibly points to some component of another graph).

The noderef, edgeref, and pathref GREs are constructed entirely by the SGM. This is no problem since each of these GREs is small and has a fixed length.

### 3.5 Representing a Graph by a Sequence of GREs

The purpose of the graph representation is to make sure that connections among the three component classes - nodes, edges, and paths - can be followed efficiently. In detail, the following connections should be supported:

(a) from a node to its outgoing edges
(b) from a node to its incoming edges
(c) from a node to the paths containing it
(d) from an edge to its source and target nodes
(e) from an edge to the paths containing it
(f) from a path to all nodes and edges in it

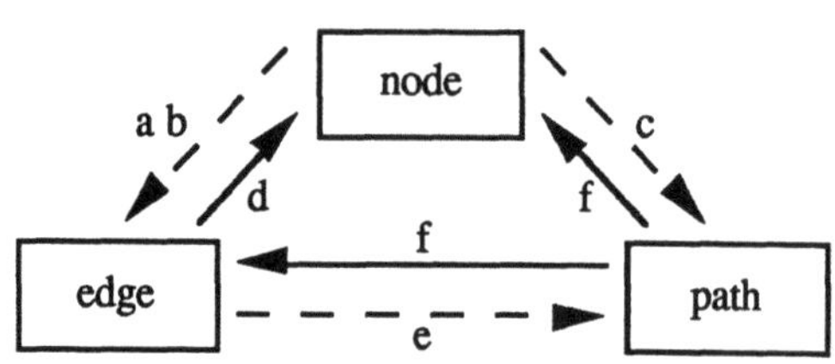


Figure 7

Some of these connections are maintained through pointers within GREs (shown as simple arrows in Figure 7). The other connections are established by placing edgeref or pathref GREs into the GRE sequence and by the ordering of the sequence (shown by dashed arrows). For these reasons the GRE sequence is layed out as follows:

<gre-sequence> ::= <node-block>*
<node-block> ::= <nodeobj GRE>
<edgeobj GRE>* <edgeref GRE>*
<pathobj GRE>* <pathref GRE>*

So a graph is represented by a "gre-sequence" which is a list of "node-blocks". This list may be empty in which case the whole graph is empty. A node block starts with a GRE representing a node object. With this node are clustered the outgoing edge objects and those path objects for which this node is the start node. Hence after the nodeobj GRE follow the edgeobj GREs of its outgoing edges. The incoming edges are obviously clustered with *their* start node, hence they are represented here by edgeref GREs. After that follow the pathobj GREs for which this node is the start node and then in the form of pathref GREs all paths containing this node elsewhere.

This scheme allows to follow all the connections mentioned above relatively efficiently. For example, given a node, one scans its node block to find all outgoing edges. All objects are stored only once. The clustering of edges and paths with their nodes minimizes the number of page accesses when some local part of a large graph is examined. Furthermore, the order of node blocks can be chosen freely. In a spatial database system, when node objects have a POINT attribute, node blocks as a whole can be clustered by location through a spatial index structure.

## 4. Conclusions

The original purpose of this paper was to describe an external data structure for the representation of graphs. To motivate the details in the design of that data structure it was necessary to give a fair amount of background on the intended data model to be supported by it. A more thorough exposition of the data model itself will be given in [Gü92]. In summary, we feel this paper presents a promising approach to move from a relational system with spatial data types to one that also adequately models and implements spatial networks and object class hierarchies.

Some interesting open issues are the following:

- Design an interface for the SimpleGraphRepresentation module that provides primitive pointer and access operations.

- Extend the next higher layer, the executable algebra (see [Gü89, BeG90]) by graph operations. For example, the A* implementation of **shortest_path** would be such an operation.
- Develop query processing strategies for the mapping of "descriptive" graph operations in the query language to executable operations. Implement these strategies by optimization rules.
- Develop design criteria that allow to decide how a large graph defined at the level of the data model should be split into parts, and thus be represented by a layered graph, at the implementation level.

**Acknowledgements.** Thanks to Heidi Wendland and Martin Erwig for many interesting discussions on the modeling of graphs in databases.

## References

[Ab89] Abel, D.J., SIRO-DBMS: A Database Tool Kit for Geographical Information Systems. *Intl. J. of Geographical Information Systems 3 (1989)*, 103-116.

[Ag87] Agrawal, R., ALPHA: An Extension of Relational Algebra to Express a Class of Recursive Queries. Proc. IEEE Data Engineering Conf. 1987, 580-590.

[BeG90] Becker, L., and R.H. Güting, Rule-Based Optimization and Query Processing in an Extensible Geometric Database System. Informatik-Bericht 97, FernUniversität Hagen, Revised Version, September 1990. To appear in *ACM Transactions on Database Systems.*

[BiRS90] Biskup, J., U. Räsch, and H. Stiefeling, An Extension of SQL for Querying Graph Relations. *Computer Languages 15 (1990)*, 65-82.

[Chan82] Chan, A., S. Danberg, S. Fox, W.K. Lin, A. Nori, and D. Ries, Storage and Access Structures to Support a Semantic Data Model, Proc. of the 8th Intl. Conf. on Very Large Data Bases, 1982, 122-130.

[CrMW87] Cruz, I.F., A.O. Mendelzon, and P.T. Wood, A Graphical Query Language Supporting Recursion. Proc. SIGMOD Conf. 1987, 323-330.

[CrMW89] Cruz, I.F., A.O. Mendelzon, and P.T. Wood, $G^+$: Recursive Queries Without Recursion. Proc. 2nd Intl. Conf. on Expert Database Systems, 1989, 355-368.

[Eg89] Egenhofer, M.J., Spatial SQL: A Spatial Query Language. Dept. of Surveying Engineering, U. of Maine, Tech. Report 103, 1989.

[EgF88] Egenhofer, M.J., and A. Frank, Towards a Spatial Query Language: User Interface Considerations. Proc. 14th Intl. Conf. on Very Large Databases, Los Angeles, 1988, 124-133.

[EgF89] Egenhofer, M.J., and A. Frank, Object-Oriented Modeling in GIS: Inheritance and Propagation. Proc. AUTO-CARTO 9, Baltimore, MD, 1989, 588-598.

[ErG91] Erwig, M., and R.H. Güting, Explicit Graphs in a Functional Model for Spatial Databases. Informatik-Bericht 110, FernUniversität Hagen, June 1991.

[Fr82] Frank, A., MAPQUERY: Data Base Query Language for Retrieval of Geometric Data and their Graphical Representation. *Computer Graphics 16 (1982)*, 199-207.

[Gü88] Güting, R.H., Geo-Relational Algebra: A Model and Query Language for Geometric Database Systems. Proc. of the Intl. Conf. on Extending Database Technology, Venice, March 1988, 506-527.

[Gü89] Güting, R.H., Gral: An Extensible Relational Database System for Geometric Applications. In: P.G. Apers and G. Wiederhold (eds.), Proc. of the 15th Intl. Conf. on Very Large Databases, Amsterdam, 1989, 33-44.

[Gü92] Güting, R.H., A Spatial Relational Data Model with Graphs, Tuple Hierarchies, and Realms. Praktische Informatik IV, FernUniversität Hagen, Manuscript in preparation, 1992.

[GüSV90] Güting, R.H., M. Scholl, and A. Voisard, Personal communication, January 1990.

[GüZC89] Güting, R.H., R. Zicari, and D.M. Choy, An Algebra for Structured Office Documents. *ACM Transactions on Office Information Systems 7 (1989)*, 123-157.

[HeSW89] Henrich, A., H.-W. Six, and P. Widmayer, The LSD Tree: Spatial Access to Multidimensional Point- and Non-Point-Objects. Proc. of the 15th Intl. Conf. on Very Large Data Bases (Amsterdam, 1989), 45-53.

[JoC88] Joseph, T., and A. Cardenas, PICQUERY: A High Level Query Language for Pictorial Database Management. *IEEE Trans. on Software Engineering 14 (1988)*, 630-638.

[LoS87] Lockemann, P.C., and J.W. Schmidt (eds.), Datenbank-Handbuch. Springer, 1987.

[LiN87] Lipeck, U., and K. Neumann, Modelling and Manipulating Objects in Geoscientific Databases. Proc. 5th Intl. Conf on the Entity-Relationship Approach (Dijon, 1986), 1987, 67-86.

[Ni80] Nilsson, N.J., Principles of Artificial Intelligence. Springer, 1982.

[OoSM89] Ooi, B.C., R. Sacks-Davis, and K.J. McDonell, Extending a DBMS for Geographic Applications. Proc. Data Engineering Conf. 1989, 590-597.

[Or86] Orenstein, J.A., Spatial Query Processing in an Object-Oriented Database System. Proc. SIGMOD Conf. 1986, 326-336.

[Or90] Orenstein, J., A Comparison of Spatial Query Processing Techniques for Native and Parameter Spaces. Proc. SIGMOD Conf., Atlantic City, 1990, 343-352.

[OrM88] Orenstein, J., and F. Manola, PROBE Spatial Data Modeling and Query Processing in an Image Database Application. *IEEE Trans. on Software Engineering 14 (1988)*, 611-629.

[Rose86] Rosenthal, A., S. Heiler, U. Dayal, and F. Manola, Traversal Recursion: A Practical Approach to Supporting Recursive Applications. Proc. SIGMOD Conf. 1986, 166-176.

[RoFS88] Rossopoulos, N., C. Faloutsos, and T. Sellis, An Efficient Pictorial Database System for PSQL. *IEEE Trans. on Software Engineering 14 (1988)*, 639-650.

[ScV89] Scholl, M., and A. Voisard, Thematic Map Modeling. Proc. 1st Intl. Symp. on Large Spatial Databases, Santa Barbara, 1989, 167-190.

[ScV91] Scholl, M., and A. Voisard, Object Oriented Database Systems for Geographic Applications: An Experience with $O_2$. To appear in: F. Bancilhon, C. Delobel, and P. Kanellakis, The $O_2$ Book. Morgan Kaufman, 1991.

[ShSN90] Shaffer, C., H. Samet, and R.C. Nelson, QUILT: A Geographic Information System Based on Quadtrees. *Intl. J. on Geographical Information Systems 4 (1990)*, 103-131.

# A Model for Complex Data: Completeness and Soundness Properties[§]

Marinella Gargano
Enrico Nardelli
Maurizio Talamo

In this paper we investigate formal properties of the logical data model for complex data introduced in [9]. We show that the *geometrical content* of the relations calculated by expressions of the extended relational algebra defined in [9] coincides with a boolean lattice. In particular we prove that (i) every relation derivable by an expression of the extended relational algebra corresponds to a subset of this lattice; (ii) every element of the lattice corresponds to an expression of the extended relational algebra.

## 1. Introduction

In these latter two decades application areas characterized by more complex and varied data with respect to traditional applications, as for example office automation, interactive computer aided design, geographic data processing, health data processing etc., have been explored.

A peculiar characteristic of these areas is the simultaneous treatment of traditional (i.e., numerals and strings) and spatial/geometric data.

In the field of data modeling, this common feature has led to focus research work on representation and manipulation of complex information, i.e. information having both a descriptive and a spatial/geometrical nature. Much work has also been done in extending traditional models to represent and manipulate spatial/geometrical information.

In particular, many researchers have worked on extending the relational model [4, 5, 10, 11, 13, 8, 9, 19]. The relational model is in fact interesting for its validated theoretical basis and its effectiveness, but is not fully able to support an effective representation and an efficient manipulation of complex information.

§ This work has been partially supported by the 'Progetto Finalizzato Sistemi Informatici e Calcolo Parallelo' of the italian National Research Council, within the research action 'MULTIDATA'.

This is mainly due to two facts: (i) attributes defined on traditional atomic domains cannot adequately represent the spatial/geometric component of the data, (ii) consequently, the relational operators used to manipulate complex information (represented by traditional domains) do not support the definition of a high level query language and cannot be translated into efficient operations at the physical level [2, 18, 16, 8, 9].

A widely followed stream of research has tackled the problem of extending the relational model by introducing abstract data types as domains [20, 11, 13, 15, 8, 9]. To our knowledge, these models are typically characterized by defining their operations in terms of some algebras, but do not deal with formal properties like, for example, their completeness and soundness with respect to the reality of interest.

In [9] we defined a logical model that allows the modeling of complex information, i.e. the modeling of information having both a descriptive and a spatial/geometrical nature.

The model rests on the relational model, which we have suitably extended with the definition of an Abstract Data Type, namely Geometry(S), for dealing with geometric nature of data. The model (i) furnishes an expressive and powerful means for integrated modeling of spatial/geometric data, (ii) provides primitives that directly model the kind of manipulation that the end-user performs on such data, (iii) constitutes a basis for the formal and correct definition of algebraic high level data manipulation and definition languages.

The main result of this paper is the precise characterization of the geometrical content of the expressions that can be written in the extended relational algebra defined in our model. In particular, starting from the geometrical content of the database, we individuate a set of basic geometrical objects that are no further decomposable. From these minimal objects we define the set of the derivable objects. This set results to be a boolean lattice. We are thus able to prove that (i) to each element of the lattice corresponds a suitable expression of the algebra that calculates a relation whose geometrical content is that element and (ii) any expression of the algebra calculates a relation whose geometrical content is a subset of the lattice.

The paper is organized as follows.

In section 2 the extended relational model and the extended relational algebra are presented. In section 3 an informal introduction has the purpose to get the reader acquainted with basic concepts and definitions. In section 4 the completeness theorems are stated and their proofs are given. Section 5 concludes the paper.

## 2. A Model for Complex Data

In order to make the paper self-contained, in this section we shortly present the model defined in [9]. An application of the model to the field of geographical databases is contained in [8].

The model rests on the definition of an Abstract Data Type, namely Geometry, to deal with geometric nature of data. This approach allows to define attributes of this type and therefore to model geometrical aspects of information on a well founded theoretical basis. It also permits to exploit the theoretical and practical knowledge acquired in the relational field by the database community. Moreover, the use of abstract data types leads to the definition of a richer and more powerful logical modeling tool [20, 7, 14, 6, 15].

In the first subsection we introduce an algebraic structure, SHAPES, defined on a set of atomic elements. In the next subsection we define our logical data model. The last subsection formally defines the manipulation operators of the introduced algebra.

### 2.1 SHAPES Algebra

Given a not empty finite set S, we denote by $H_S$ the set $H_S = P(P(S))$ where P(S) is the powerset of S (hence $H_S$ is the set of all sets of sets of elements of S). The elements of S are called *atoms*.

**Definition 1** (*) : Given a not empty finite set S, we call SHAPES on S the 6-tuple $\Psi_S = (H_S, \cup, \cap, \cap^*, geo, compl)$, where $\cap^*$, *geo*, and *compl* are three operators defined on $H_S$, for all A, B $\in H_S$, as it follows:

$A \cap^* B = \{ c \in P(S) : (\exists a)(\exists b)\ a \in A \ \wedge\ b \in B \ \wedge\ c = a \cap b \}$.

$geo(A) = \{ X \}$, where $X = \{ x \in S : a \in A \ \wedge\ x \in a \}$.

$compl(A,B) = \{ Y \}$, where $Y = \{ y \in S : a \in A \cup B \wedge y \notin a \}$.

We shall denote $compl(A,\emptyset)$ by $compl(A)$.

The operators $\cup$ and $\cap$ respectively denote the union and the intersection operations in the way they are used in set theory, i.e. :

$A \cup B = \{ c \in H_S : c \in A \ \vee\ c \in B \}$.

$A \cap B = \{ c \in H_S : c \in A \ \wedge\ c \in B \}$.

**Proposition 1.** $\Psi_S$ is an algebraic structure (briefly, an algebra).

**proof:** $\Psi_S$ is clearly closed under the operations $\cup$, $\cap$, $\cap^*$, *compl* and *geo*.

---

(*) The symbol $\cap$ is used both on the right and left side of the formulas to denote the usual operator of intersection. Though on the left side the elements are belonging to P(P(S)) and on the right one to P(S) we hope the difference is clear from the context.

An element A of $\Psi_S$ is called a Shape. If A contains exactly one element (i.e. a single set of atoms) it is called simple. Otherwise, each of its element (i.e. each one of the set of atoms it is made up by) is called A-component. Informally speaking, when the $\cap^*$ operator is applied to two Shapes A and B it returns a Shape C whose components are the result of the intersection of every A-component with every B-component. When the *geo* operator is applied to a Shape A containing more than one component it returns a simple Shape whose single component is the union of all the A-components. If *geo* is applied to a simple Shape A it returns Shape A itself. When *compl* operator is applied to two Shapes A and B it returns a Shape C whose unique component is made up by all atoms belonging neither to either components of A nor components of B. Figures 1a, 1b, ...,1f show in informal way the operators of the algebra $\Psi_S$ introduced in this section.

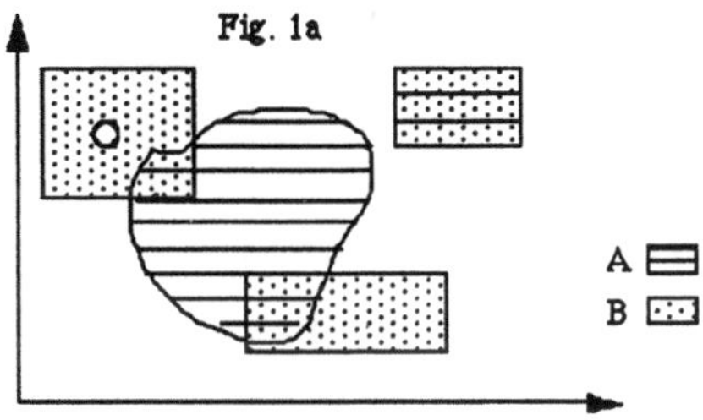

Fig. 1a

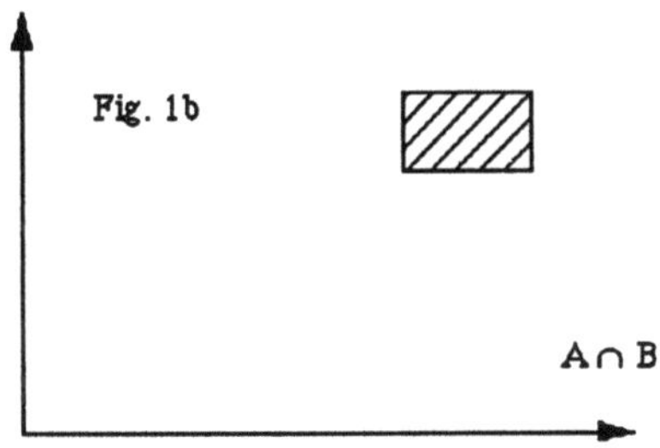

Fig. 1b

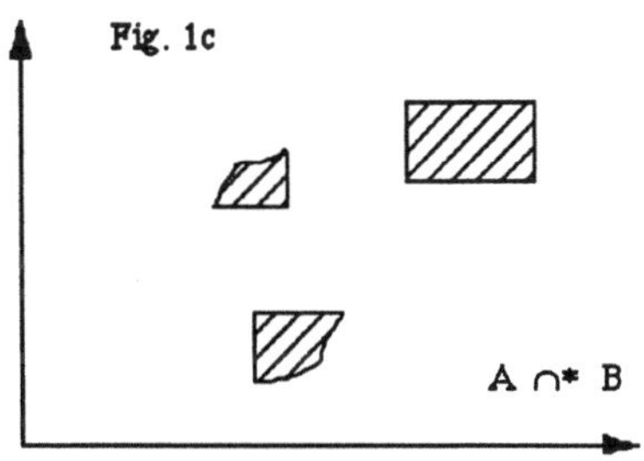

Fig. 1c

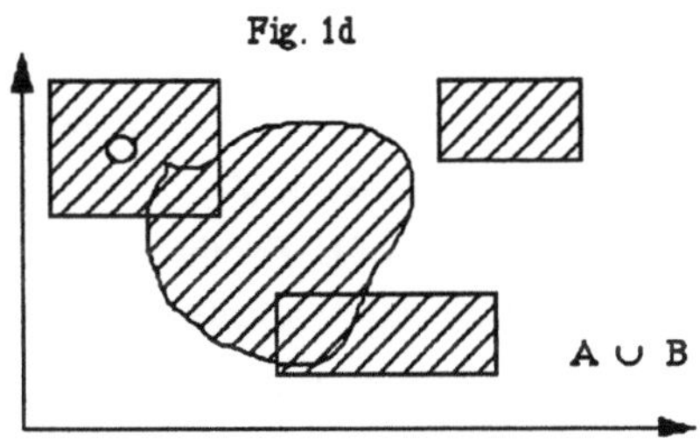

Fig. 1d

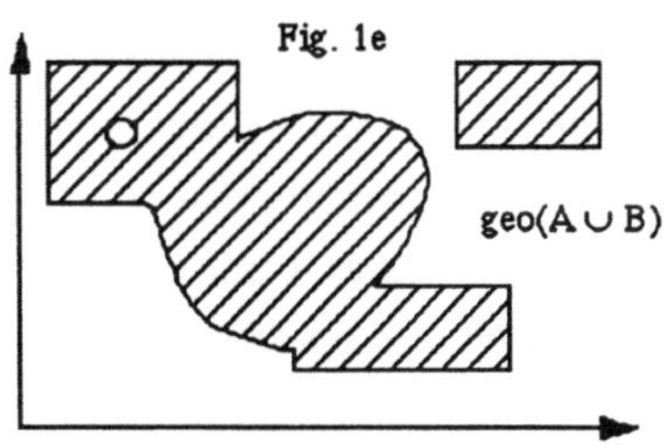

Fig. 1e

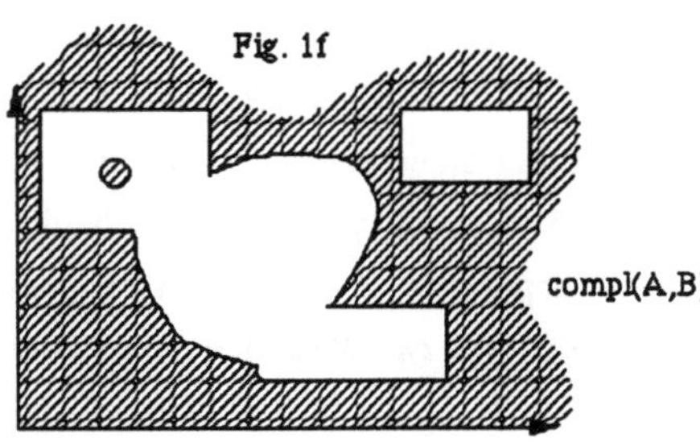

Fig. 1f

The introduction of an algebra corresponds to the definition of an ADT whose syntactical part is concerned with the signature of the algebra (operational syntax) and whose semantical part with its interpretation (operational semantics) [12]. The ADT we introduce for representing SHAPES Algebra is called *Geometry*(S) [9].

## 2.2 The Logical Data Model

In this subsection our logical data model is presented. The model supports the representation of SHAPES valued attributes (i.e. attributes whose value is an instance of the ADT Geometry).

Let U be a given set of names, let $S_1, \ldots, S_p$ be p distinct finite sets of elements such that $S_i \cap S_j = \varnothing$ $(i \neq j)$ ($S_i$ can be a finite subset of integers, or a finite set of strings, a set of floats or a finite subset of points of a n-dimensional space, etc.) and let T be a finite set of type symbols $T=\{T_1, \ldots, T_m, G_{S_1}, \ldots, G_{S_p}\}$. The elements of U are called "attributes". A function *type*: $U \rightarrow T$ is defined that associates to each attribute $A \in U$ a unique symbol $t=$*type*$(A) \in T$. To any attribute $A \in U$ a totally ordered domain, called the domain of A, of values of type $t=$*type*$(A) \in T$ is associated and it is denoted by *dom*(A). For all $A \in U$, *type*$(A)=G_{S_i}$ if and only if *dom*$(A)=H_{S_i}$ when $S_i$ is associated with the algebra $\Psi_{S_i}$, i.e. if the elements of *dom*(A) are instances of the ADT Geometry($S_i$). Two attributes $A_i$, $A_j$ such that *dom*$(A_i)=$*dom*$(A_j)$ are called *type consistent* if *type*$(A_i)=$*type*$(A_j)$.

We shall assume that for each attribute $A_i$, *type*$(A_i)$ and hence *dom*$(A_i)$ have been formally specified by means of an ADT definition. This implies that $T_i \equiv$*type*$(A_i)$ denotes the ADT whose domain is *dom*$(A_i)$ on which manipulations have to be carried on only using the operation defined in the ADT specification.

A relation schema $R_n = (A_1, \ldots, A_n)$ (where n is called the *degree* of the relation schema $R_n$) is a subset of U. For the sake of simplicity we call $A_i$ a *descriptive attribute* if *type*$(A_i)=T_i$ and we call $A_i$ a *geometric attribute* if *type*$(A_i)=G_{S_i}$. We briefly write $R_n$ to denote a relation schema $R_n$ with attributes $A_1, \ldots, A_n$, or simply R when no ambiguity is possible. An instance of a relation schema $R_n$ is a set $\{v_1, \ldots, v_h\}$ where each $v_i$, called "tuple", belongs to $\{dom(A_1) \times \ldots \times dom(A_n)\}$. An instance of $R_n$ is denoted by $R_n(I)$.

Let X be an attribute. We denote with $f_X$ an associative and commutative binary function that takes as source data two values of type *type*(X) and returns a unique computed value of type *type*(X). Thanks to the associative property that allows the iterative application of the function it is possible to extend it to take as source data a subset of values of type *type*(X) and to return a unique computed value of *type*(X). Therefore we write $f_X\{A, B, C, \ldots\}$ instead of $((A\ f_X\ B)\ f_X\ C)\ldots$, and we call it a *fusion function* on (*dom*(X)).

Notice that the domain *dom*(X) of a fusion function is individuated through the ADT $T_i$=*type*(X) associated to X. The ADT $T_i$ in fact uniquely identifies the domain of elements *dom*(X) and the set of the operations allowed on such a set. Consequently, $f_x$ is a typed function, i.e it takes a subset of elements of a given type $T_i$ and returns a single element of the same type. Fusion functions are employed when computations on sets of attributes values are needed to answer extended relational queries (see 9b of the following section).

## 2.3 The Extended Relational Algebra

In this subsection, the set **E** of the expressions of the extended relational algebra is formally defined and the syntax and the semantics of each element of E are given. Advantage is taken of the formal framework introduced in section 2.1 in defining the primitive operations of the extended relational algebra. Borrowing the approach and the notation from [17], let us consider a database schema $D=(R_1,..,R_M)$, where $R_i$ is a relation schema.

(1) **Literals.** For any $c \in \cup_{A \in U} dom(A)$, $\{c\} \in$ **E**, it has degree 1 and $\{c\}(I)=\{c\}$.

(2) **Relations.** For each $R_j$ in D, $R_j \in$ **E** and $R_j(I) = r_j$.

(3) **Projection.** Let $a \in$ **E**, and let X be a subset of attributes of a. Then $a[X] \in$ **E**, and its degree is equal to card(X), where card(X) is the function that returns the number of elements in X, and $a[X](I) = \{t[X] \mid t \in a(I)\}$. We assume that $card(\emptyset)=0$ and $a[\emptyset](I)=\emptyset$ denotes the empty relation.

(4) **Cross Product.** Let $a_n, b_m \in$ **E**. Then $(a_n \times b_m) \in$ **E**, its degree is n+m, and $(a_n \times b_m)(I) = \{t_1 \circ t_2 \mid t_1 \in a_n(I) \wedge t_2 \in b_m(I)\}$, where $\circ$ denotes concatenation.

(5) **Restriction.** Let $a \in$ **E** and let X and Y be two type consistent attributes of a. If

$$\Theta_d = \{=, <\}$$
$$\Theta_g = \{\equiv, \subset, \cap, \cap^*\},$$

then $a[X\theta Y] \in$ **E**, where $\theta \in \Theta_d$ (if X and Y are both descriptive attributes) or $\theta \in \Theta_g$ (if X and Y are both geometric ones), and $a[X \theta Y](I) = \{t \mid t \in a(I) \wedge t[X] \theta t[Y]\}$.

(6) **Union, difference.** Let $a_n, b_n \in$ **E**, and such that their corresponding attributes are type consistent. Then $a_n \cup b_n$ and $a_n - b_n$ belong to **E** and both have degree n. They are defined as it follows:

$(a_n \cup b_n)(I) = \{ t \mid t \in a_n(I) \vee t \in b_n(I) \}$,

$(a_n - b_n)(I) = \{ t \mid t \in a_n(I) \wedge t \notin b_n(I) \}$.

(7) **G-Compose.** Let $R_n \in \mathbf{E}$, let X and G be two subsets of attributes of $R_n$, $X \cap G = \varnothing$ and $card(G) \neq \varnothing$. Then $\text{G-Compose}_X(G)(R_n) \in \mathbf{E}$ and it has degree equal to card(X)+card(G). Semantically, for each tuple $p \in R_n(I)$, let $v_p$ be the tuple defined as follows(**) :

$$v_p[X] = p[X]$$

$$v_p[G] = \begin{cases} \cup_{t \in R_n(I) \wedge t[X]=p[X]} t[G] & \text{if G is a not empty subset of geometrical attributes} \\ p[G] & \text{otherwise} \end{cases}$$

Then+

$$\text{G-Compose}_X(G)(R_n)(I) = \{ v_p \mid p \in R_n(I) \}$$

(8) **G-Decompose.** Let $R_n \in \mathbf{E}$, X and G be two subsets of attributes of $R_n$, $X \cap G = \varnothing$, and card(G)=1. Then $\text{G-Decompose}_X(G)(R_n) \in \mathbf{E}$ and it has degree card(X)+1. Semantically, for each tuple $p \in R_n(I)$, let $U_p$ be the set defined as it follows:

$$U_p = \begin{cases} \{t \mid (\exists y)\ y \in p[G] \wedge t[G]=\{y\} \wedge t[X]=p[X]\} & \text{if G is a geometrical attribute} \\ \{p\} & \text{otherwise} \end{cases}$$

Then++

$$\text{G-Decompose}_X(G)(R_n)(I) = \cup_{p \in R_n(I)} U_p.$$

(9a) **G-Fusion** (without fusion function). Let $R_n \in \mathbf{E}$, let X and G be two subsets of attributes of $R_n$, $X \cap G = \varnothing$ and $card(G) \neq \varnothing$. Then $\text{G-Fusion}_X(G)(R_n) \in \mathbf{E}$, and it has degree equal to card(X)+card(G). Semantically, for each tuple $p \in R_n(I)$, let $v_p$ be the tuple defined as follows:

$$v_p[X] = p[X]$$

$$v_p[G] = \begin{cases} geo(\cup_{t \in R_n(I) \wedge t[X]=p[X]} t[G]) & \text{if G is a not empty subset of geometrical attributes} \\ p[G] & \text{otherwise} \end{cases}$$

---

(**) We shall assume that $p[\varnothing]$ is defined as the identity element with respect to concatenation between tuples.

+ The effect of $\text{G-Compose}_\varnothing(G)$ is that all the tuples of $R_n$ (projected on G) are "composed" in a single one.

++ The effect of $\text{G-Decompose}_\varnothing(G)$ is that all tuples of $R_n$ (projected on G) are "decomposed".

Then[+++],

$$G\text{-}Fusion_X(G)(R_n)(I) = \{ v_p \mid p \in R_n(I) \}$$

(9b) **G-Fusion** (with fusion function). Let $R_n \in \mathbf{E}$, let X, Y and G be mutually not intersecting subsets of attributes of $R_n$, card(Y)=1, and let $f_y$ be a fusion function defined on *dom*(Y). Then $G\text{-}Fusion_X((f_y,Y),G)(R_n) \in \mathbf{E}$ and it has degree equal to card(X)+card(G)+1. Semantically, for each tuple $p \in R_n(I)$, let $v_p$ be the tuple defined as follows:

$$v_p[X] = p[X]$$

$$v_p[G] = \begin{cases} geo(\cup_{t \in R_n(I) \wedge t[X]=p[X]} t[G]) & \text{if G is a not empty subset of geometrical attributes} \\ p[G] & \text{otherwise} \end{cases}$$

$$v_p[Y] = \begin{cases} f_y\{t[Y] \mid t \in R_n(I) \wedge t[X]=p[X]\} & \text{if G is a not empty subset of geometrical attributes} \\ p[Y] & \text{otherwise} \end{cases}$$

Then[++++],

$$G\text{-}Fusion_X((f_y,Y),G)(R_n)(I) = \{ v_p \mid p \in R_n(I) \}$$

# 3. Basic Issues

## 3.1 Basic Concepts

To derive completeness and soundness properties of our model we need some concepts. This paragraph is supposed to introduce them by an informal approach.

Given a set of type-consistent attributes, the *evaluation* of a relation R is the set of values taken by those attributes in the relation R.

When the set of all the attributes of a given type T in a given relation R is specified, the evaluation of R is called the *content* of relation R with respect to T.

We are interested to investigate on completeness properties for what regards the modeling of geometrical nature of data. Thus we only concentrate on the *geometrical*

---

+++ The effect of $G\text{-}Fusion_{\emptyset}(G)$ (without fusion function) is that all tuples of $R_n$ (projected on G) are "fused" in a single one.

++++ The effect of $G\text{-}Fusion_{\emptyset}(G)$ (with fusion function) is that all tuples of $R_n$ (projected on YG) are "fused" in a single one whose Y value is generated by the application of the fusion function. On the other side, the effect of $G\text{-}Fusion_X(\emptyset)$ is the same as the operator "Aggregate Formation" of [17].

*content* of the database and on how this content is modified from the extended relational algebra. In particular we are interested:

- to individuate the set of the geometrical objects generated from the expressions of the extended relational algebra
- to verify that this set is an algebraic structure, and to identify its combinatorial properties
- to understand what formal properties of the model derive from these combinatorial properties

In this paper we answer the first two points by the completeness theorems of section 4.

By *geometrical completeness* we mean that each element (geometrical object) in the given algebraic structure is derivable (as an entity) by a suitable expression of extended relational algebra. By *geometrical soundness* we mean that the geometrical content of each entity derivable by expressions of the extended relational algebra is a subset of the algebraic structure.

When we say that the geometrical content of a relation R is the set of shapes $\{ A_1, A_2, ..., A_n \}$ we mean that each shape $A_i$ is the geometrical value of at least one tuple and that the geometrical value of each tuple is one shape from the given set. The notion of geometrical content plays a central role in the proofs of the theorems because it permits to denote all the geometrical values of a given relation at a given time. We are interested on the geometrical content of a given database. From the definition of "geometrical content of a database" (see next paragraph) it follows that we can focus on the schema of a single relation (corresponding to the cartesian product of the relation schemas of the considered database). An expression of the extended relational algebra can be thus interpreted as a function taking a relation R and returning a relation R'. We analyze how this function manipulates the relation R by studying how the geometrical content of the relation is modified.

### 3.2 Informal Description of Theorems

The basic idea of our completeness theorems is to characterize precisely the geometrical content of expressions that can be written in the extended relational algebra.

Starting from the geometrical content of the database we individuate a set of basic geometrical objects that are no further decomposable. From these minimal objects we define the set of the derivable objects. This set results to be a boolean lattice.

In the completeness theorem (Theorem 1) we prove that to each element of the lattice corresponds a suitable expression of the algebra that calculates a relation whose geometrical content is that element. In the soundness theorem (Theorem 2) we prove that any expression of the algebra calculates a relation whose geometrical content is a subset of the lattice.

We first prove the completeness of our model when the geometrical content of the database is made up by only two shapes. Then we show how to extend our results to a database whose geometrical content is a collection (of cardinality greater than two) of shapes.

# 4. The Completeness Theorems

## 4.1 Definitions and Notations

**Def.1** (active domains) [1]

Given an instance r of a relation schema $R(A_1...A_n)$, let $D_i = dom(A_i)$, $1 \leq i \leq n$. We call *active domain* of $A_i$ the set $\delta_i = \{d \in D_i \mid \exists\, t \in r \wedge t[A_i] = d\}$.

**Def.2** (evaluation V of a relation with respect to a subset of attributes)

Given a relation schema R and a subset $A_1...A_n$ of its type consistent attributes, let us denote by $V_{A_1...A_n}(R)$ the union of active domains of attributes $A_i$ (recall that all attributes $A_i$ have the same domain). We shall denote by $V_G(R)$ the evaluation of relation R with respect to the set G of its geometrical attributes.

**Def. 3** (geometrical content of a database)

Let $R_1,...,R_m$ a database schema. Let us denote by G the set of the names of the geometrical attributes in $R_1,...,R_m$. We say that the geometrical content of the database is the set $\{s_1,...,s_n\}$ iff $V_G(R_1 \times ... \times R_m) = \{s_1,...,s_n\}$, where $R_1 \times ... \times R_m$ denotes the schema of the cartesian product of the relations $R_1,...,R_m$. Analogously it is possible to define the content with respect to a subset X of attribute of type *type*(X).

We shall focus now on a database whose geometrical content is {A,B}, where A and B denote two sets belonging to $H_S$.

**Def.4** (Atoms(A,B))

Let us consider the following sets belonging to $H_S$:

$A \cap^* B = \{ c \in P(S) : (\exists a)(\exists b)\ a \in A \wedge b \in B \wedge c = a \cap b \}$.

*compl*(A,B) = { Y }, where $Y = \{ y \in S : (\forall a)\ a \in A \cup B,\ y \notin a \}$.

$A \cap^*$*compl*(B) $= \{ c \in P(S) : (\exists a)(\exists b)\ a \in A \wedge b \in$ *compl*(B) $\wedge\, c = a \cap b\}$.

$B \cap^*$*compl*(A) $= \{ c \in P(S) : (\exists a)(\exists b)\ a \in B \wedge b \in$ *compl*(A) $\wedge\, c = a \cap b\}$.

Let us now denote by Atoms(A,B) the set whose elements are the components of the sets just defined, i.e.

Atoms(A,B)={c∈P(S): c ∈ A∩*B ∨ c ∈ A∩**compl*(B) ∨ c ∈ B∩**compl*(A) ∨ c∈*compl*(A,B)}

Let $L_1$(A,B) be the lattice defined on Atoms(A,B) with respect to the union (∪) and intersection (∩) operators. Let {$L_1$(A,B)} denote the set {{c}: c ∈ $L_1$ (A,B)}.

Let Expand(A,B) denote the lattice built on the powerset of $L_1$(A,B).

Given l and m belonging to $L_1$(A,B), we denote by $lub_1$(l,m) the least upper bound of l and m (l∪m), and by $glb_1$(l,m) the greatest lower bound of l and m (l∩m). Analogously we shall use $lub_2$(l,m) and $glb_2$(l,m) when l and m belong to Expand(A,B).

Since the elements of the minimum antichain of both the lattices are mutually not divisible (they are not intersecting each other) the following fact holds [3]:

**Fact 1.** Each element p of the lattice $L_1$(A,B) (Expand(A,B)) can be written, in a unique way, as greatest lower bound of a subset of elements of its minimum antichain, i.e. it exists and it is unique a subset $p_1..p_K$ of Atoms(A,B) ({$L_1$(A,B)}) so that p=$glb_1$($p_1..p_K$) (p=$glb_2$($p_1..p_K$)).

**Def.5** (mapping function $F_{A,B}$)

Let $F_{A,B}$: Expand(A,B) → ERA be the function that to each element l of Expand(A,B) associates the relation R with only one attribute such that $V_G$(R) = l.

## 4.2 Theorems

**Lemma 1.** Expand(A,B) contains both A and B.

**Proof.** We show that A∈ Expand(A,B). Let A={$A_1$,...,$A_n$} and B={$B_1$,...,$B_m$}. For each $B_j$∈ B we have $A_i \cap B_j$ ∈ A ∩* B. Let b∈*compl*(B): since $A_i$∈ A then $A_i \cap$ b∈ A ∩**compl*(B). It follows that, for each $A_i$:

$$A_i \cap b \in \text{Atoms(A,B)},$$
$$A_i \cap B_j \in \text{Atoms(A,B)}, \quad j = 1, 2, ..., m;$$

Therefore,

$$A_i = lub_1(A_i \cap b\ , A_i \cap B_1, A_i \cap B_2, ..., A_i \cap B_m) \in L_1(A,B)$$

and hence {$A_i$} ∈ {$L_1$(A,B)}.

Finally, from Fact 1,

$$A = lub_2(\{A_1\},...,\{A_n\}) \in \text{Expand(A,B)}$$

**THEOREM 1 (geometrical completeness).** For each element $l \in \text{Expand}(A,B)$ it exists a sequence of operators of the extended relational algebra calculating $F_{A,B}(l)$.

**Proof:** It will be done in two steps:

**(i)** we show that for each element l belonging to the minimum antichain $\{L_1(A,B)\}$ of Expand(A,B) it exists an expression of ERA that computes $F_{A,B}(l)$;

**(ii)** we show that to each element l of Expand(A,B) it corresponds a relation R such that $V_G(R) = l$

Proof of **(i)**:

We notice that each element belonging to the minimum antichain $\{L_1(A,B)\}$ of Expand(A,B) has exactly one component. Also notice that Lemma 1 implies that $F_{A,B}(A)$ and $F_{A,B}(B)$ are definable.

The hypothesis that the geometrical content of the database is $\{A, B\}$ implies that there exist two instances of relation $R_1$, $R_2$ and two tuples $t_1 \in R_1$, $t_2 \in R_2$ such that

$t_1[G] = A$ and $t_2[G] = B$

We prove this step by first showing that the statement holds for an element of the minimum antichain of Expand(A,B) (i.e. $\{L_1(A,B)\}$) whose component belongs to one of the four sets $A \cap^* B$, *compl*(A,B), $A \cap^*$*compl*(B), and $B \cap^*$*compl*(A).

**1 - for each {l} such that $l \in A \cap^* B$**

Let

$l \in A \cap^* B$,

$R_3 = (\text{Select}_{t[G]=A} R_1) \cup (\text{Select}_{t[G]=B} R_2)$,

$R_4 = \text{G-Fusion}_\varnothing\ (\cap^*, G)\ (\varnothing)\ R_3$.

Then

$$\Pi_G(\text{Select}_{t[G]=\{l\}}(\text{G-Decompose}_\varnothing(G)(R_4)) = F_{A,B}(\{l\}) \quad .$$

**2 - for each {l} such that $l \in A \cap^*$ *compl*(B)**

Let

$l \in A \cap^* compl(B)$,

$R_1^1 = \Pi_G(\text{Select}_{t[G]=A} R_1)$,

$R_2^1 = \Pi_G(\text{Select}_{t[G]=B} R_2)$,

$R_2^2 = \text{G-Fusion}_\varnothing\ (compl, G)\ (\varnothing)\ R_2^1$.

Since $l \in A \cap^* compl(B)$ it follows that

$$\text{Select}_{t[G]=\{l\}}(\text{G-Decompose}_\varnothing G(\text{G-Fusion}_\varnothing\ (\cap^*, G)\ (\varnothing)\ (R_2^2 \cup R_1^1))) =$$

$$F_{A,B}(\{l\}) .$$

**3 - for each {l} such that l∈ B ∩* *compl*(A)**

Analogous to previous case, inverting the role of A and B.

**4 - for each {l} such that l∈ *compl*(A,B)**

Let us first observe that:

(a) l ∈ *compl*(A,B) if and only if {l} = *compl*(A,B),

(b) *compl*(A,B) = *compl*(A) ∩**compl*(B).

Let now

l ∈ *compl*(A,B)

and

$R_1^2$ = (G-Fusion$_\varnothing$ (*compl*, G) (∅) $R_1^1$).

Then

$$\text{G-Fusion}_\varnothing\ (\cap^*, G)\ (\varnothing)\ (R_1^2 \cup R_2^2) = F_{A,B}(\{l\}).$$

We now complete the proof by showing that the statement holds for any other element in the minimum antichain of Expand(A,B).

**5 - for each other {l} ∈ {L1(A,B)}**

In this case, it follows from Fact 1 that there exist j elements $a_1...a_j$ belonging to Atoms(A,B) (i.e. the minimum antichain of the lattice $L_1$(A,B)) such that

$l = glb_1(a_1...a_j) \equiv a_1 \cup ... \cup a_j$

Since each $a_i$ is an element of Atoms(A,B), it follows that for {$a_i$} it holds one among 1), 2), 3) and 4). Thus j sequences of ERA operators exist that return j relations $R_1...R_j$ such that

$R_h = F_{A,B}(\{a_h\})$, h=1, 2, ..., j.

It follows that

$$\text{G-Fusion}_\varnothing(G)(R_1 \cup ... \cup R_j) = F_{A,B}(\{l\}).$$

Proof of **(ii)**:

Let l ∈ Expand(A,B). From Fact 1 it follows that there exist k element $l_1...l_k$ of {$L_1$(A,B)} s.t. $l = l_1 \cup ... \cup l_k \equiv glb_2(l_1...l_k)$. From 5) we have that there exist k expressions of the extended relational algebra calculating the following k relations:

$$R_h = F_{A,B}(\{l_h\}) \qquad h=1, 2, ..., k.$$

Then

$$\text{G-Compose}_\varnothing(G)(R_1 \cup ... \cup R_k) = F_{A,B}(l)$$

and the theorem is completely proved.

**THEOREM 2 (geometrical soundness).** Let {A,B} be the geometrical content of the database. Then any legal sequence of operators $op_1...op_n$ calculates a relation R whose geometrical content is a subset of Expand(A,B).

**Proof:** It will be done in three steps:

(i): *a single operator* of the algebra applied to a relation whose geometrical content is {A,B} computes a relation whose content is a subset of Expand(A,B),

(ii): *any sequence of operators* applied to a relation whose geometrical content is {A,B} computes a relation whose content is a subset of Expand(A,B),

(iii): *any sequence of operators* applied to any relation containing elements of Expand(A,B) computes a relation whose content is a subset of Expand(A,B).

Proof of (i):

Let R be a relation of the database and p any consistent predicate defined on the attributes of R. We suppose that $A=\{A_1,...,A_n\}$, $B=\{B_1,...,B_m\}$ and the geometrical content of R is a subset of {A,B}. Let G denote the set of the geometrical attributes contained in the schema of R.

Then

$$V_G(R) \subset \{A,B\}$$

We first observe that

$$V_G(\Pi_p R) \subset V_G(R)$$
$$V_G(\text{Select}_p R) \subset V_G(R)$$

Beside, if R' is an other relation of the database s.t. $V_{G'}(R') \subset \{A,B\}$ (where G' denotes the set of the geometrical attributes of R') then

$$V_{GG'}(R \cup R') \subset \{A,B\}$$
$$V_{GG'}(R - R') \subset \{A,B\}$$
$$V_{GG'}(R' - R) \subset \{A,B\}$$
$$V_{GG'}(R \times R') \subset \{A,B\}$$

Therefore, any sequence of algebra operators not containing G-Compose or G-Decompose or G-Fusion operators, does not add geometrical content to the database. Hence, let us concentrate on G-Compose, G-Decompose and G-Fusion operators.

**i.1 - G-Decompose$_X$Y (R)**

If this operator does not really decompose the geometrical attributes of R (i.e. when Y is a set of descriptive attributes or when Y contains both descriptive and geometrical attributes), then

$$V_G(R) = V_G(\text{G-Decompose}_X Y\ (R))$$

for any attribute X.

Otherwise

$$V_G(\text{G-Decompose}_X Y\ (R)) \subset \{A\ ,\ B,\ \{A_1\}\ ,...,\ \{A_n\}\ ,\ \{B_1\}\ ,...,\ \{B_m\}\}\ .$$

In any case

$$V_G(\text{G-Decompose}_X Y\ (R)) \subset \text{Expand}(A,B).$$

**i.2 - G-Compose$_X$Y (R)**

If this operator does not really compose the geometrical attributes of R (i.e. when Y is a set of descriptive attributes or when Y contains both descriptive and geometrical attributes), then

$$V_G(R) = V_G(\text{G-Compose}_X Y\ (R))$$

for any attribute X.

Otherwise

$$A \cup B \equiv \text{lub}_2(\text{lub}_2(\{A_1\},\ ..,\{A_n\}),\ \text{lub}_2(\{B_1\},\ ..,\{B_n\})) \in \text{Expand}(A,B),$$

and

$$V_G(\text{G-Compose}_X Y\ (R)) \subset \{A\ ,\ B\ ,\ A \cup B\}.$$

In any case it is $V_G(\text{G-Compose}_X Y\ (R)) \subset \text{Expand}(A,B)$.

**i.3 - G-Fusion$_X$Y (R)**

If this operator does not really fuse the geometrical attributes of R (i.e. when Y is a set of descriptive attributes or when Y contains both descriptive and geometrical attributes), then

$$V_G(R) = V_G(\text{G-Fusion}_X Y\ (R))$$

for any attribute X.

Otherwise let A' be

$$A' = \text{lub}_1((\text{lub}_1(A_1,\ ..,A_n),\ \text{lub}_1(B_1,\ ..,B_n)) \in L_1(A,B)$$

then $\{A'\} \in \text{Expand}(A,B)$ and

$$V_G(\text{G-Fusion}_X Y\ (R)) \subset \{A\ ,\ B\ ,\ \{A'\}\ \}.$$

In any case it is $V_G(\text{G-Fusion}_X Y\ (R)) \subset \text{Expand}(A,B)$.

**i.4 - G-Fusion$_X$((f$_Y$ , Y), G) (R)**

If this operator does not effectively fuse the geometrical attributes of R (i.e. when Y is a descriptive attribute of R and G contains only descriptive or both geometrical and descriptive attributes), then

$$V_G(R) = V_G(\text{G-Fusion}_X Y\ (R)).$$

When Y is a descriptive attribute and G is a not empty subset of geometrical attributes, we can easily extend point (i.3) above to show that the G-Fusion operator (with fusion function) does not increase the geometrical content of the database.

Then let us analyze the case when Y is a geometrical attribute of R. In this case the fusion function $f_Y$ can be any one among " $\cup, \cap, \cap^*$, *compl* ". Notice that we have already shown that the G-Fusion operator (without fusion function) does not increase the geometrical content of the database. Hence a detailed analysis only concerns $V_Y(R)$.

### i.4.1 - $f_Y \equiv \cap$

Since $(A \cap B) \subset (A \cap^* B)$ then

$$c_i \in (A \cap B) \Rightarrow c_i \in \text{Atoms}(A,B) \subset L_1(A,B).$$

Then

$$\{c_1, .., c_h\} \equiv A \cap B \in \text{Expand}(A,B),$$

and finally

$$V_{XYG}(\text{G-Fusion}_X((\cap, Y), G)(R)) \subset \text{Expand}(A,B).$$

### i.4.2 - $f_Y \equiv \cup$

Lemma 1 and Fact 1 imply that there exist n+m elements $\{A_1\}...\{A_n\}$, $\{B_1\}...\{B_m\}$ of $\{L_1(A,B)\}$ s.t. $A = \{A_1\} \cup ... \cup \{A_n\} \equiv glb_2(\{A_1\}...\{A_n\})$ and $B = \{B_1\} \cup ... \cup \{B_m\} \equiv glb_2(\{B_1\}...\{B_m\})$.

Since

$$A \cup B = \{A_1\} \cup ... \cup \{A_n\} \cup \{B_1\} \cup ... \cup \{B_m\} = lub_2(A,B) \in \text{Expand}(A,B)$$

then

$$V_{XYG}(\text{G-Fusion}_X((\cup, Y), G)(R)) \subset \{A \cup B, A, B\} \subset \text{Expand}(A,B)$$

### i.4.3 - $f_Y \equiv \cap^*$

From the definition it follows that $A \cap^* B \subset \text{Atoms}(A,B)$, hence we have

$$A \cap^* B \in \text{Expand}(A,B),$$

and finally

$$V_{XYG}(\text{G-Fusion}_X((\cap^*, Y), G)(R)) \subset \{A \cap^* B, A, B\} \subset \text{Expand}(A,B)$$

### i.4.4 - $f_Y \equiv$ *compl*

Let us observe that the fusion function *compl* computes *compl*(A) and *compl*(B) and/or *compl*($A \cup B$), and that $A = \{A_1,...,A_n\}$ where each $A_i \in L_1(A,B)$. From the definitions we have

$$compl(A) = \{\{ y \in S : (\forall i)\ y \notin A_i \}\}.$$

Then let $a^i_1,\ldots,a^i_{k_i}$ be the $k_i$ elements of Atoms(A,B) s.t.

$$A_i = lub_1(a^i_1,\ldots,a^i_{k_i}) = a_1 \cup \ldots \cup a^i_{k_i}.$$

Therefore

$$compl(A) = \{ y \in S : (\forall i)(\forall j)\ i=1..n,\ j=1..k_i,\ \ y \notin a^i_j \}.$$

Since

$$S = lub_1(at_1,...,at_S), \text{ where } at_i \in Atom(A,B),$$

if $A^c$ denotes the set

$$A^c = \{\ at \in Atoms(A,B) : (\forall i)(\forall j)\ i=1..n, j=1..k_i\ \ at \neq a^i_j\ \},$$

and l denotes the cardinality of $A^c$ then $lub_1(at_1,...,at_l)$, where $at_i \in A^c$, belongs to $L_1(A,B)$ and hence

$$\{lub_1(at_1,...,at_l),\ at_i \in A^c\} \equiv \{\{y \in S : (\forall i)(\forall j)\ i=1..n, j=1..k_i,\ \ y \notin a^i_j\ \}\} \equiv$$

$$compl(A)$$

is an element of Expand(A,B). A similar proof show that $compl(A \cup B) \in$ Expand(A,B).

Finally,

$V_{XYG}$(G-Fusion$_X$((*compl*, Y), G) (R)) $\subset$ {*compl*(A), *compl*(B), *compl*(A$\cup$B), A, B} $\subset$ Expand(A,B).

Proof of **(ii)**:

We prove that $V_G(op_1....op_n(R)) \subset$ Expand(A,B) by induction on the length of the sequence of the operators.

The thesis holds for n=1 as it follows from the proof of previous point **(i)**. Assuming that the thesis holds for i=1,...,n, let us now prove that it holds for i=n+1.

Given any sequence $op_1....op_{n+1}$ we have that

$$op_1....op_{n+1}(R) = op_{n+1}(....op_1(R)) = op_{n+1}(R'),$$

where, by the induction hypothesis, R' is a relation containing only the elements of Expand(A,B), formally

$$V_G(R') \subset Expand(A,B).$$

Since any relational operator cannot modify the geometrical content of the database, we focus on the case when $op_{n+1}$ is any one among G-Compose, G-Decompose and G-Fusion operators.

**ii.1 - $op_{n+1}$= G-Compose$_X$Y (R')**

If Y is an empty subset of attributes of R' or it is not entirely made up by geometrical attributes, then the geometrical content of R' is not modified from $op_{n+1}$.

Otherwise, let $A^i_1...A^i_k \in V_Y(R')$ be the k elements, corresponding to the value (or the r-tuple of values if card(X)=r) $x_i \in V_X(R')$, to be composed by G-Compose operator. Since $A^i_j \in$ Expand(A,B) (by the induction hypothesis) then

$$A_i = lub_2(A^i_1...A^i_k) = A^i_1 \cup...\cup A^i_k \in Expand(A,B)$$

for each i=1..s where s is the cardinality of $V_X(R')$.

Let $X_G$ denotes the set of the geometric attributes of X and $V_{X_G}(R')=\{X_{G_1},...,X_{G_p}\}$. Since the schema of the resulting relation is XY and the set of its geometrical attributes is $X_GY$ then

$$V_{X_GY}(\text{G-Compose}_XY\ (R')) \subset \{X_{G_1},...,X_{G_p}, A_1,...,A_s\} \subset \text{Expand}(A,B).$$

**ii.2 - $op_{n+1}$= G-Decompose$_X$Y(R')**

Let $A_i \in$ Extend(A,B) an element of $V_Y(R')$. **Fact 1** implies that there exist $A_1^i,...,A_{k_i}^i$ belonging to $\{L_1(A,B)\} \subset$ Expand(A,B) s.t.

$$A_i = lub_2(A_1^i...A_k^i) = A_1^i \cup...\cup A_{k_i}^i$$

Since each $A_j^i$ contains only one component, if s is the cardinality of $V_Y(R')$ then

$$V_Y(\text{G-Decompose}_XY\ (R')) \subset \{A_1^1...A_{k_1}^1,...,A_1^s...A_{k_s}^s\} \subset \text{Expand}(A,B).$$

If $X_{G_1},...,X_{G_p}$ and $X_G$ are defined as in the previous (ii.1) then

$$V_{X_GY}(\text{G-Decompose}_XY\ (R')) \subset \{X_{G_1},...,X_{G_p},...,A_1^1...A_{k_1}^1,...,A_1^s...A_{k_s}^s\} \subset \text{Expand}(A,B).$$

**ii.3 - $op_{n+1}$= G-Fusion$_X$Y (R')**

If Y is an empty subset of attributes of R' or it is not entirely made up by geometrical attributes, then the geometrical content of R' is not modified from $op_{n+1}$.

Otherwise, let $A_1^i,...,A_k^i \in V_Y(R')$ be the k elements, corresponding to the value (or the n-tuple of values if card(X)=n) $x_i \in V_X(R')$, to be fused by the G-Fusion operator. Fixed $A_j^i$, let $a_{j_1}^i...a_{j_p}^i$ be its p components. Since each $a_{j_h}^i \in L_1(A,B)$ and also

$$a_j^i \equiv lub_1(a_{j_1}^i,...,a_{j_p}^i) = a_{j_1}^i \cup...\cup a_{j_p}^i \in L_1(A,B).$$

Since each $a_j^i \in L_1(A,B)$ then

$$a^i \equiv lub_1(a_1^i,...,a_k^i) = a_1^i \cup...\cup a_k^i \in L_1(A,B),$$

and hence $\{a^i\} \in$ Expand(A,B) for each i=1..t where t is the cardinality of $V_X(R')$. If $X_{G_1},...,X_{G_p}$) and $X_G$ are defined as in (ii.1) then

$$V_{X_GY}(\text{G-Fusion}_XY\ (R')) \subset \{X_{G_1},...,X_{G_p}, a_1,...,a_t\} \subset \text{Expand}(A,B).$$

**ii.4 - $op_{n+1}$= G-Fusion$_X$(($f_Y$,Y),G) (R')**

If this operator does not really fuse the geometrical attributes of R' (i.e. when Y is a descriptive attribute of R' and G contains only descriptive or both geometrical and descriptive attributes), then

$$V_G(R') = V_G(\text{G-Fusion}_X Y\ (R')).$$

When Y is a descriptive attribute and G is not empty subset of geometrical attributes, we can easily extend (ii.3) to show that the G-Fusion operator (with fusion function) does not increase the geometrical content of the database.

Then let us analyze the case when Y is a geometrical attribute of R'. In this case the fusion function $f_Y$ can be anyone among " $\cup$, $\cap$, $\cap^*$, *compl* ". Since we have shown in (ii.3) that the G-Fusion operator does not increase the geometrical content of the database, a detailed analysis only concerns $V_Y(R)$.

**ii.4.1 - $f_Y \equiv \cup$**

Arguments similar to those of (ii.1) hold.

**ii.4.2 - $f_Y \equiv \cap$**

Let $A^i_1 \ldots A^i_k \in V_Y(R')$ be the k elements, corresponding to the value (or the r-tuple of values if card(X)=r) $x_i \in V_X(R')$, to be intersected by the fusion function $\cap$. Since $A^i_j \in$ Expand(A,B) (by the induction hypothesis) then

$$A_i = glb_2(A^i_1 \ldots A^i_k) = A^i_1 \cap \ldots \cap A^i_k \in \text{Expand(A,B)}$$

for each i=1..s where s is the cardinality of $V_X(R')$. If $X_{G_1},\ldots,X_{G_p}$ and $X_G$ are defined as in (ii.1) and $V_G(\text{G-Fusion}_X G\ (R')) = \{G_1,\ldots,G_q\}$ then

$$V_{X_G Y G}(\text{G-Fusion}_X((\cap,Y),G)\ (R')) \subset \{X_{G_1},\ldots,X_{G_n}, A_1,\ldots,A_s, G_1,\ldots,G_q\} \subset \text{Expand(A,B)}.$$

**ii.4.3 - $f_Y \equiv \cap^*$**

Let $A^i_1 \ldots A^i_k \in V_Y(R')$ be the k elements, corresponding to the value (or the r-tuple of values if card(X)=r) $x_i \in V_X(R')$, to which the fusion function $\cap^*$ must be applied. Let us first concentrate on $A^i_1$ and $A^i_2$. We first show that $A^i_1 \cap^* A^i_k$ belongs to Expand(A,B). Successively we extend the result to the k sets $A^i_1,\ldots,A^i_k$.

Let $a1^i_1,\ldots,a1^i_n$ and $a2^i_1 \ldots a2^i_m$ be respectively the components of $A^i_1$ and $A^i_2$. Each of these components belongs to $L_1$(A,B). It follows that, for each j=1..n and h=1..m,

$$glb_1(a^i_{1_j}, a^i_{2_h}) \equiv a^i_{1_j} \cap a^i_{2_h} \in L_1(A,B),$$

and hence

$$A_1 \cap^* A_2 \equiv \{a^i_{1_j} \cap a^i_{2_h}\}_{j=1..n\ h=1..m} \in \text{Expand}(A,B).$$

By renaming $A_1$ and $A_2$ applying the formula

$$A_i = A_i \cap^* A_{i+1} \text{ and } A_{i+1} = A_{i+2} \ ,$$

and iterating the proof for i=1,..,(n-2) we obtain that

$$A_j = \cap^*_{i=1..k} A^i_j \in \text{Expand}(A,B)$$

for each j=1..s where s is the cardinality of $V_X(R')$. If $X_{G_1},...,X_{G_p}$ and $X_G$ are defined as in (ii.1) and $V_G(\text{G-Fusion}_X G\ (R')) = \{G_1,...,G_q\}$ then

$$V_{X_G Y G}(\text{G-Fusion}_X((\cap^*,Y),G)\ (R')) \subset \{X_{G_1},...,X_{G_p}, A_1,...,A_s, G_1,...,G_q\} \subset \text{Expand}(A,B)$$

**ii.4.4 - $f_y \equiv$ *compl***

Let us first prove that, for any X,

$$X \in \text{Expand}(A,B) \Rightarrow \text{compl}(X) \in \text{Expand}(A,B).$$

We have

$$X \in \text{Expand}(A,B) \Rightarrow X = \{X_1,...,X_n\},\ X_i \in L_1(A,B) \text{ for } i=1..n.$$

Then let $X^i_1,\dots,X^i_{k_i}$ be the k elements of Atoms(A,B) so that

$$X_i = lub_1(X^i_1,\dots,X^i_{k_i}) = X_1 \cup \dots \cup X^i_{k_i} \ .$$

Denoting by $X^c$ the set

$$X^c = \{\ at \in \text{Atoms}(A,B) : (\forall i)(\forall j)\ i=1..n,\ j=1..k_i\ \ at \neq X^i_j\ \} \subset \text{Atoms}(A,B),$$

we have that $lub_1(at_1,...,at_l)$, where $at_i \in X^c$, belongs to $L_1(A,B)$ and hence

$$\{lub_1(at_1,...,at_l),\ at_i \in X^c\} \equiv \{\{x \in S : (\forall i)(\forall j)\ i=1..n,\ j=1..k_i,\ \ y \notin a^i_j\ \}\} \equiv \mathit{compl}(X)$$

belongs to Expand(A,B).

Let now $A^i_1 \dots A^i_k \in V_Y(R')$ be the k elements, corresponding to the value (or the r-tuple of values if card(X)=r) $x_i \in V_X(R')$, to which the fusion function *compl* must be applied. We must show that for each i $\mathit{compl}(\cup_j A^i_j) \in$ Expand(A,B). Since $\cup_j A^i_j \in$ Expand(A,B), from the previous proof it follows that

$$\cup_j A^i_j \in \text{Expand}(A,B)$$

for each j=1..s where s is the cardinality of $V_X(R')$. If $X_{G_1},...,X_{G_p}$ and $X_G$ are defined as in (ii.1) and $V_G(\text{G-Fusion}_X G\ (R')) = \{G_1,...,G_q\}$ then

$$V_{X_G Y G}(\text{G-Fusion}_X((compl,Y),G)\ (R')) \subset \{X_{G_1},...,X_{G_n}, A_1,...,A_s, G_1,...,G_q\} \subset \text{Expand}(A,B).$$

Proof of **(iii)**:

Given a relation R whose geometrical content is a subset of Expand(A,B), we prove that $V_G(op_1....op_n(R)) \subset$ Expand(A,B) by induction on the length of the sequence of the operators. The thesis immediately follows from (ii).

### 4.2.1 Generalization to n Sets

Let $\{A_1...A_n\}$ is the geometrical content of the database. Let

$\cap_i^* A_i = \{\ c\in P(S): (\exists i_1...i_n))\ a_{i_1}\in A_1 \wedge ... \wedge a_{i_n}\in A_n \wedge c = \cap_j a_{i_j}\ \}$.

$compl(\cap_i A_i) = \{\ Y\ \}$, where $Y = \{\ y\in S : (\forall a)\ a\in\cap_i A_i,\ y\notin a\ \}$.

$A_1 \cap^* compl(A_2,...,A_n) = \{c\in P(S) : (\exists a)(\exists b)\ a\in A \wedge b\in compl(A_2,...,A_n) \wedge c = a \cap b\}$.

$A_2 \cap^* compl(A_1,A_3,...,A_n) = \{\ c\in P(S) : (\exists a)(\exists b)\ a\in B \wedge b\in compl(A_1,A_3,...,A_n) \wedge c = a \cap b\}$.

.

.

$A_n \cap^* compl(A_1,...,A_{n-1}) = \{\ c\in P(S) : (\exists a)(\exists b)\ a\in B \wedge b\in compl(A_1,...,A_{n-1}) \wedge c = a \cap b\}$.

Starting from these basic sets we can define the set $Atoms(A_1,...,A_n)$, $L_1(A_1,...,A_n)$ and finally $Expand(A_1,...,A_n)$. Finally, Lemma 1, Theorem 1 and Theorem 2 can be easily extended on the basis of the definition of these sets.

## 5. Conclusions

In this paper we have investigated formal properties of the logical data model for complex data introduced in [9].

The main result of this paper is the precise characterization of the *geometrical content* of the expressions that can be written in the extended relational algebra defined in our model. We have shown that this *geometrical content* coincides with a boolean lattice. In particular we have proved that (i) every relation derivable by an expression of the extended relational algebra corresponds to a subset of this lattice; (ii) every element of the lattice corresponds to an expression of the extended relational algebra.

To our knowledge, this is a novel result in the field studying how to extend the relational model by introducing abstract data types as domains [20, 11, 13, 15]. These approaches are typically characterized by defining operations of the extended models in terms of some algebras, but do not deal with formal properties like, for example, their completeness and soundness with respect to the reality of interest.

We are currently investigating on the formal properties of the model derivable from the combinatorial properties of the universe of the derivable objects, since we think that query processing and optimization in our model are strongly dependent on these properties.

## References

[1] Atzeni P., Batini C., De Antonellis V., La Teoria Relazionale dei Dati, Editrice Boringhieri, Torino, 1985.

[2] Abiteboul S., Scholl M., Gardarin G., Simon E., "Towards DBMSs for Supporting New Applications" Proc. VLDB, Kyoto, August, 1986.

[3] Birkhoff G., Lattice Theory, American Mathematical Society, Providence, RI, 3rd edition, 1979.

[4] Chang N.S. and K.S. Fu, "A Relational Database System for Images". In: S.K. Chang and K.S. Fu (eds.), Pictorial Information Systems, Springer, 1980, 288-321.

[5] Chang N.S. and K.S. Fu, "Picture Query Languages for Pictorial Data-Base Systems". Computer 11 (1981), 23-33.

[6] Proceedings of International Conference on Extending Data Base Technology, Venice 1988.

[7] G. Gardarin, J.P. Cheiney, G. Kirnan, D. Pastre, H. Stora, "Managing Complex Objects in an Extensible Relational DBMS", Proc. 15th VLDB, Amsterdam, 1989.

[8] M. Gargano, E. Nardelli, "A Logical Data Model for Integrated Geographical Databases", Proc. of ACM-IEEE Int. Conference on Systems Integration, 1990.

[9] M. Gargano, E. Nardelli, M. Talamo, "Abstract Data Types for the Logical Modeling of Complex Data", Information Systems, Vol. 16, No.6, 1991.

[10] Grosky W.I., "Toward a Data Model for Integrated Pictorial Databases", Computer Vision, Graphics, and Image Processing 25, 371-382 (1984).

[11] R.H. Guting, "Gral: An Estensible Relational Database System for Geometric Application", Proc. 15th VLDB, Amsterdam, 1989.

[12] S. Kamin, "Final Data Types and their Specification", ACM Trans. Programming Lang. and Sys. N.1, Vol.5, 1983.

[13] Lorentzos N. A. , and Johnson R.G., "An Extension of the Relational Model to Support Generic Intervals", In Conference on Extending Database Technology (EDBT 88).

[14] B. Meyer, "Reusability: the Case for Object-Oriented Design", IEEE Software, March 1987.

[15] S.L. Osborn, T.H. Heaven, "The Design of a Relational Database System with Abstract Data Types for Domains", ACM Trans. on Database Systems, 1986.

[16] Orestein J.A. and Manola F.A, "Probe Spatial Data Modeling and Query Processing in an Image Database Application", IEEE Trans. on Software Engineering, Vol 14, No. 5, May 1988, 611-629.

[17] Ozsoyoglu G., Ozsoyoglu M. and Matos V., "Extending Relational Algebra and Relational Calculus with Set-Valued Attributes and Aggregate Functions", ACM Trans. on Database Systems, Vol 12, No. 4, December 1987, 566-592.

[18] Roussopoulos N., Faloutsos C., Sellis T. G., "An Efficient pictorial Database System for PSQL", IEEE Trans. on Software Engineering, Vol 14, No. 5, May 1988, 639-651.

[19] Scholl, M., Voisard A., "Modeling of thematic maps: an application to geographic databases", Int. Symp. on the Design and Implementation of Large Spatial Databases, Santa Barbara, July, 1989.

[20] M. Stonebraker, B. Rubenstein, A. Guttman, "Application of Abstract Data Types and Abstract Indices to CAD Bases", IEEE Trans. on Software Engineering, 1983.

# Object-Based Data Model and Deductive Language for Spatio-Temporal Database Applications

Terence R. Smith
Raghu Ramakrishnan
Agnès Voisard

**Abstract:** We discuss data models and languages in relation to computational environments in which data access and tools for analysis are integrated in order to support data-intensive and numerically intensive modelling activities in a class of scientific database systems. We describe a simple model of scientific activity in order to motivate the development of a new conceptual data model for complex, spatio-temporal objects and a deductive, object-oriented language to support the definition and manipulation of such objects. We provide three examples of the application of both the model and the language.

# 1 Introduction

A database system and its conceptual schema provide a *model* of phenomena that are represented in the database [Ull88]. Many non-standard database applications may be viewed as *system modeling activities.* For such applications, the database system must provide a means to represent and manipulate classes of complex objects; it must also provide a means to manipulate information about objects and their interrelationships; and it must also provide a facility for deriving various properties of objects and their interrelationships from information that is represented in the database in implicit as well as explicit form [LN87].

In *standard* database applications, modeling support has generally assumed discrete domains and has focussed on structural relationships between individual values [Cod70, Che76, HM81, HK87]. A panel that studied the achievements and opportunities for database systems [SSU90] indicated that the next generation of applications, such as data-intensive scientific applications, will require more sophisticated support. Such support relates to data access and type management for large and internally complex objects, capabilities for processing large numbers of rules, and the ability to handle new concepts such as spatial data, time and uncertainty. A second panel that examined scientific database management [FJP90] observed that: "*To manipulate data and produce information, a scientist needs to access data and apply analysis tools in concert. Failure to integrate the data management and analysis environments restricts the productivity of the scientist... extant systems are not integrated...due to the fact that the data management environment was created by a computer scientist and the analytic environment... by a discipline specialist.*"

In the remainder of the chapter, we discuss several issues relating to the integration of data access and analysis tools for scientific database and modelling support. We first describe a simple model of scientific activity, and use this model to provide motivation for the development of a conceptual data model for complex, spatio-temporal objects and a deductive, object-oriented language to support the definition and manipulation of such objects. We then provide three examples of the application of both the model and the language.

# 2 Data-Intensive Scientific Investigations

In order to examine issues relating to the integration of data access and analysis tools indicated above, we briefly describe a simple model of an important class of scientific investigations relating in particular to earth system science. We employ this model in an informal manner in order to elucidate critical issues relating to computational support for data and modeling in this class of scientific investigations.

The core of many scientific investigations typically involves: (1) a conceptualization of some relatively complex domain of investigation, (2) a formal representation of the conceptualization, and (3) an inference process based in part on the formal representation and a (frequently large) database of information in both explicit and implicit form. A key modeling activity in such investigations is the definition of semantically important classes of entities and relationships among the entities. The complexity of the systems that are currently being modeled increasingly forces investigators to adopt computational approaches. For example, in many studies of complex environmental systems, both the conceptualization of the domain and the formal representation of the domain typically involve many classes of complex, spatio-temporal objects and relatively complex relationships among the classes. The inference procedures relating to the objects and their relationships frequently require access to large databases of objects, relationships, and observed facts. It is increasingly evident that there is a need for computational environments that provide a uniform approach to the definition and manipulation of entities and entity classes that contain objects ranging from raw dataset objects to high-level conceptual objects.

We conceptualize the *computational modeling* of complex natural systems in terms of a two dimensional space (Figure 1). The vertical axis, which we term the "extraction" axis, characterizes the scientist's representation of the world in terms of the degree to which the objects and relationships of the scientist's conceptualization have been made explicit. The horizontal axis, which we term the "abstraction" axis, characterizes that degree of abstraction in the computational representation of the data of interest to the scientist. To the extent that these dimensions are orthogonal, it is our contention that the scientist needs to operate in an environment that allows full access to all levels of the extraction axis, but that hides many of the issues

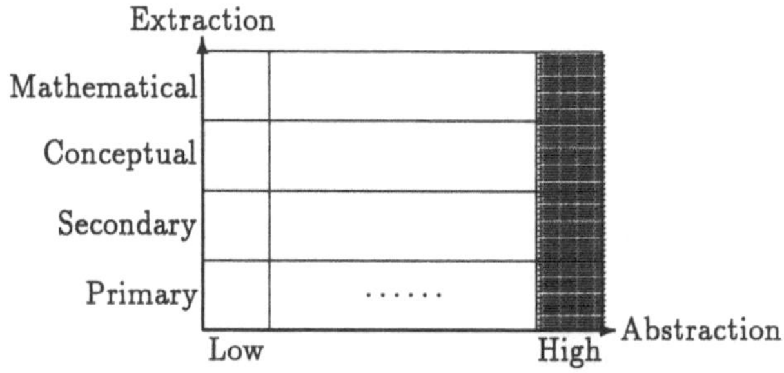


Figure 1: Two Dimensions of A Scientific Database System

relating to lower levels of the abstraction axis. This region of operation is illustrated in the shaded area of Figure 1.

The extraction axis represents the key dimension with respect to scientific modeling activities. It is possible to identify four broad levels of the extraction axis with respect to datasets:

(1) A *primary* database that contains information about objects in largely implicit form. In many applications, such a database may contain sets of images or digitized maps and sets of data obtained from measurements taken over some set of locations.

(2) A *secondary* database that contains datasets that are typically derived from procedures applied to various datasets from the primary database and to other information. These are not necessarily the final products of computation. A database of river valleys derived from a database of surface elevations provides an example of such a dataset [SZP90].

(3) A *tertiary* database (or *conceptual representation*) containing complex spatio-temporal objects derived from the primary and secondary databases. Such objects and their interrelationships represent models of the phenomena in the application domain. An example of such a complex object in a database containing hydrological information might be a connected set of river links and the nested areas that they drain.

(4) A *quaternary* level in which relationships between complex objects are precisely represented. Typically this is the domain of equations of various classes, and especially differential/difference equations, and equations relating to statistical inference.

The abstraction axis, on the other hand, represents the level of details and abstractions at which the scientist wishes to operate. In particular, issues along this axis includes data storage, operation execution, and collaboration. Storage abstraction enables the scientists to access information without being hindered by the location, distribution, and heterogeneity of the database. Operation abstraction allows a scientist to execute operations or programs concurrently without being concerned about the possibility of interference and, hence, inconsistency in the database. Finally, collaboration permits teams of scientists to accomplish a single complex task in a tightly coupled environment. The significance of Figure 1 rests on our strong beliefs that any useful system should attempt to support all levels of the extraction axis *in the most explicit manner possible*, as well as the representations and processing that relate to all but the highest levels of the abstraction axis *in the most transparent manner possible.*

With respect to the extractive axis, important issues that require investigation, particularly in the area of large, spatial databases [SF90], include:

1. the design of adequate data models for complex spatio-temporal objects and object classes;

2. the design of (deductive- and object-based) languages to support applications involving complex spatio-temporal objects and their manipulation, including support for mathematical and statistical modeling activities that require data integration;

Although there are many other important issues relating to both the extractive and abstractive axes that require investigation, we focus our attention in this chapter on these two issues, since we believe that they are of central importance for database systems that support scientific modelling activities.

## 2.1 Domains of Application

We believe it is critical to develop new database technology on the basis of substantive investigations of specific domains. While no single domain of scientific endeavor can serve as a basis for designing general purpose modeling support systems, an appropriate choice of specific domains is important in providing significant insights relating to requirements for database and modeling support. Two important criteria in choosing such domains are that they involve phenomena of major importance from a scientific point of view and that they involve data-intensive and numerically-intensive modeling activities that require significant computational and database support.

An application domain that meets these criteria and serves our present purposes involves large-scale, hydrologic networks. Hydrological systems are of scientific importance because they characterize the surface of much of the earth, and are deeply related to many fundamental research issues in several areas of science, including hydrology, geology, ecology, climatology, and soil science. Hydrological systems involve data-intensive and numerically-intensive modelling activities since they involve phenomena that are naturally modeled in terms of complex spatio-temporal objects and interrelationships between these objects; such modeling typically involves sophisticated mathematical and statistical techniques, as in the case of the modeling of the flow of water through the various subsystems of hydrologic network [Dun88]. Increasingly large amounts of data are necessary for such modelling.

While one of the most important classes of queries about such systems relate to the runoff that occurs in various parts of the channel and drainage area system at different times before, during and after a rainfall event, there are a large number of other classes of queries that are of scientific interest. These include the computation of functions of the geometric structure of the system, including the subnetworks defined by the set of channels, or about aggregate areas drained by given channels; the static attribute characteristics of the basin, such as the distribution of soils; and the state of the dynamic attributes of the system, such as the proportion of the basin that may be saturated at any time or whether the discharge at a given channel will exceed some critical amount under a variety of initial and boundary conditions. Such queries require major computational support for the data-intensive and numerically-intensive modeling activities. In relation to data-intensive computation, for example, the amount of data that has accumulated and will accumulate over the next few decades about such systems is massive, and involves large collections of images, digitized maps and experimental datasets. In relation to both data-intensive and numerically-intensive computation, there is a growing realization that the mode for hydrologic system investigations must change from one in which relatively simple, lumped-system modeling is dominant to one in which distributed system modeling based on detailed physical process modeling is the standard approach.

There are currently no DBMS that could support the data and modeling requirements of such distributed systems. Apart from the problems of handling large volumes of heterogeneous, distributed and spatially-indexed datasets, computational issues relating to such support include: the definition and manipulation of the complex objects that arise in part from the nature of the domain and from the fact that modeling is occurring over several scales of spatial and temporal resolution; the integration of modelling of activities that span the whole of the extractive axis; and the manipulation of mathematical models consisting, for example, of sets of algebraic and partial differentiation equations.

# 3 The Data Model

We now present a simple data model that allows us to define and manipulate both spatio-temporal objects and spatio-temporal object classes. Our constructs are based on the general requirements outlined above, and involve a data model that, while simple, is highly expressive with respect to a broad range of spatial and spatio-temporal objects that we believe are necessary in many applications. The key ideas concerning spatial objects are that they possess an identity; that they have a structure that contains an essential component defined over some spatial domain and an optional component defined over some non-spatial domain; that they possess values that depend on time; and that they may be organized into a partially ordered class structure in terms of their structure and constraints upon their structure.

Our presentation of the data model begins with a stepwise development of the domain of objects of interest. It is important to note that the distinction between data model and data manipulation language is blurred by the fact that we allow intensional definitions of *classes*. We defer discussion of these rule-oriented aspects of the data model until we have introduced the data manipulation language. Hence, this part of the definition of the data model is only focussed on the domain of spatial objects, as defined in terms of their basic structures, and not in terms of the constraints upon the structures, which we develop below.

## 3.1 Domains and Types

Domains of values form the basis for constructing spatial object types, spatial objects and spatial object classes. We introduce several preliminary domains before defining our *domain of discourse* $\mathcal{D}$. In particular, $\mathcal{N}$ is a domain of non-spatial attributes of spatial objects; $\mathcal{S}$ is a domain of spatial attributes of objects; $\mathcal{I}$ is a domain of object identities; $\mathcal{T}$ is a domain of time; $\mathcal{O}$ is a domain of spatial objects; and $\mathcal{TO}$ is a domain of spatio-temporal objects. We employ the concept of *type* to indicate the structure of an element in some domain, and allow a simple naming convention to facilitate the definition of types. Our notion of type equivalence is structural. The structure of an object is of critical importance in our development, since one of our principal mechanisms for creating semantically important classes of objects is to place constraints on the values that a structure may assume.

$\mathcal{N}$ is the domain of non-spatial values or *terms*. The domain $\mathcal{N}$ is similar to the domains of typical CO-DATALOG languages such as COL [AG91] or LDL [NT89]. We treat this domain as a single entity, since our focus is on spatial objects, and assume for simplicity that the elements of $\mathcal{N}$ are of a single type **N_type**. In constructing $\mathcal{N}$, we assume that sets of constants, variables and function symbols are available. A term is a constant, variable or an n-ary function symbol applied to n terms. A *ground term* is a term that does not contain any variables. We assume that the set of function symbols includes the special arity-less set-constructor $\{\}$, which can be applied to any number of terms to obtain a set containing exactly these terms. We also assume that the set of function symbols contains the special k-ary tuple constructor $[]_k$ for each positive integer value k. $[]_k$ applied to a sequence of k terms yields a tuple value.

$\mathcal{S}$ is the domain of spatial values or *terms*. This domain is the *distinguishing domain* of spatial objects, and serves three main roles of providing values for the spatial components of complex spatial objects; defining a special subclass of spatial objects; and providing values for a special set of functions defined on spatial objects, namely the *S_projections* of the objects. As in the case of $\mathcal{N}$, we assume that sets of constants, variables and function symbols are available; that a term is a constant, variable or an n-ary function symbol applied to n terms; that a *ground term* is a term that does not contain any variables; and that the set of function symbols includes the special arity-less set-constructor {} and the special k-ary tuple constructor $[]_k$.

The construction of $\mathcal{S}$ is based upon a predomain of points $\mathcal{P} \subseteq R^n$ and a collection of **S_types** in $\mathcal{S}$. Basic *S_types* are of two forms, *Point* and *Pointset*. The elements of *S_type Point* are the values in $\mathcal{P}$. The elements of *S_type Pointset* are the values in $\mathbf{P}^{\mathcal{P}}$, which denotes the powerset of $\mathcal{P}$. [1] Tuple *S_types* are constructed as follows. If $T_1, \ldots, T_k$ are *S_types*, then $[T_1, \ldots, T_k]$ is also a *S_type*. Set *S_types* are of the form $\{T\}$ if $T$ is also a *S_type*. The domain $\mathcal{S}$ is defined to contain all elements of any basic, set or tuple *S_type*. Clearly the domain $\mathcal{S}$ may in general be extremely large, and will typically contain elements that are uncountable. These characteristics raise two issues. First, the domain may be much larger than is needed for many applications and second, we require finite representations of elements that are infinite. Hence a key idea that we refine below involves defining special cases of $\mathcal{S}$ that satisfy these criteria.

$\mathcal{I}$ is the domain of *object identities* of spatial objects.

$\mathcal{T}$ is the domain of *time*.

$\mathcal{O}$ is the domain of spatial objects. This domain is the main focus of the current paper and is developed in in the next section. In general, a spatial object is an object that is defined in terms of $\mathcal{N}$ and $\mathcal{S}$, and that possesses an essential projection into $\mathcal{S}$. A spatial object may also have projections into the non-spatial domain $\mathcal{N}$. We assign to each spatial object a unique identity, and use identity-based equality extensively. The construction of $\mathcal{O}$ is based upon a collection of **O_types**.

$\mathcal{TO}$ is the domain of spatio-temporal objects. In general, a spatio-temporal object is an object that is defined in terms of $\mathcal{N}$, $\mathcal{S}$ and $\mathcal{T}$, and that possesses an essential projection into $\mathcal{S}$. We assign to each spatio-temporal object a unique identity, and use identity-based equality extensively. The construction of $\mathcal{TO}$ is based upon a collection of **TO_types**.

Finally, D is our overall domain of discourse. This domain may be constructed as follows: Every term of $\mathcal{N}$, every term of $\mathcal{S}$, every element of the spatial object domain $\mathcal{O}$ and every element of the spatial object domain $\mathcal{TO}$ is a term of $\mathcal{D}$. Additionally, terms can be constructed by applying function symbols to terms as in the definition of $\mathcal{N}$.

## 3.2 The Domains of Spatial and Spatio-Temporal Objects Revisited

In this section, we develop in greater detail the domains of spatial and spatio-temporal objects $\mathcal{O}$ and $\mathcal{TO}$. The goal is to develop a relatively small set of "primitive" types that are sufficiently expressive to cover a range of objects along the

[1]Pointset can also be defined using Point and the set-type constructor defined below, but it is useful to distinguish this special *S_type*.

extractive axis. We first develop the concept of spatial objects and their types and then the concept of spatio-temporal objects and their types; we next develop the idea of the *S_projections* of objects; and finally we develop the concept of spatial object classes.

**Definition 3.1** *A* **point atomic spatial object** *is a tuple [id, t] where id* $\in \mathcal{I}$ *and* $t \in \mathcal{P} \times \ \mathcal{X}_1 \times ... \times \ \mathcal{X}_n,\ n \geq 1$ *and* $\mathcal{X}_i = \mathcal{P}$ *or* $\mathcal{N},\ i = 1,\ n$. *The O_type of a point atomic spatial object is defined to be* **pointatomic**.

**Definition 3.2** *A* **pointset atomic spatial object** *is a tuple [id, t] where id* $\in \mathcal{I}$ *and* $t \in \mathbf{P}^{\mathcal{P}} \times \mathcal{X}_1 \times ... \times \ \mathcal{X}_n,\ n \geq 0$ *and* $\mathcal{X} = \mathbf{P}^{\mathcal{P}}$ *or* $\mathcal{N},\ i = 1, n$. *The O_type of a point atomic spatial object is defined to be* **pointsetatomic**.

**Definition 3.3** *A* **point relation spatial object** *is a tuple [id, R] where id* $\in \mathcal{I}$ *and* $R \subseteq \mathcal{P}_1 \times \ \mathcal{X}_1 \times ... \times \ \mathcal{X}_n,\ n \geq 1$ *and* $\mathcal{X}_i = \mathcal{P}$ *or* $\mathcal{N},\ i = 1,\ n$. *The O_type of a point atomic spatial object is defined to be* **pointrelation**.

**Definition 3.4** *A* **pointset relation spatial object** *is a tuple [id R] where id* $\in \mathcal{I}$ *and* $R \subseteq \mathbf{P}^{\mathcal{P}} \times \mathcal{X}_1 \times ... \times \ \mathcal{X}_n,\ n \geq 0$ *and* $\mathcal{X}_i = \mathbf{P}^{\mathcal{P}}$ *or* $\mathcal{N},\ i = 1,\ n$. *The O_type of a point atomic spatial object is defined to be* **pointsetrelation**.

As we note below in our examples, the first two of our four primitive types are of value in representing objects that are high on the extractive axis while the second two are of value for representing objects that are low on this axis. We now complete our basic data model by applying tuple and set constructors to our basic types.

**Definition 3.5** *A* tuple spatial object *is a tuple* $[id,\ [o_1, o_2, ..., o_k]]$, *where for* $1 \leq i \leq k$, $o_i$ *is a spatial object of O_type* $T_i$ *and id* $\in \mathcal{I}$. *The O_type of the tuple spatial object is defined to be* $[T_1, \ldots, T_k]$.

**Definition 3.6** *A* set spatial object *is a tuple [id,* $\{o\}$*], where* $\{o\}$ *is a set of spatial objects of O_type T and id* $\in \mathcal{I}$. *The O_type of a set spatial object is defined to be* $\{T\}$.

The values associated with the two types of atomic objects are both tuples whose first element **must** be an element of the domain $\mathcal{S}$. In the case of point atomic objects, it must be a point in $\mathcal{P}$ and in the case of pointset atomic objects, it must a pointset in $\mathcal{P}$ or, equivalently, an element of $\mathbf{P}^{\mathcal{P}}$. In the case of point atomic objects, the tuple must contain at least two elements. In the case of pointset atomic objects, the tuple may contain only one element, which must be a pointset. A point atomic object may contain any finite positive number of points, and any positive finite number of elements from $\mathcal{N}$, while a pointset atomic object may contain any positive finite number of pointsets and any finite number of elements of $\mathcal{N}$. The point and pointset relation objects take *sets* of the tuples employed in the definition of atomic objects as the values of objects. It is easy to see that these four type of objects are disjoint in terms of their respective domains of values. We note, for example, that pointsets cannot be obtained from point relation objects.

In many applications of importance, spatial objects vary as a function of time. It is therefore important to define the concept of a spatio-temporal object as a spatial object for which a *subset* of its values vary as a function of time and to denote such objects as being of **TO_type**. It should be noted that this definition limits the spatio-temporal objects with which we deal to be those whose structure (and hence type) remains invariant as a function of time, although the associated values may change. The definitions of spatio-temporal objects are similar to those of spatial objects, with the difference that we allow a "slot" for time in the four primitive object types. For example, we may define a **point atomic spatio-temporal object** as a tuple [id, t] where $id \in \mathcal{I}$ and $t \in \mathcal{T} \times \mathcal{P} \times \mathcal{X}_1 \times \ldots \times \mathcal{X}_n$, $n \geq 1$ and $\mathcal{X}_i = \mathcal{P}$ or $\mathcal{N}$, $i = 1, n$. The other definitions are similar, and in general, a *spatio-temporal object* is a point atomic spatio-temporal object, a pointset atomic spatio-temporal object, a point relation spatio-temporal object, a pointset relation spatio-temporal object, a tuple spatio-temporal object or a set spatio-temporal object.

## 3.3 Properties of Spatial Object Types

We take it as given that the focus of much computation over spatial databases involves various properties of complex spatial objects and object classes. Typically such properties are functions defined on the values of the structure of a spatial object or class, and hence depend on the *O_type*. The most fundamental property defined on the structure of a spatial object is the *S_projection*. In particular, we will see that a spatial object is an object that possess a non-empty *S_projection*, under a reasonable definition of *S_projection*.

In general the *S_projection* of a spatial object is a mapping from objects to the domain $\mathcal{S}$. There is a natural class of spatial projections from which one may compute many classes of useful projections. Essentially, the *natural S_projection* of an object possesses the same structure as the object itself. If we let $\pi_S^{nat}(o)$ denote the natural *S_projection* of the object $o$, then we have the

**Definition 3.7** *The natural S_projection of a spatial object is:*

1. $\pi_S^{nat}(o) = [p_1, \ldots, p_k],\ k \geq 0,\ p_i \in \mathcal{P}$
   *for a point atomic spatial object o. Each point $p_i$, $i = 1, k$ occurs in the S_projection iff it occurs in the spatial object.*

2. $\pi_S^{nat}(o) = [\{p\}_1, \ldots, \{p\}_k],\ k \geq 0,\ p \in \mathcal{P}$
   *for a pointset atomic spatial object o. Each pointset $\{p\}_i$, $i = 1, k$ occurs in the S_projection iff it occurs in the spatial object.*

3. $\pi_S^{nat}(o) = \{[p_1, \ldots, p_k]\},\ k \geq 0,\ p_i \in \mathcal{P}$
   *for a point relation spatial object o. Each tuple of points occurs in the S_projection iff it occurs in the spatial object.*

*4.* $\pi_S^{nat}(o) = \{[\{p\}_1, \ldots, \{p\}_k]\},\ k \geq 0,\ p \in \mathcal{P}$
*for a pointset relation spatial object o. Each tuple of pointsets occurs in the S_projection iff it occurs in the spatial object.*

The natural projections of tuple objects are defined recursively as
$\pi_S^{nat}(o) = [\pi_S^{nat}(o_1), \ldots, \pi_S^{nat}(o_n)]$
while for set objects, we have
$\pi_S^{nat}(o) = \{\pi_S^{nat}(o)\}$.
We denote by $\mathcal{S}_\pi^{nat}$ the domain of *S_projections* over all objects in the domain $\mathcal{O}$.

There are some strong relationships between the domains $\mathcal{S}$, $\mathcal{O}$ and $\mathcal{S}_\pi^{nat}$. Clearly, the natural *S_projection* of any spatial object contains at least one point $p \in \mathcal{P}$. Since every spatial object has a non-null natural *S_projection*, we may define a set of equivalence classes over $\mathcal{O}$ in terms of their natural *S_projections*. We term this set the *spatial equivalence classes* of domain $\mathcal{O}$. It is clear that $\mathcal{S}_\pi^{nat} = \mathcal{S}$. Hence it follows that there is a one-one correspondence between the elements of the domain $\mathcal{S}$ and the set of spatial equivalence classes of domain $\mathcal{O}$. In many applications, it is often convenient to define other *S_projections* of spatial objects. Hence we define a *general S_projection* to be some mapping $\mathcal{M}$ from the set of natural *S_projections* into $\mathcal{S}$: $\mathcal{M} : \mathcal{S}_\pi^{nat} \rightarrow \mathcal{S}$. In particular, it is often of value to conceptualize the *S_projection* of a spatial object as a pointset that is an element of $\mathbf{P}^{\mathcal{P}} \subset \mathcal{S}$. We term this class of projections the class of *spatial projections*.

## 3.4 Spatial Object Classes

As noted above, an core activity in scientific investigations involves the characterization of important classes of objects in terms of their properties, behavior and interrelationships. While a class of spatial objects may, in general, be conceptualized simply as a collection of spatial objects, it is much more fruitful to view a class as a set of objects that possess common properties, particularly in terms of their structure and their behavior. Hence we will typically associate a given *O_type* with all members of a given class and also associate a set of *methods* with all members of the class, and define a class as a tuple [*classname*, *O_type*, *methods*]. Methods can be defined for a general spatial object class, and can be inherited by specialized classes of spatial objects.

# 4 A Language for Defining and Manipulating Complex Spatial Objects and Object Classes

Having presented a conceptual data model, we now describe the elements of a formal language for defining and manipulating spatial objects and classes. We term the language CSOL (complex spatial object language). For present purposes, it is convenient to base our language for spatial objects on current logic-based complex object (or CO-DATALOG) languages such as COL [AG91] or LDL [NT89]. We provide a brief synopsis of the main elements of our language, which we subsequently extend with additional object-oriented features for dealing with spatial objects.

As usual, the language possesses an alphabet of constants, variables, function symbols, constant symbols, logical symbols and punctuation symbols; we also define terms, atoms, literals, rules and programs in the usual manner. In particular, a query is expressed using a collection of rules of the form **head** $:-$ **body**, where the head is a *positive literal* and the body is a *conjunction of literals*. As in LDL, the construction of set terms is supported through a special kind of rule in which the head literal is allowed to contain the special *set-generation term* $< X >$, where X is a variable that appears in the body and nowhere else in the head.

The language also possesses a relatively standard model-theoretic semantics, although the use of set-generation ($< ... >$) and negation ($\neg$) is constrained to be *stratified*. That is, if predicate p is defined in terms of q using either set-generation or negation, then q cannot be defined in terms of p.

## 4.1 A Set of CSOL Primitives

In addition to general rules introduced in the previous section, we require some primitives for defining and manipulating spatial objects and spatial object classes easily. First, it is of value to have a small number of primitives that may be used to define elements of domain $\mathcal{S}$. With respect to domains $\mathcal{O}$ and $\mathcal{OT}$, we need to be able to define new object *types* and to access the structure of a *type*; we need to be able to define new objects in terms of their *type* and values and to access the value, *type* and *S_projection* of an object; we need to be able to define new classes of objects and be able to obtain their *type* and associated methods, and to assign subclass membership; finally, we need to be able to assign an object to a class and to obtain its class membership. Our classification rules permit both structural and semantic information to be used in specifying the class to which a spatial object belongs.

The following primitives are useful in defining types, objects and classes in the domain $\mathcal{O}$. Similar primitives may be defined for the domains $\mathcal{S}$ and $\mathcal{TO}$.

1. **O_type** $T =$ *O_type_definition*
   defines T as an *O_type*.

2. **O_class** $= \{$*classname*, *O_type*, *methods*$\}$
   defines C as an *O_class*.

3. *makeo*([*id*, *V*]) $\Leftarrow$ *new*(*id*), *conditions*
   The conditions define the object to to have value V. A unique new id-value

is generated by the use of the special primitive *new*, which can only be used in the body of a *makeo* rule. This is an imperative construct, in contrast to the declarative rules of the basic language. Its result depends upon when it is executed and it is not repeatable (since different id values are generated each time it is used). In database terminology, this construct belongs in the data definition and update language, and not the query language.

4. *oval(id, V)*
   This is a predicate which takes an object (with identity id) and returns its value.

5. *otype(id, T)*
   This is a predicate which takes an object (with identity id) and returns its *O_type* (and thus implicitly its structure). It can be used to check whether the structure matches some pattern through unification.

6. *S_proj(id, P)*
   Given an object id, this is a primitive predicate, analogous to *oval*, that enables us to obtain the spatial component of the object. We assume that this predicate returns the natural *S_projection* of an object defined above. However, as noted above, a different representation of the spatial component is frequently desirable. We allow a user to define a different projection function for each class, overriding the default:
   $S\_proj(C, P) \leftarrow$ *conditions*
   This is taken to be the *S_projection* function for all objects in class C.

7. $classof(id,\ C) \leftarrow$ *conditions*
   This states that the object with identity id is a member of class C if the conditions in the body hold. It is the main mechanism for defining the conditions that an arbitrary object must satisfy if it is to belong to the class. The body is allowed to contain any literal that appears in the body of a rule, and such literals may be viewed as constraints on the objects, and as constraints on the structure of the objects in particular. This construct is again part of the data definition language, but we permit its use in the body of an ordinary rule with an overloaded useage.

8. *subclass*$(C1,\ C2)$
   This is a predicate that declares class C2 to be a subclass of class C1. We require that every object of class C2 must also be of class C1. To this end, the conditions associated with class C2 (by an *class* assertion) are taken to include all the conditions associated with class C1. This is part of the data definition language, but again we permit an overloaded usage in the bodies of rules.

9. $makeoin([id,\ V,\ C]) \Leftarrow new(id)$, *conditions*
   This construct is a variant of *makeo* which creates the object and declares it to

be in class C. It is intended to be a concise notation for a frequently occurring pattern of *classof* followed by *makeo* commands.

Finally, we require methods for accessing particular values in some structure. If the structure S is a tuple [X, Y], then we employ the notation S.X to denote the term X; if the structure is a set, we employ the predicate $X \in S$ to denote the element X.

# 5 Defining Special Subdomains of $\mathcal{S}$ and of $\mathcal{O}$ in CSOL

The general domain $\mathcal{S}$ of spatial attribute values, which results from unlimited applications of the set and tuple constructors to predomain $\mathcal{P}$, is extremely large. In many important classes of applications of spatial database systems it is unnecessary to employ such a large domain, and it is therefore of advantage to consider special subdomains of $\mathcal{S}$. The resulting simplifications have important computational consequences, particularly when manipulations of interest are largely concerned with the spatial component of objects. For example, in the case of the drainage basin application, an adequate domain would be one sufficient to represent the spatial projections of channel links, drainage areas and storms. Similar remarks apply to subdomains of the domain $\mathcal{O}$ of spatial objects, in which one is only interested in objects defined over special subdomains of $\mathcal{S}$.

We now describe applications of CSOL in which we first construct special subdomains of $\mathcal{S}$ and then use these subdomains to create special subdomains of $\mathcal{O}$. In particular, we create subdomains of $\mathcal{O}$ that consist of *pointset* objects.

## 5.1 A Procedure for Defining Subdomains

We employ the following general strategy in defining a set of first order theories of special subdomains of $\mathcal{S}$. These theories may be used to define special subdomains of $\mathcal{O}$ of pointset objects:

1. choose a **finite** number of points $\mathcal{P}_0 \subseteq \mathcal{S}$;

2. apply the $\{\}$ and $[]_k$ constructors a relatively small number of times to the set $\mathcal{P}_0$;

3. apply a (generally small) subset of special *pointset constructors* $\{C_1, C_2, C_3, ...\}$ to the elements of relevant subdomains of values created in the first two steps, in order to create new pointsets in $\mathbf{P}^{\mathcal{P}}$. Simple examples of such constructors are: the *line* constructor, which operates on 2-tuples of points from $\mathcal{P}_0$ and constructs a set of points that lies on a line segment connecting the two points;

the *rectangle* constructor, which takes a 2-tuple of points and constructs a rectangular set of points in the natural way; and the *ngon* constructor, which operates on n-tuples of points from $\mathcal{P}_0$ and constructs a set that is an n-sided polygon;

4. compute new elements in this set of pointsets under a set of pointset operators, such as $\cup$ and $\cap$;

5. compute the closure of this domain under $[]_k$ and $\{\}$ and hence determine an appropriate subdomain of values $\mathcal{S}_0 \in \mathcal{S}$;

6. create particular object types, objects and object classes based on the subdomain $\mathcal{S}_0$.

Since many of the elements of $\mathcal{S}_0$ and many of the pointset objects so constructed involve an uncountable number of points from $\mathcal{P}$, an important problem is to find a satisfactory finite representation of such elements. We may, however, "solve" this problem by both viewing and employing the constructors in terms of a finite set of *constraints* on the points that may belong to a pointset. Since these constructors are applied to a finite set of points, we may view the combination of the finite set of points and the finite set of constraints as a finite representation of a typically infinite set of points.

We now illustrate this approach by constructing, in terms of CSOL, a sequence of first order theories $T_{P_1} \subset T_{P_2} \subset \ldots \subset T_{P_k}$. A small number of special cases of these theories are adequate for many current applications. In developing this and later applications, we assume the existence of certain basic types, such as *reals* and *integers*. We present these applications in terms of the six steps listed above. Before proceeding, it is of value to make some basic stype definitions:
**S_type** $point = [real, ..., real]$
**S_type** $pointset = \{point\}$
**S_type** $kpointtuple = [point, ..., point]_k,\ k = 1,\ N$
and to define a few basic predicates:
$point(P)\ :-\ stype(P,\ point)$
$pointset(PS)\ :-\ stype(PS,\ pointset)$
$kpointtuple(PKT)\ :-\ stype(PKT,\ kpointtuple)$

## 5.2 Pointset Theories $T_{P_1}$, $T_{P_2}$,...,$T_{P_n}$

The pointset theory $T_{P_1}$ is simple and does not require the use of any pointset constructors. Applying our six steps:

1. we first choose n points $p_1, ..., p_n$ from $\mathcal{P}$.

2. We apply the set constructor {} once to each of the n points to produce the singleton pointset elements $\{p_i\} \in \mathcal{S},\ i = 1, n$

3. There are no pointset constructors required in this example.

4. We compute any required pointset elements of $\mathcal{S}_0$ by applying the $\cup$ and $\cap$ operators, using the following rules:
$pointset(PS)\ :-\ pointset(PS_1),\ pointset(PS_2),\ \cup(PS_1,\ PS_2,\ PS)$
$pointset(PS)\ :-\ pointset(PS_1),\ pointset(PS_2),\ \cap(PS_1,\ PS_2,\ PS)$

5. We do not need to apply the [] and {} constructors further in the present example in order to define $\mathcal{S}_0$.

6. In order to create objects from the domain of values defined above, we first create an appropriate class for such objects
**O_class** = (*pointsetclass*, *pointsetatomic*, *pointsetmethods*)
and, for convenience, define the predicate:
$pointset(P)\ :-\ classof(P,\ pointsetclass)$
on the domain $\mathcal{O}$. We now use our initial set of points to construct singleton set atomic objects
$makeoin([id,\ ps_i,\ pointsetclass]) \Leftarrow new(id),\ pointset(ps_i)$
for specific pointsets $ps_i$, $i = 1, n$. The *S_projection* of such a pointset object is defined to be the pointset itself. By employing the predicates $\cup$ and $\cap$ we may now define the union or intersection of any two given pointsets objects to be a pointset object using
$makeoin([id,\ V,\ pointsetclass]) \Leftarrow new(id),\ pointset(PS_1),\ pointset(PS_2),$
$oval(PS_1,\ V_1),\ oval(PS_2,\ V_2),\ \cup(V_1,\ V_2,\ V)$
and
$makeoin([id,\ V,\ pointsetclass]) \Leftarrow new(id),\ pointset(PS_1),\ pointset(PS_2),$
$oval(PS_1,\ V_1),\ oval(PS_2,\ V_2),\ \cap(V_1,\ V_2,\ V)$

The preceding discussion defines $T_{P_1}$. We may, however, easily extend the theory and, in particular, create more structured pointsets in $\mathcal{S}_0$ by placing constraints on the points in a pointset, or on the coordinates of the points. For example, we may create *latticepointsets*, in which each point lies at the vertex of a rectangular lattice, by placing constraints on the relative locations of points.

In defining the pointset theory $T_{P_2}$, we first assume $T_{P_1}$. We wish to define a class of *line* pointsets and objects in terms of a 2tuple of points in $\mathcal{S}_0$ that serve as the endpoints of the line and the pointset formed from all convex combinations of the two points. It is therefore appropriate to define a line pointset to be the pointset that is defined in terms of some pointset constructor defined on the tuple. Hence we define the following predicate
$linetuple(LT)\ :-\ stype(LT,\ 2pointtuple)$
as a basis for defining a similar predicate for line pointsets.

We again follow the six steps:

1. Step one is identical to that of the preceding example.

2. In step two, we apply the 2pointtuple constructor to points in $\mathcal{P}$ in order to obtain tuples $[P_1,\ P_2]$.

3. We apply the following linepointset constructor $C_{lineset}$ that is a predicate defined by the following rule:
$C_{lineset}(LT,\ <[Y_1, ..., Y_n]>)\ :-\ linetuple(LT)), real(\alpha),\ real(\beta),$
$\geq(\alpha,\ 0),\ \geq(\beta,\ 0),\ +(\alpha,\ \beta,\ 1),\ \times(\alpha,\ LT.P_1.X_1,\ PX_{11}), ...,\ \times(\alpha,\ LT.P_1.X_n,\ PX_{1n}),$
$\times(\beta,\ LT.P_2.X_1,\ PX_{21}), ...,\ \times(\beta,\ LT.P_2.X_n,\ PX_{2n}),$
$+(PX_{11},\ PX_{21},\ Y_1), ...,\ +(PX_{1n},\ PX_{2n},\ Y_n),\ point([Y_1, ..., Y_n])$
We note how the constructor is expressed in terms of a finite number of points and constraints.

4. We may apply the union and intersection operators as in the preceding case in order to define new pointsets, although they will in general not be simple linesets.

5. We do not need to apply the set and tuple constructors in the present case in order to define further $\mathcal{S}_0$.

6. We may now create various object types, object classes and objects. We focus on objects that are linetuple objects and line pointset objects. Although a linetuple is of type *pointatomic*, it is convenient, for later applications to define a special type for such objects
**O_type** *linetuple* = [*point*, *point*]
We now define appropriate classes of objects
**O_class** = (*linetupleclass*, *linetuple*, *linetuplemethods*)
**O_class** = (*lineclass*, *setatomic*, *linemethods*)
We may employ the simplifying rules
*linetuple*$(LT)\ :-\ classof(LT,\ linetupleclass)$
*line*$(L)\ :-\ classof(L,\ lineclass)$.

For theory $T_{P_n}$, we assume $T_{P_{n-1}}$. In a procedure that is exactly analogous to the procedure for defining linetuple and line elements in $\mathcal{S}$ and linetuple and line pointset objects in $\mathcal{O}$, we may define *ngontuples* and *ngon pointset* elements in $\mathcal{S}$ and ngontuple and ngon object types, objects and objects classes in $\mathcal{O}$. In particular, we employ the pointset constructor $C_{ngonset}$ over ntuples to construct ngon pointsets, using convex combinations of the points in the ntuple. It is clear that the theory $T_{P_n}$, for sufficiently large n and specialized to a 2-dimensional Euclidean space, may be viewed as the appropriate geometrical theory that underlies the set of **vector** (or point, line and polygon) models of spatial objects.

## 5.3 Pointset Theories: Their Significance and Extensions

The simple theories developed above may be viewed as constructive approaches to defining special cases of the spatial domain $S$, as well as to creating special classes of objects in $\mathcal{O}$. In many applications, theories similar to those described above are quite sufficient for defining relevant subdomains of both $S$ and $\mathcal{O}$, as we noted in the cases of $T_{P_1}$, $T_{P_2}$ and $T_{P_n}$.

In constructing the theories $T_{P_i}$ for i=1,n, the basic approach involved the construction of a class of (typically uncountable) pointsets from an initial number of tuples of points and a set of constructors that were expressed in terms of a finite number of constraints. The various constraints that were employed in these constructions were linear constraints relating to the points and/or their coordinates. It would appear that in a large number of applications of interest, it is sufficient to define pointsets and objects of relevance in terms of finite tuples of points and a finite set of polynomial constraints or finite order on the coordinates of the points, expressed in the form of CSOL rules.

# 6 Defining Dataset Objects in CSOL

We now exemplify applications of our data model and CSOL with respect to an important class of data objects that represent data sets. While such objects are low on the extractive axis, it is frequently convenient to define and manipulate such data sets as objects in the same environment as objects that are higher on the extractive axis. The point and pointset relation objects are of particular value in such applications, in which defining and manipulating sets of spatially indexed data is of great significance.

We noted above the importance in many scientific investigations of datasets in which measured values of certain attributes were associated with some finite set of points. It is clear that such data sets may be represented naturally as a *point relation spatial object*. We may, for example, define special types, classes and predicates for such datasets

**O_class** = (*pointdatasetclass*, *pointspatialrelation*, *methods*)

$pointdataset(PDS)\ :-\ classof(PDS,\ pointdatasetclass)$

and create specific objects using

$makeoin(id,\ \{[p_1,\ a_1],\ ...,\ [p_n,\ a_n]\},\ pointdatasetclass)\ \Leftarrow\ new(id)$

If the associated attribute values are each measured over some area (ie a pointset), then we may treat a set of such observation as a *pointset relation spatial object* and define special types and classes for pointset datasets

**O_class** = (*setdatasetclass*, *pointsetspatialrelation*, *methods*)

and create them using

$makeoin(id,\ \{[\{p\}_1,\ a_1],\ ...,\ [\{p\}_n,\ a_n]\},\ setdataclass \Leftarrow new(id).$

Many dataset objects, such as images or certain classes of digitized maps, are defined over a set of points that have regular structure, such as a lattice. As noted in the previous section, we may define special classes of *rasterdatasetclass* objects that satisfy given geometrical constraints.

It is relatively straightforward to model more complicated dataset objects in terms of tuple and set constructors and constraints on class membership. It is frequently advantageous to treat several datasets as a "named" unit. For example, many databases contain such layers, and the objects of ultimate interest in system modelling are typically represented *implicitly and jointly* in several of the layers. Hence one must apply procedures to several layers jointly in order to make the objects explicit. Particular domains of objects may require particular subsets of layers, and in such cases the ability to treat any given subset as a unit with its own object identity is an obvious advantage. As an example, one may wish to collect certain raster dataset objects, whose *S_projections* intersect a given rectangular geographical window, and treat them as an object that can be accessed and processed as a unit.

# 7 Modelling Spatio-Temporal Objects in a Complex System

We now exemplify applications of our data model and CSOL to the complex domain of application mentioned above, namely a large-scale hydrologic network, which involves objects that occur across the whole of the extractive dimension. We exemplify the definition and manipulation of complex spatial objects; the use of deductive rules; and the use of class hierarchies. In particular, we construct, using CSOL, a simple model of a drainage basin and its flow dynamics during a storm event. It is to be emphasized that, from phenomenological point of view, the conceptual model on which the CSOL program is based is simplistic. This construction is used only to exemplify the main features of our data model and deductive language in a relatively complex *domain* of application, although we would like it to be viewed as a simple prototype of much more realistic models that might be constructed in a similar manner.

In the *simplest* of terms, a drainage basin is a system of stream channels that serve to drain some area of land, together with the associated areas of land that are being drained and the relevant properties of both the channels and the associated areas of land. The system of stream channels may be viewed as a binary tree composed of a set of linear links. Each link is a maximal length section of a stream to which no other streams are tributary. Each link has an associated area of the *total* drainage basin that drains *directly and only* into that link. This drainage area may be characterized in terms of the proportions of different rock, soil and vegetation types that occur within it. These characteristics determining the amount of rainfall that flows overland from the drainage area into the stream during and between storm events. Such runoff occurs in a storm when the soil becomes saturated. For example, runoff from the saturated area occurs when the difference between the rainfall intensity and the amount of water that can be absorbed by the soil, is positive. The amount of discharge into a link at a given point in time is the discharge from the two parents of the link and the runoff from the saturated portion of the drainage area of the link.

For current purposes, we model both a storm event and the drainage basin response in relatively simple terms. With respect to a storm event, we assume a simple *S_projection* and simple dynamics, in the sense that a storm moves in a straight-line at a constant-speed while the rainfall in the area of the *S_projection* of the storm is assumed to possess a uniform intensity at any time, although this area and the rainfall intensity change over time in a proportionately decreasing manner. We characterize the flow of water in an element of the basin corresponding to a link as follows. We assume that if the storm intersects some part of the area associated with the link and if the rainfall intensity exceeds a certain value, dependent on the geology, then the rainfall will flow over this saturated area into the link. We further assume that the water flow in the link in a given time interval is the sum of the runoff from the saturated area, generated during that time interval, and of the flow from the parent links that was generated in the *previous* interval of time.

## 7.1 Spatial Object Classes

We now proceed to define the spatial object classes in $\mathcal{O}$ that we use to model a drainage basin. We first assume that the modelling of a complex drainage basin requires data objects that are low on the extractive axis. As an example, we assume the existence of a *pointdatasetobj*, as defined above, in which the ground surface elevations at a sufficient number of points in the area underlying the drainage basin are represented in an object of type *pointspatialrelation* defined by
$makeoin([id,\ \{[p_1,\ e_1], ...,\ [p_n,\ e_n]\},\ pointdatasetclass]) \Leftarrow new(id)$
We further assume that, for this object, $id =$ **elev**.

In terms of objects that are high on the extractive axis, the most fundamental object class is a *linkclass*, which we employ in representing the links in a drainage network. We define link objects to be a subclass of linetuple objects, defined above, in which the defining endpoints are ordered in terms of their elevation, so that the first point is the upstream endpoint. Hence we have the following class definition
**O_class** = (*linkclass*, *pointatomic*, *linkmethods*)
*subclass*(*linkclass*, *linetupleclass*).
Since the elevation of the upstream point of a link must be greater than the downstream elevation of a link, we may access and use the data set **elev** in order to define a legal link. We therefore have the defining rule
$classof(LT,\ linkclass) \leftarrow linetuple(L),\ oval(L,\ V),\ pointdataset($**elev**$),$

*oval*(**elev**, $V_E$), $\in$ ($T_1$, $V_E$), $=$ ($V.P_1$, $T_1.P$), $\in$ ($T_2$, $V_E$), $=$ ($V.P_2$, $T_2.P$), $>$ ($T_1.E$, $T_2.E$)

A *basin element*, or *basel*, characterizes the static characteristics of the smallest integral areal unit of a basin. A basel includes a single link, together with its related parent and child links, the area that it drains, its order and the associated geological characteristics. We define the structure of a basel in terms of a tuple type
**O_type** *basel* = [*linetuple*, *linetuple*, *linetuple*, *linetuple*, *pointsetatomic*]
The first four elements are linetuple types representing links which are ordered as follows: the first is the link of interest; the second is the right parent link (if any); the third is the left parent link (if any); and the fourth is the child link. The fifth element is a pointset atomic type having the form
[*id*, [*pointset*, *integer*, *integer*]]
in which the pointset is the area drained by the link, the first integer denotes the geological type of the area drained by the link and the second integer denotes the order of the link draining the area. Hence we now define a basel class of objects
**O_class** = (*baselclass*, *basel*, *baselmethods*)
and define an object to be of basel class if the following rule is satisfied
*classof*(*B*, *baselclass*) : − *otype*(*B*, *basel*), *oval*(*B*, *V*), *joins*(*V*.2, *V*.1), *joins*(*V*.3, *V*.1), *joins*(*V*.1, *V*.4)
The predicate *joins* is true for two linetuple objects if the second point of the first line equals the first point of the second line. Finally, we may define the useful predicate
*basel*(*B*) : − *classof*(*B*, *baselclass*)

It is possible in this domain to define a large number of other classes of important objects in a similar manner. We may, for example, define a class that contains sets of basels or *baselset*, and a class of *basins*. Any object in the class of basins denotes, for a given link the entire set of basels whose links drain into the link, or equivalently, the transitive closure of a set of basels under link ancestry. It is easy to express such transitive closures in our base language

## 7.2 Spatio-Temporal Object Classes

We define two major classes of spatio-temporal objects, namely a *storm* and *link flow element*. We commence with the useful type and class definitions
**TO_class** = (*stormclass*, *pointsetatomic*, *methods*)
The definition for a storm is
*classof*(*S*, *stormclass*) $\leftarrow$ *totype*(*S*, *storm*), *toval*(*S*, [$\tau$, *SPS*, *F*]), *stype*(*SPS*, *pointset*), *real*(*F*)
We have the useful predicate
*storm*(*S*) : − *classof*(*S*, *stormclass*)
In this definition, $\tau$ represents time, and indicates that a storm is an element of a spatio-temporal object class; the set atomic object type of the type definition is a

tuple in which the first element is the area covered by the storm (ie its *S_projection*) while the second element is the intensity of the rainfall over the area of the storm.

A *link flow element*, or a *flowel*, is a basel together with an object of type *tpointrelation* whose values characterize channel flow at the downstream point of a link at different times $[id, \{[\tau_1, P, F_1], ..., [\tau_n, P, F_n]\}]$. We first have a type definition
**TO_type** $flowel = [basel, tpointrelation]$
**TO_class** $= (flowelclass, flowel, methods)$
The definition of a flowel is
$classof(FL, flowelclass) \leftarrow totype(FL, flowel), toval(FL, V), otype(V.1, basel),$
$oval(V.1, V_1), oval(V_1.1, LT_1), oval(LT_1.2, P), point(P),$
$totype(V.2, tpointrelation), toval(V.2, V_2), \in ([\tau, P, F]], V_2), < (\tau, T), real(F)$
with the useful predicate
$flowel(S) \; :- \; classof(S, flowelclass)$
In this definition, the point $P$ must be the same as the downstream point of the linetuple defining the link, the real number $F$ is the link flow at this point.

In analogy with the definition of a basin in terms of a basel and the baselset of ancestors of the basel, we may construct a spatio-temporal equivalent of a basin in terms of a flowel and the set of its flowel ancestors, which we term a a *flowbasin*.

## 7.3 Deductive Rules Governing System Behavior

We now define a set of rules that permit us to infer properties relating to the time-dependent behavior of spatio-temporal classes of objects. We may view these rules as *methods* that apply to particular classes of objects. Such rules are based on the use of finite difference schemas over time to infer information about the temporal variation of complex object classes and their interrelationships. We implement this approach in the present context in terms of special inference rules that change the values (ie *tovals*) associated with a set of spatio-temporal objects as a function of time.

It is frequently the case that the order of computation is important in relation to computing changes in state values, and difficult semantic issues arise when one is attempting to formulate such a problem in declarative terms. When necessary, therefore, we use a left to right ordering convention with respect to the literals in the body of a rule that is applied in order to infer changes in object values as a function of time.

We first define the rule that governs the behavior of *storm* objects, and then the rule that governs the behavior of a drainage basin system during the passage of a given storm in terms of the *flow at the downstream end of a given link and at a given time*. Using the conceptual model of a storm presented above, the sets of literals in this rule are ordered in terms of computing: an updated value of time, and a safety condition on time; accessing values of the storm object; a new *S_projection*

for the storm; a new rainfall intensity for the storm. Since a storm is a single spatio-temporal atomic object, we need only the following rule
$toval(S,\ [\tau^+,\ [SPS',\ R']])\ :-$
$\leq (V.1\ ,\ T),\ +(V.1,\ 1,\ \tau^+),$
$storm(S),\ toval(S,\ [\tau,\ [SPS,\ R]]),$
$geoscale(SPS'',\alpha,\ SPS'),\ translate(SPS,\ [\delta x,\ \delta y],\ SPS''),$
$attrscale(R,\ \beta,\ R')$
where the special predicates *geoscale*, *translate* and *attrscale* respectively scale the spatial projection of the storm by a factor $\alpha$, translate the spatial projection of the storm by an amount $\delta x$, $\delta y$ and scale the intensity of the rainfall by a factor $\beta$, in accordance with the model of a storm outlined above. We may view this rule as a method associated with the class of storm objects.

The rule for determining changes in the state of the link flow elements is a little more complex, since a flowel has two relatively complex objects in its definition. We first compute the overland flow from the saturated area as a result of the storm intersecting the area of the flowel. The sets of literals in the rule indicate respectively: an updated value of time, and a safety condition on time; the flowbasin defined by the flowel in question, so that the appropriate set of ancestor flowels are available; access to appropriate information about the *flowel* in question; access to appropriate information about the *flowel* of the right parent link, if any, of the link in the flowel of interest; access to appropriate information about the *flowel* of the left parent link, if any, of the link in the flowel of interest; access to appropriate information about the storm; the intersection of the storm with the area of the link in order to compute the saturated area; whether the rainfall, if any, over this area exceeds the critical value given the geology; and the addition of a new tuple relating to flow in the link to the tpointrelation object of the flowel. We assume a left to right order of evaluation in this rule:
$toval(FL,\ [X,\ Y])\ :-$
$\leq (\tau,\ T),\ +(\tau,\ 1,\ \tau^+),$
$flowbasin(FB),\ toval(FB,\ [FL,\ FLS]),\ oval(FLS,\ V_{FLS})$
$flowel(FL),\ (toval(FL,\ [X,\ Y]),$
$toval(Y,\ S_Y),\ \in ([\tau,\ P,\ F],\ S_Y),$
$oval(X.5,\ [BPS,\ I_1,\ I_2]),$
$flowel(FLR),\ \in (FLR,\ V_{FLS}),\ toval(FLR, [X_R,\ Y_R]),\ = (X.2,\ X_R.1),\ toval(Y_R,\ S_{Y_R}),$
$in([\tau,\ P,\ F_{RL}],\ S_{Y_R}),$
$flowel(FLL),\ \in (FLL,\ V_{FLS}),\ toval(FLL,\ [X_L,\ Y_L]),\ = (X.3,\ X_R.1),\ toval(Y_L,\ S_{Y_L}),$
$in([\tau,\ P,\ F_{LL}],\ S_{Y_L}),$
$storm(S),\ toval(S,\ [\tau^+,\ [SPS,\ R]]),$
$\cap(SPS,\ BPS,\ WPS),\ crit(I_1,\ CR),\ > (R,\ CR), -(R,\ CR,\ E),\ \times(E,\ WPS,\ F_A),$
$+(F_A,\ F_{RL},\ F_{LL},\ F),$
$\cup([\tau^+,\ P,\ F],\ S_Y,\ S_Y^+),\ toval(Y,\ S_Y^+).$
where *crit* denotes a critical rainfall value. We note that this rule is recursive in

both time and space with respect to the values of flowels. Furthermore, we note that the rule involves two classes of objects, but that we may interpret it as a method associated with the class of flowel objects.

# 8 Conclusions

The data model that we have presented together with the deductive, object-based language CSOL currently appear to be promising as an initial candidate for defining and manipulating complex, spatio-temporal objects that arise over the spectrum of the extractive axis in a variety of scientific applications. We judge the degree of appropriateness in terms of the expressiveness of the model and the language with respect to our three examples, and particularly the drainage basin example. In particular, we believe that in one particularly important domain of application, we have shown the model and language to lead to compact representations of complex modelling situations. In future research we plan to explore alternative data models, to refine the language and to examine its semantics in greater depth and to examine issues that relate in particular to the representation of the spatial projections of objects and efficient support for the associated geometrical operations.

# 9 Acknowledgements

Support for the research on which this chapter is based was provided under NASA NAGW-1743 and NSF SES-8810917.

# References

[AG91] S. Abiteboul and S. Grumbach. A rule-based language with functions and sets. *ACM Trans on Database Systems*, 16(1):1–30, 1991.

[Che76] P.P.S. Chen. The entity relationship model — towards a unified view of data. *ACM Trans on Database Systems*, 1(1):9–36, 1976.

[Cod70] E.F. Codd. A relational model of data for large shared data banks. *Communications of the ACM*, 13(6):377–387, 1970.

[Dun88] T. Dunne. Models of runoff processes and their significance. *Scientific Basis of Water Management, National Academy of Sciences, Washington, D.C., 17–30, 1982. Reproduced in Perspectives on Water (eds. D.H. Speidel, L.C. Ruedisili and A.F. Agnew), Oxford Univ. Press, NY*, pages 66–72, 1988.

[FJP90] J.C. French, A. K. Jones, and J. L. Pfaltz. Scientific database management. *Dept. of CS, U. of Virginia*, 1990.

[HK87] R. Hull and R. King. Semantic data modeling: Survey, applications, and research issues. *ACM Computing Surveys*, 19(3):201–260, 1987.

[HM81] M. Hammer and D. McLeod. Database description with sdm: A semantic database model. *ACM Transactions on Database Systems*, 6(3):351–386, 1981.

[LN87] U.W. Lipeck and K. Neumann. Modelling and manipulating objects in geoscientific databases. *S. Spaccapletra, North-Holland*, pages 67–83, 1987.

[NT89] S. Naqvi and S. Tsur. A logical language for data and knowledge bases. *Computer Science Press, New York*, 1989.

[SF90] T. R. Smith and A. Frank. Report on workshop on very large spatial databases. *Journal of Visual Languages and Computing*, 1:291–309, 1990.

[SSU90] A. Silberschatz, M. Stonebraker, and J. D. Ullman. Database systems: Achievements and opportunities. *ACM SIGMOD Record*, 19(4):6–22, 1990.

[SZP90] T. Smith, C. Zhan, and G. Peng. A knowledge-based, two-step procedure for extracting channel networks from noisy dem data. *Computers and Geoscience*, 16(6):777–786, 1990.

[Ull88] J. D. Ullman. Principles of database and knowledge-dase systems. *Computer Science Press, Maryland*, 1988.

# Object-Oriented Database Systems for Geographic Applications: an Experiment with $O_2$*

Michel Scholl
Agnès Voisard

**Abstract**
In this paper, we study the implementation of a spatial relational-like query language for geographic applications in an object-oriented environment. A prototype has been developed on top of the object-oriented database system $O_2$. We report on our design choices, describe the current status of implementation and conclude on the suitability of the object-oriented approach and more specifically of the $O_2$ system for developing such applications.

## 1 Introduction

Data intensive geographic applications such as cartography, urban planning and natural resource management are built using Geographic Information Systems (GIS). GISs are database systems that allow the manipulation, storage and retrieval of geographic data and the display of data in the form of maps.

Conventional database management systems (DBMS) do not address correctly the needs of geographic applications: the underlying data models, query languages and access methods were designed to deal with simple data types such as integers and strings. GISs must handle spatial data and other forms of data having both complex structure and semantics. Furthermore, conventional DBMSs do not efficiently support geometric operations required by geographic applications. And, last but not least, user interfaces in conventional DBMSs are not particularly suitable for the display of geographic data. For a comprehensive study on the modeling and design of GISs, see [Fra84, SMSE87].

Thus, designing a GIS includes the design of new data models and query languages (or the extension of existing ones). Such a design might be complex. Indeed, GISs cover a wide range of applications dealing with objects for which, neither a common definition nor a common set of operations is known. Several propos-

*This work has been partially supported by the French Programme de Recherche Coordonnée BD3 and the ESPRIT BRA working group "Basic GOODS".

als for query languages, called spatial query languages have recently been made [BB81, Fra82, SMO87, CJ88, Dav88, Gut88, Abe89, Ege89b, GNT91, ODD89].

A spatial query language basically has the following functionalities:

1. managing, manipulating and querying large amounts of objects whose value includes geometric as well as alphanumeric data,

2. representing arbitrarily complex structures for these objects,

3. performing application dependent geometric operations on these objects.

Clearly, (extended) relational systems are good candidates for fulfilling our first requirement. There exists a large number of proposals in the literature for modeling and designing GISs using extended relational query languages *à la* SQL (see for example [SMO87, WH87, Gut89, BDQV90]).

Within such languages, geometric objects are represented as tuples of flat (normalized) relations. Regular relational operations can be performed on these objects as well as application dependent geometric operations.

The model proposed in [SV89] differs from these approaches in the following aspects:

- it allows the direct representation of arbitrarily complex structures (in the pure relational model, only records of atomic values are allowed) and the expression of more complex operations on objects such as nest and unnest [AFS88]. It uses the Abiteboul-Beeri complex objects algebra [AB88].

- it allows to extend general purpose retrieval and manipulation operations with application/user dependent geometric operations through the use of a functional *apply* similar to the *replace* operator of [AB88], the *filter* operator of FAD [BBKV87], or the $\lambda$ operator of [Gut88].

The main objective of this paper is not to propose another spatial query language but to study the implementation of a relational-like spatial query language by means of an object-oriented database system. Therefore, in the sequel, we do not allow the representation of arbitrarily complex objects and restrict our attention to the use of flat relations to model geometric objects, since this approach is adopted by all proposals in the literature mentioned above. Although these languages do not rely on a sound theoretical framework, they all more or less distinguish two levels, the relational or map level and the spatial or geometric level:

1. At the *map level*, geographic information is represented by geographic relations (maps) whose only particularity is the existence of at least one geometric attribute whose (abstract data) type is defined at the second level. Maps are queried and manipulated through a relational like query language which possibly calls for functions on the geometric attributes. These functions are implemented at the second level.

2. The *geometric level* consists of geometric abstract data types (ADT) with the following characteristics:

- The ADTs correspond to a given data model or representation of geometry. Recent geometric data models (see for example [Ege89a, Smy90]) use geometric models derived from Euclidean geometry and algebraic topology.
- The internal ADTs structure is hidden from the map level, i.e., from the user who sees only functions (name and signature) on objects with such types.
- Application dependent functions on these types are available. Such functions are supposed to be very general. In particular, it is not assumed that objects of geometric type are closed under these functions[1] and the signature of such functions may use any type (geometric or not). Because of this generality, the power of expression of such languages is not well defined, as it was with standard relational languages.

Some general purpose extended relational systems such as Postgres [SR86] or Sabrina-Lisp [GCK*89] are based on this approach. This provides the user with some programming facility linked to the use of high level general purpose database operations.

In order to study an object-oriented implementation of spatial query languages, we chose to develop a prototype of GIS on top of the $O_2$ system developed by Altaïr [O290].

The remainder of the paper describes our design choices and the current state of implementation. A definition of the map model is given in Section 2 as well as a sketch of a spatial query language. Section 3 gives the main features of the $O_2$ environment and the architecture of the prototype. Section 4 is dedicated to the implementation of a spatial query language on top of the database system $O_2$. Section 5 addresses the implementation of geometric data types. This requires the definition of classes and methods for handling geometry: the corresponding object-oriented schema is given in Appendix 1.

One of the most challenging tasks of geographic applications is the design of user interfaces whose characteristics is the visualization of (and interaction with) geometric objects. This is shortly discussed in Section 6 which reports on our design choices. Some conclusions on the suitability of the object-oriented approach and more specifically of the $O_2$ system for such applications are given in Section 7.

## 2 Map Model and Query Language

It is not our goal to be exhaustive in the description of a map model and query language. Instead, we give a few examples to illustrate some of the most common features and operations of the languages one can find in the literature, and express

[1]As an example of function, consider line intersection. It takes two lines as arguments and returns their intersection point: lines are not closed under intersection.

a few queries with a syntax *à la* SQL. We first need to specify a *map model* and to introduce a few definitions.

We believe that looking for a model general enough to encompass a large variety of geographical applications is bound to fail. Instead, any map model will strongly depend on the geographic application, on the structure of the geometry visible by the user and on the geometric functions to be applied to these objects. Of course, part of the structure is hidden to the user as well as the implementation of the functions (encapsulation). The more encapsulated the geometry, the cleaner the separation between application independent database operations and application dependent geometric operations. Encapsulation will allow modularity, i.e., extensibility and customization of the geometric operations independently of the general purpose database operations.

## 2.1 Map model

Any relation with at least one geometric attribute is called a *map*. Non geometric attributes are called *descriptive* attributes. They are defined as usual, i.e., they have a name and a definition domain [Ull88]. We do not require in a map the existence of descriptive attributes: a map can have only geometric attributes. These are specified by a (geometric) data type. Any application requires the specification of one or more types. We chose for our examples three data types: *Points*, *Lines* and *Regions*.

A *geographic database* consists of a set of maps. A map is a set of *tuples*. With each tuple of a map is associated a subset $S$ of the 2-dimensional plane $R^2$ with common alphanumeric information. The geometric attributes of the tuple represent the (possibly infinite) set $S$ of points[2], the descriptive attributes represent the alphanumeric information common to all points in $S$. We give below an informal specification of geometric types necessary for the examples of queries to be studied.

### Informal specification of the geometric types

The geometric attribute of a map may have one among three types depending on whether $S$ topology is 0-dimensional, or 1-dimensional or 2-dimensional:

1. A 0-dimensional geometric object of type *Points* is defined as a finite set of points in $R^2$.

2. Let an *arc* be the infinite set of points of a linear segment in $R^2$. Whether the *end points* of the segment belong or not to the arc is of no concern for our examples. Given a finite set of arcs $A$, let $N$ be the union of end points of arcs in $A$. With $A$ is associated a graph $G=(N,E)$ where to each arc in $R^2$ corresponds an edge in $E$. $A$ is a *line* iff $G$ is connected. A 1-dimensional object of type *Lines* is defined as a finite set of lines.

[2]We assume here that there is only one geometric attribute in any map. If there are $p > 1$ geometric attributes, with sets of points $S_i$, $i = 1,p$, the interpretation will depend on the model of geographic data of the application. An interpretation would be to associate with each tuple the set of points $S$, intersection of the $S_i$'s. Another interpretation would be to consider $S$ as the union of the $S_i$'s.

3. A 2-dimensional object of type *Regions* is defined as a finite set of regions, where a *region* is the infinite set of points in $R^2$ included in and possibly on the boundary of a simple polygon.

We represent geometric objects (of types *Regions*, *Lines*, *Points*) as finite sets of more elementary objects (*region*, *line*, *point*), because the former objects are closed under operations such as intersection, union, difference.

Note that, at the map level, one does not see the internal structure of geometric types (arc, line, region,...) which was given here only for illustrative purposes. One sees only functions, such as the following:

- *included-in* with signature *Points* $\times$ *Regions* $\rightarrow$ *Boolean* is defined as set inclusion in $R^2$: let p be of type Points and r of type Regions; *included-in(p, r)* is true if all points in p are in r,
- *inside* with signature *Lines* $\times$ *Regions* $\rightarrow$ *Lines* corresponds to intersection in $R^2$: let l be of type Lines and r of type Regions, then *inside(l, r)* represents all points in l which are in r,
- *inter* with signature *Lines* $\times$ *Regions* $\rightarrow$ *Boolean* tests whether a 1-dimensional object crosses a 2-dimensional object: let l be of type Lines and r of type Regions, *inter(l, r)* is true iff the intersection of points in l with points in r is non empty.
- *border* with signature *Regions* $\rightarrow$ *Lines* gives the 1-dimensional boundary of a 2-dimensional object.
- *Radjacent* (signature *Regions* $\times$ *Regions* $\rightarrow$ *Boolean*) and *Ladjacent* (signature *Lines* $\times$ *Lines* $\rightarrow$ *Boolean*) test respectively adjacency of 2-dimensional objects and adjacency of 1-dimensional objects. Informally speaking, two 2-dimensional objects are adjacent if their intersection is non empty and is on their boundaries and on their boundaries only; two 1-dimensional objects $l_1$ and $l_2$ are said to be adjacent if there exists one arc of $l_1$ colinear with one arc of $l_2$ (Figure 1).

Fig. 1: Arcs a and b are colinear

Note that the semantics of the functions is more or less dependent on the topology of the geometric objects: this is particularly true for the last three functions. A formal specification of geometric types and in particular of the above types should use geometric modeling. [Ege89c] is a step in this direction.

**Examples of tuples**

1. *[Name: fontainebleau, Vegetation: beech, R: r]* represents a forest whose name is "Fontainebleau" and whose geometric attribute has for value "r". r identifies the set of two polygons displayed in Figure 2.

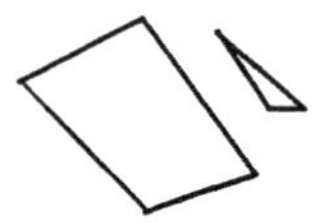

Fig. **2**: The value of a geometric attribute

This tuple belongs to relation FORESTS*(Name: String, Vegetation: String, R: Regions)*.

2. *[Number: a6, L: l]* represents highway number "A6" whose 1-dimensional geometry value "l" identifies a set including the single line of Figure 3:

Fig. **3**: A line

This tuple belongs to relation HIGHWAYS*(Number: String, L: Lines)*.

3. *[Name: petit-clamart, P: p]* represents a (highway) intersection point whose name is "Petit-Clamart" and whose geometry is a set including a single point. This tuple belongs to relation CROSSINGS(*Name: String, P: Points*).

Such a representation of geographic objects is simple. It has however a limitation due to the relational model: we forbid the use of non atomic domains for descriptive attributes. We already addressed this issue in the introduction. One way of getting around this restriction is to allow complex values as in nested relational models [AFS88], or in complex object models [AB88]. Object-oriented models allow an even more powerful representation of complex domains. As a simple example, the domain of a highway number is chosen as "String". Since any highway is described by an alphabetic character string followed by an integer, it would be more appropriate to define the domain of attribute "Number" in relation HIGHWAYS by a type "highway-type" whose definition in terms of atomic domains is *[Alpha: String, Num: Integer]*. As another example, geometric attributes are always seen as attributes whose value is atomic. This is of course a strong limitation.

## 2.2 Examples of queries on maps

There is a very large number of geometric queries which vary from one paper to the other in the literature. The reader is referred for example to [Ege89b, Gut88, QDDA89] for a thorough description of queries and geometric functions used at the map level.

We will use the following map schemas in our examples:

CITIES(*Name: String, Population: Integer, R: Regions*)
CROSSINGS(*Name: String, P: Points*)
HIGHWAYS(*Number: String, L: Lines*)
DISTRICTS(*Name: String, R: Regions*)
FORESTS(*Name: String, Vegetation: String, R: Regions*)

For each query, we give the English formulation and the query itself in a language whose syntax is close to SQL. All queries have a SELECT - FROM - WHERE (SFW) form. The arguments of the SELECT clause can be descriptive attributes drawn from the source maps, as well as values computed by the use of a function of the abstract geometric data type. In general, the arguments of the SELECT give the schema of the target relation. But the result of a query is not necessarily a map: it can be any relation (whose attributes are descriptive attributes); it can also be the result of any function such as length, distance, etc. The arguments of the FROM clause are the source map(s) and the arguments of the WHERE clause are a conjunct of functions whose range is boolean. These functions are either regular SQL predicates or boolean functions on geometric attributes.

1. Highway intersections located within the city of Versailles?

   SELECT CROSSINGS.Name
      FROM CITIES, CROSSINGS
      WHERE CITIES.Name = "Versailles"
      AND CROSSINGS.P *included-in* CITIES.R

2. Highway portions inside district "Var"?

   SELECT HIGHWAYS.L *inside* DISTRICTS.R
      FROM HIGHWAYS, DISTRICTS
      WHERE DISTRICTS.Name = "Var"
      AND HIGHWAYS.L *inter* DISTRICTS.R

3. Cities adjacent to Versailles?

   SELECT C1.Name, C1.R
   FROM CITIES C1, CITIES C2
   WHERE C2.Name = "Versailles"
   AND C1.R *Radjacent* C2.R

4. Forests bordered by highway A7?

   SELECT FORESTS.Name, FORESTS.Vegetation, FORESTS.R
   FROM FORESTS, HIGHWAYS
   WHERE HIGHWAYS.Number=A7
   AND FORESTS.R *border Ladjacent* HIGHWAYS.L

In Query 3, observe that the last predicate (C1.R Radjacent C2.R) could be expressed as: C1.R *border Ladjacent* C2.R *border* testing whether the two border lines have one or more colinear segments. One could also have defined a function, say RLadjacent, testing adjacency between a 2-dimensional object and a 1-dimensional object. This example illustrates that the query language strongly depends on the definition of the geometric types, i.e., on the spatial semantics visible at the user level. Other typical queries are given in Section 5.

# 3 Architecture of the Prototype

A prototype of a spatial query language has been developed on top of the Version V1 of the $O_2$ system. We do not describe in detail the $O_2$ environment, and give only the main features necessary to describe the prototype. For more information about $O_2$, see [BDK91, AK89, LR89]. This prototype allows queries on maps through a graphical user interface.

## 3.1 The $O_2$ environment

We give a rapid sketch of the main features of the $O_2$ model and data manipulation facilities. We assume the reader familiar with object oriented database concepts.

### 3.1.1 The $O_2$ Data Model

$O_2$ distinguishes *values* and *objects*. Values and objects have a given *type* (or structure). Types are either *atomic types* such as *strings*, *integers*, *reals*, *bitmaps*, *object identifiers*, ... , or complex types built from atomic types by using the *tuple* ([ ]), *list* (<>) and *set* ({}) *constructors*.

An *object* has a *value* and an *identifier* independent of its value. Objects are encapsulated into *classes* whose only visible interfaces are the names and signatures of the *methods* to be applied to objects of that class. The invisible part is made of

(i) the structure common to all objects of the class and (ii) the body of the methods. Values are not encapsulated and user defined functions cannot be applied to values. There are two ways of executing a function on objects and values: (i) by applying one of the few built-in functions on values, (ii) by sending to an object a *message* including a method to be applied to this object. The visible part of all classes of the database of the application are described in a *schema*.

### 3.1.2 Programming in $CO_2$

Programming, i.e., defining the schema and writing the body of the methods is done in $CO_2$. This language is an extension of the C language: the schema is specified in the $O_2$ language, while methods are written in $CO_2$ and precompiled in C. For further information, the reader is referred to [LR89].
As an example we define a schema (with one class and one method) and give the body of a method:

```
add class Person
    type tuple (last-name : string,
                first-name : string,
                age : integer,
                photo : bitmap ,
                children : list(Person));


/* a person is a tuple with 4 atomic attributes and an attribute whose
structure is a list of objects of class Person. The following is the
value of an object of class Person (note that the value of attribute
children is a list of object identifiers):
[last-name:Brown,first-name:John,age:25,photo:101..,children:<i1,...,in>]*/

add method is-adult (): boolean /* in class Person */
    body (
      if (self → age >= 18)
      return(true);
      else return(false))


  /* self is the receiver of the method,i.e., an object of class Person,
     self->age is the value of attribute age of the object */
```

As another example, consider the implementation of the intersection of two sets of lines, the result being the set of their intersection points. We give below the corresponding schema with one class (*Lines*) and one method (*p-intersection*) whose body is explained.

```
add class Lines
    type set (l: Line);
```

```
/* an object of  class Lines is structured as a set of objects of
class Line. An object of class Line represents a single line */


/* The following method is sent to an object of class Lines and
has for an argument, another object of class Lines. The result is
an object of class Points: the set of intersection points between
lines (objects of class Line) of the receiver and lines of the
argument */

add method pintersection(l:Lines): Points /* in class Lines */
      body(
        Points v,s
        Line t,u
        s=new(Points)
        for each t in *self
          for each u in *l
            v=new(Points)
            v=t→pinter(u)
            s+=v
        return(s))


/* v and s (t, u) are object variables declared of type Points (Line).
An object of class Points is created by the built-in construct ''new''
All lines (object of class Line)  of the receiver are scanned. *self
(*l) is the value of the receiver (argument): it is a set of objects
of class Line. For each line of the receiver, all lines of the
arguments are scanned. An object v of class Points is created and is
assigned the result of the intersection of two lines (objects t and u
of class Line): method pinter of class Line is sent to t with argument
u. A set union (+=) is performed between the result of the
intersection (set of points v) and the accumulator to be returned (set
of points s) */
```

### 3.1.3 Data manipulation

$O_2$ provides three interfaces for manipulating data:

- The programming language $CO_2$.
- A graphical user interface constructed with *Looks* [PCL*90]. The *Looks* system is implemented on top of the Aïda toolkit [DD88]. *Looks* provides capabilities to browse and manipulate an $O_2$ database, such as display the value of an object, run methods on it, manipulate interactively objects, ...
- The query language $O_2$Query [BCD89] implemented in C and interfacing with the Object Manager of $O_2$

### 3.2 Functionalities

The compiler of the SQL-like language of Section 2 was not available at implementation time. Thus we decided to query maps using an algebraic language. Querying maps is done through algebraic operations such as:

- standard relational selection/projection/join,
- algebraic operations with geometric functions. The simplest operation is *geometric selection*: select regions (or lines): *included in*, or *containing*, or *intersecting,...*, a constant window chosen on the screen. As an example, *clipping* selects all regions (lines) intersecting a rectangle drawn by the user on the screen. *Windowing* is another example which selects all geometric objects inside a rectangle drawn on the screen.

For a more complete definition of these algebraic operations, see [SV89]. In the prototype, a map is an object and a query is implemented as a method call. The user interface is provided by the $O_2$ interface tool *Looks* [PCL*90] extended to deal with maps. The user first selects a map he/she wishes to query. He/she displays the map and can perform the following operations:

1. displaying the description (alphanumeric attributes values) of objects (regions, lines or points) containing a point selected with the mouse on the screen,
2. projecting or selecting on descriptive attributes,
3. clipping and windowing.

### 3.3 Architecture

The implementation of such functionalities corresponds to the development in parallel of three independent modules:

1. *User interface*: it uses the *Looks* system which has been extended in Aïda [DD88] to allow the display of geometric vectorized objects and a geometric interaction with the user (see Section 6).
2. *Map level* [Sou90]: a map is represented by an object. Algebraic operations are implemented by methods on map objects. This module corresponds to the programming in the $CO_2$ programming language [LR89] of all the methods implementing algebraic operations on maps as well as methods for creating maps and entering actual map data. Most of this module could have been implemented by using the $O_2$ query language $O_2$Query. We explain in Section 4 why this was not possible, i.e., why we had to program these algebraic operations in $CO_2$ .
3. *Geometric level*: this module corresponds to the implementation in $CO_2$ of the geometric ADTs (*Regions, Lines, Points*). A tuple of a map is represented as an object. Each tuple object is composed of a subobject which represents the geometry of this object. As an example, the object whose value is *[Name: Versailles, R: r]* is composed of the object with identifier "r" of class "Regions".

When executing a database query, e.g., "Cities neighboring Versailles" (Query 3, Section 2) a method is sent to an object representing the map of CITIES. The execution of a geometric operation may be necessary (here Radjacent) which corresponds to calling a method with receiver an object of class Regions. The design of the schema (classes and methods) corresponding to the internal representation of the geometry (Regions) and the implementation of geometric operations (e.g., Radjacent) is considered in Section 5.

# 4 Implementation of the Map Level

In this section, we are more specifically interested in the implementation of the map level, i.e., in the design of methods on classes representing maps. These methods implement algebraic operations for answering database queries.

Two approaches can be taken:

1. either these methods on map objects are implemented in the programming language of the object-oriented system (here $CO_2$),

2. or a query language is available, powerful enough to express the queries of the application.

We start by the second alternative, namely the use of $O_2$Query for implementing extended database queries.

## 4.1 Expressive power of $O_2$Query

$O_2$Query is powerful enough to express database queries such as those described in Section 2. To illustrate this we take the example of Query 2: *Display the highways portions inside district "Var"*. This query can be expressed with $O_2$Query as:

```
SELECT x.L inside (y.R)
FROM   x in HIGHWAYS
       y in DISTRICTS
WHERE  y.Name=''Var'' and
       x.L inter (y.R)
```

A map, say HIGHWAYS, is represented by a set of objects representing map tuples. These objects belong to a class with structure *[Name: String, L: Lines]*. The geometric attribute (L) has for a value an object of a geometric class: Lines. Geometric functions (*inside, inter*) are implemented as methods (*inside, inter*) of class Lines.

The FROM arguments indicate that x (respectively y), is an object to be taken in a set of HIGHWAYS (respectively DISTRICTS) objects. The SELECT argument expresses that method *inside* is sent to object x.L (x.L is an object representing the highway geometry) of class Lines, with argument object y.R of class Regions (y.R is the geometry of a district). The second WHERE predicate is true if method *inter* sent to highway geometry x.L with argument district geometry y.R returns "true".

The result is a set including objects of class Lines. With the $O_2$ current interface, we can output the value of this result, but we cannot visualize it as a map.

## 4.2 Integration of $O_2$Query

The use of $O_2$Query through a graphical user interface requires its full integration into the programming language. By full integration, we mean that the query language is a subset of $CO_2$ and in particular any SFW query can be used in a $CO_2$ method.

To illustrate the use of $O_2$Query integrated in $CO_2$ for implementing algebraic operations, let us take for example the standard *relational projection*. Tuples of a relation are represented by objects with a tuple structure while the extension of the relation is represented by the set of objects identifiers stored in a named value: persistent objects (values) are represented in the $O_2$ model by named objects (values). To represent relation R(*A1: D1,..., An: Dn*), we define a class *R* with structure *[A1: D1,..., An: Dn]* and a named value *Rext* with type $\{R\}$. Relation R is represented by the set *Rext* of tuple objects of class *R*. Projection $\pi_{Ai}(R)$ is implemented as a method with the following body:

```
V = select[x.Ai] from x in Rext
for each v in V
  a=new(T)
  a.Ai=v
  Text+={a}
```

The target relation is represented by a set value *Text* of type $\{T\}$ where *T* is a class with type *[Ai: Di]*. The first line performs the projection: for each tuple x in relation Rext, the value of attribute Ai is kept (x.Ai). The result is a set of tuple *values*[3] assigned to *V*. But we would like the result to be a set of *objects* (a relation) stored in *Text*. Then for each tuple value *v* of *V*, an object *a* of class *T* is created (a=new(T)). The value of *a* is assigned *v* (a.Ai=v) and *a* is added to *Text*. This implementation assumes the existence of class *T* and named value *Text*.

This example illustrates the use of $O_2$Query as an ad-hoc query facility embedded in $CO_2$ for extracting information in the database. One may wonder whether it is necessary after projection (first line of the program above) to sequentially scan the result *V* and thus loose part of the efficiency of the optimized query facility. In other words, is it enough to represent any relation by a set of values? This implies that algebraic operations be implemented by functions on (set) values. But user defined functions on values are not yet provided by $CO_2$. Then the algebraic operations must be implemented as methods sent to objects. This is why tuples of a relation are represented by objects. Then

- either the result of an $O_2$Query filter query is a set of objects, which is not the case in the current version,

[3]Recall that the $O_2$ model distinguishes (encapsulated) objects with methods from non encapsulated values (with no possibility of defining user functions on these values).

- or it is a set of values. Then, as in the above example, this set must be scanned and each element of the set is assigned as a value of a new object representing a tuple of the target relation.

But, at the time of implementation, $O_2$Query was only available as an interactive tool. It was not fully integrated in $CO_2$. The alternate approach to the use of $O_2$Query consists then in directly implementing algebraic operations on maps in $CO_2$.

As relational operations, operations on maps have various parameters such as source and target maps, attributes to be projected,... We show in the following sections that the implementation of such parameterized operations is not simple if we want to take advantage of the typing power of the database programming language.

## 4.3 Implementing generic relational algebra operations

In relational DBMSs, relational algebra interpreters have the capability of run-time access to the relation type definitions (schema). This is in contradiction with database programming languages using static type checking. This problem is illustrated in [SFSS90] where the integration in the *ADABTPL* language of generic projection, selection, join and nest is studied. It is shown that a solution exists for the two first operations but that the two last ones are problematic. Another approach is studied in Machiavelli [OBB89]. Strongly typed object-oriented database programming languages such as $CO_2$ have an insufficient level of polymorphism to handle query algebra capabilities, i.e., to allow the implementation of generic relational operations. In particular, it is impossible to implement a generic projection with $CO_2$. To show this, consider the above example. One would like to implement a method "Projection" on class *R* taking as arguments attribute names in {A1,...,An}. We hit upon two problems:

- projecting tuple *[A1: a1,...,An: an]* on Ai where attribute name Ai is given as a *parameter* is not possible: run time access to type information is impossible. Analogous problems arise for selection, join,...
- dynamic class creation: after projection, the target relation may have one among $2^n - 1$ schemas. This suggests the existence of as many classes for representing target relations since it is impossible (i) to create a class with parametric type, and (ii) to create a new class at run time with $CO_2$.

It should be emphasized, however, that such a limit is the price to be paid to strong type checking. This is not particular to $O_2$ but should be expected from any typed language such as Modula, Pascal or Eiffel [Mey88].

## 4.4 Design of a query language

$O_2$Query was designed as an ad-hoc optimized query facility, i.e., a short cut in the programming language [BCD89] meant to extract information in the database. The system designer who would like, instead of using $O_2$Query, to develop its own query language for performance reasons[4] has to face the problem of dynamic class creation. Assuming this solved, it is necessary to afford the capability of writing a query language

- either in the programming language of the database system,
- or directly on lower levels of the system (object manager). Then a more user friendly interface to the object manager should be provided to the system designer.

In Appendix 2, we sketch a possible extension of $CO_2$ with a few constructs and the simulation of genericity by inheritance [Mey86, Mey88] in order to allow the implementation of generic relational algebra operations. These extensions solve the problem of run time access to type information, by manipulating parametric attribute names as strings. However, they are of limited use if dynamic class (type) creation is not provided by the system.

## 4.5 Current implementation

In order to solve the above problems and to actually implement filter queries of the SFW form in $CO_2$, an alternate approach, also suggested in [SFSS90] was chosen. It is far from elegant and complicates the implementation of sets of tuples (relations or maps). But it does not require any additional features to the type mechanism, it does not require dynamic class creation nor run time access to type information.

The approach consists in encapsulating the attribute name with its value so that type information is manipulated by the programmer as values. All maps are implemented as objects of a single class *Map*. All algebraic operations on maps are implemented by methods of this single class and give as an output an object (a map) of that class (there is no more need for creating a new class for a target map).

Each map tuple is modeled as a list of tuples *[Attribute-name, Attribute-value]*. Such a schema sketched in Figure 4 loses part of the typing power of $O_2$. The enforcement of constraints such as all tuples of a single map have same number of attributes and same attribute names is under the responsibility of the programmer. Access to the attribute value and attribute name is explicitly done by the programmer. Thanks to this subterfuge, some genericity is obtained.

More precisely, all maps have a single tuple structure with two attributes: a Name and an Extension which is an object of class *Geotuples*. An object of class *Geotuples* has a set value. An element of this set represents a map tuple: it is an object of class *Geotuple* and has two attributes: one is a list of descriptive attributes, the other one is the geometric attribute, which is an object of class *Geos*. There

[4]Specific applications require specific optimizations dependent on the physical organization of data. This is particularly true for geographic applications where a significant effort is conducted towards efficient data structures [Sam89, SW88, GSS91] or efficient implementation of query languages (see for example [Gut88]).

are three classes, one for each geometric type: *Points*, *Lines*, *Regions* with common superclass *Geos*. This will be detailed in the following section.

Each descriptive attribute has a name and an atomic value (integer or string or real,...). This is represented by a class hierarchy with root *Att* of type a tuple with a single ($O_2$) attribute whose value is the name of the descriptive attribute of the map. The descriptive attribute value is specified in the subclass corresponding to its domain (*IntAtt*, or *StringAtt*,...). Method *compare* of this subclass tests whether the descriptive attribute value is equal to the value in argument. Similarly, other methods are provided for other comparators $(<, >, \ldots)$.

```
class Map
    type tuple(Name: String,
               Extension: Geotuples)

class Geotuples
    type set(Geotuple)

class Geotuple
    type tuple(Attributes: list(Att)
               Geometry: Geos)

class Att
    type tuple(Name: String)
    method compare(x: Att): boolean

class IntAtt
    inherits Att
    type tuple(Value: Integer)
    method compare(x: IntAtt): boolean

class StringAtt
    inherits Att
    type tuple(Value: String)
    method compare(x: StringAtt): boolean
```

Fig. 4: $O_2$ schema for map representation

For example, consider map FORESTS(*Name, Vegetation,R*) and tuple

*[Name: fontainebleau, Vegetation: beech, R: g].*

Map FORESTS is represented by object *[Name: forests, Extension: e]* of class *Map*. *e* is an object of class *Geotuples* with value *{f,...}* where *f* represents Fontainebleau forest and is an object of class *Geotuple* with value: *[list(n,v), g]*. *n* and *v* are objects of class *StringAtt* and *g* (geometry) is an object of class *Geos*. *n* and *v* have respectively for value:

*[Name: name, Value: fontainebleau]* and *[Name: vegetation, Value: beech].*

# 5 Implementation of Geometric Functions: Geometric Level

We illustrate the object-oriented approach and describe the $O_2$ schema for implementing a few geometric functions appearing in the following queries.

## 5.1 Queries

1. Which are the cities adjacent to Versailles? (Query 3 of Section 2.2)

   SELECT C1.Name, C1.R
   FROM CITIES C1, CITIES C2
   WHERE C2.Name = "Versailles"
   AND C1.R *Radjacent* C2.R

2. Which forests are bordered by highway A7? (Query 4 of Section 2.2)

   SELECT FORESTS.Name, FORESTS.Vegetation, FORESTS.R
   FROM FORESTS, HIGHWAYS
   WHERE HIGHWAYS.Number=A7
   AND FORESTS.R *border Ladjacent* HIGHWAYS.L

3. Display the intersection between European countries and Leman lake.

   SELECT LAKES.R *Rintersection* COUNTRIES.R
   FROM LAKES, COUNTRIES
   WHERE LAKES.Name= "Leman"

4. Select parts of trails along (colinear to) highways.

   SELECT HIGHWAYS.L *Lintersection* TRAILS.L
   FROM HIGHWAYS, TRAILS

5. Select bridges (i.e. intersection points of highways and rivers).

   SELECT HIGHWAYS.L *Pintersection* RIVERS.L
   FROM HIGHWAYS, RIVERS

Geometric functions are implemented by methods on geometric classes the schema of which is now given.

## 5.2 Structure

Figure 5 presents the geometric classes hierarchy.

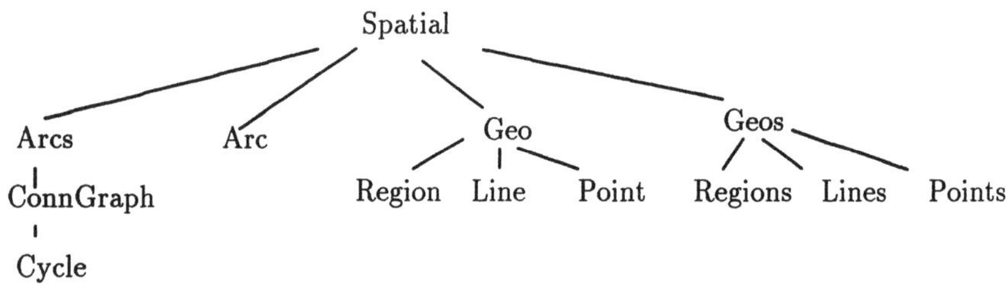


Fig. 5: Inheritance tree for the geometric classes

All classes have a common root *Spatial* that represents geometric information common to all maps (e.g., coordinates system, scale,...). Appendix 1 gives for each class its structure and its methods. For the time being, we are only interested in the structure of the classes. The hierarchy of geometric classes includes four subtrees with roots Geos, Arc, Arcs and Geo:

1. *Geos*
   Recall a tuple of a map has at least one geometric attribute of type Points or Lines or Regions. Values of geometric attributes are implemented as objects of a class *Geos* with subclasses *Regions, Lines* and *Points.* Objects of class Regions (respectively Lines and Points) represent geometric values of type Regions (respectively Lines and Points). The value of an object of class *Geos* (or Regions, Lines, Points) is a set of objects of class *Geo* (or *Region, Line, Point*). An object of class Region represents a region, i.e., a polygon (Figure 6.1). An object of class Line represents a line (a finite set of linear segments or arcs whose associated graph is connected, Figure 6.3). An object of class Point represents a point.

   For example, tuple: *[Administrative-region: brittany, Roads: l]* represents the roads of Brittany (*l* is an object of class Lines, and each road is represented by an object of class Line) and tuple: *[Administrative-region: brittany, Crop: corn, Counties: r]* represents the counties of Brittany where corn is grown (*r* is an object of class Regions). Each county is represented by an object of class Region.

   Classes Region, Line and Point have class Geo as superclass.

2. *Arc*
   An object of class Arc represents an arc, i.e., a linear segment represented by its end points (objects of class Point). The distinction between a begin point and an end point allows to represent oriented arcs.

3. *Arcs*
   The value of an object of class *Arcs* is a list of objects of class *Arc* (it represents a list of linear segments in $R^2$). Class *ConnGraph* is a subclass of *Arcs.* It

inherits Arcs structure and methods and has an additional method *connected* of type boolean which checks whether the set of arcs is a line (its associated graph is connected). Class *Cycle* is a subclass of ConnGraph with an additional method *cyclic*, of type boolean, that checks whether the line is a boundary of a simple polygon: in particular, its associated connected graph is a simple cycle. Non simple polygons (Figure 6.1 c) are not considered in this paper.

4. *Geo*
This hierarchy of classes implements "atomic" objects (Figure 6). A point (class Point) is represented by its two coordinates (relative to the coordinates system). A line is represented by an object of class Line: it has a tuple structure with one attribute whose value is of class *ConnGraph* (a line is *composed* of a set of connected arcs). Similarly, a region object (class Region) consists of a cycle (boundary of the polygon, Figure 6.1): although a region is a 2-dimensional object, it is represented by its 1-dimensional boundary. If regions with holes were to be represented (Figure 6.2), one could as well add a subclass of *Region* called *RegionWithHoles* with an additional attribute for the holes whose type is: set (Cycle). The extension of class *Geo* is the union of the extensions of *Point*, *Line* and *Region*: an "atomic" object is either a point or a line or a region.

a) Convex Polygon b) Simple Polygon c) Non Simple Polygon

Fig. **6.1**: Objects of class Region

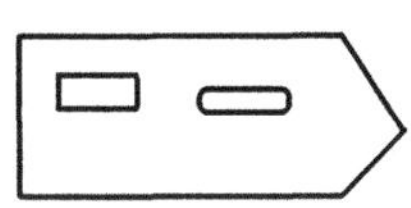

Fig. **6.2**: Region with holes

Fig. **6.3**: Objects of class Line

Fig. **6**: Atomic objects

Modeling *Regions* (*Lines*, *Points*) as a set of objects of class *Region* (*Line*, *Point*) is motivated by the fact that sets of homogeneous geometric objects are closed under operations such as intersection, union or difference. But this is somewhat arbitrary. For some applications, *Regions* could be structured as a set of sets of objects of class *Region*. For example, assume that administrative regions are split into districts which in turn are split into counties. Then, one might decide to keep for the geometry of tuple "Brittany" in map "France", the borders of all counties of Brittany, grouped by district, instead of either the set of district borders, or the outer border of the administrative region itself. Then, depending on how much of the semantics of the geometry should be visible at the map level, one has the choice between two extremes:

1. render the structure visible at the map level: use types Region, Line and Point. Thanks to an extended relational model using complex objects through the set and tuple constructors, the map level sees atomic attributes whose type is Line or Polygon. With such an approach, in the above example, counties are visible in tuple "Brittany" of map "France":

   *[Administrative-region: brittany,*
   *Crop: seaweed,*
   *Districts:* {
   *[District: morbihan,*
   *Counties:* {
   *[County: groixisland,*
   *Geometry: r],*
   ...
   } *]* ... } *],*

   where r is an object of class Region,

2. hide the structure of geometric objects: at the map level, it is an "atomic" attribute (of type Regions,...) on which some operations are performed and whose hidden structure is any complex structure whose atoms are lines or polygons. The choice of this structure would then depend not only on the application semantics, but also on implementation, optimization and performance choices:

   [*Administrative-region: brittany, Geometry: r*],
   where r is an object of a class of structure:

   [*ARouterborder: Region,*
   *Districts:* {
   *[Districtborder:Region,*
   *Counties:* {*Countyborder: Region*}
   *]* } *].*

Although the first approach is more appropriate, in this paper we chose the latter alternative in order to stick to the flat relational model assumed by spatial query languages proposed in the literature. We further assumed the simplest structure for representing complex geometric objects (Geos hierarchy) in terms of "atomic" geometric objects (Geo hierarchy).

### 5.3 Methods (see Appendix 1)

We now present the implementation of the geometric functions appearing in the five queries of Section 5.1: Radjacent, Ladjacent, border, Rintersection, Lintersection and Pintersection. Each function is implemented by a method of a given class (see Figure 7).

| Function | Function signature | Method | Class |
|---|---|---|---|
| Radjacent | Regions × Regions → Bool | adjacent | Geos |
| Ladjacent | Lines × Lines → Bool | adjacent | Geos |
| border | Regions → Lines | border | Regions |
| Rintersection | Regions × Regions→ Regions | intersection | Geos |
| Lintersection | Lines × Lines → Lines | intersection | Geos |
| Pintersection | Lines × Lines → Points | pintersection | Lines |

Fig. 7: Implementation of geometric functions by methods

#### 5.3.1 Adjacency: Queries 1 and 2 of Section 5.1

Note in Figure 7 that the class name (Geos) does not match the first argument of the function signature (Regions or Lines) when a generic operation can be implemented by a single method defined in superclass Geos: *Radjacent* (adjacency of 2-dimensional objects) and *Ladjacent* (adjacency of 1-dimensional objects) are implemented by a *single* method *adjacent* defined in class Geos. This is a typical advantage of object-oriented programming. As a consequence, at the query language level, it is not necessary to distinguish between *Radjacent* and *Ladjacent.* Independently of the type of the geometric attribute (Regions or Lines), method *adjacent* of class Geos is executed. In its body, method *adj* of class *Geo* appears. This method is redefined in subclasses *Region* and *Line.* At run time, the system chooses which code is to be executed according to the class (Region or Line) of the receiver of method *adj* (overloading). Method *adj* of class Geo should never be executed.

To complete the implementation, let us look at method *adj* redefined in class Region (adjacency test for two polygons): two regions are adjacent if their boundaries are colinear. Method *border* is sent to an object of class Region. It returns an object of class Cycle to which method *colinear* (inherited from class *Arcs*) is sent. This method checks whether two lists of arcs are colinear and therefore "calls" method *colinear* of class *Arc.*

#### 5.3.2 Intersection: Queries 3, 4 and 5 of Section 5.1

To illustrate the limits of the above mechanism for implementing generic operations, we chose operation intersection, because its semantics varies not only according to the objects type, but also within a given type. The intersection of two sets of polygons is either the empty set or a set of polygons. A naive algorithm consists in iterating on each set of polygons, using as a primitive the intersection of two elementary polygons.

The intersection of two sets of lines may have at least two interpretations (Figure 8):

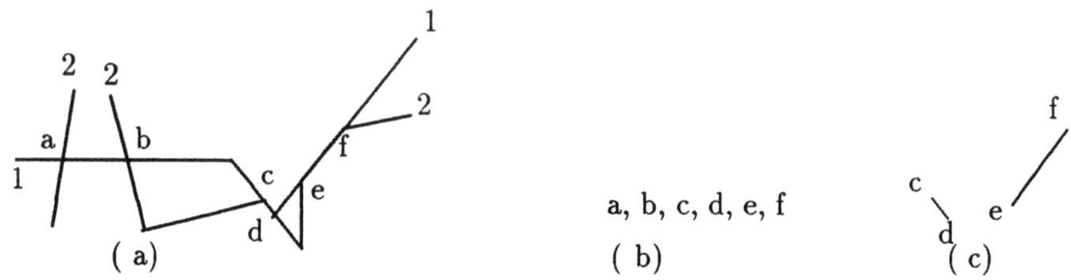


Fig. 8: Intersection of sets of lines

Either it is a set of points (Figure 8b), or it is a set of lines (Figure 8c). In the first case, it uses as a primitive the intersection of two lines which in turn utilizes the intersection of two arcs; in the second case it uses as a primitive another operation on two lines which utilizes the colinearity of two arcs.

Functions *Rintersection* and *Lintersection* (Queries 3 and 4) are implemented by a single method *intersection* of class *Geos*: without loss of generality, we chose to interpret intersection as either intersection of sets of regions or intersection of sets of lines whose result is a set of lines (second interpretation of lines intersection) and implemented a generic operation as follows:

Method intersection of class Geos is redefined in classes Regions and Lines. Take the example of class Regions. Its body consists in (i) enforcing method *intersection* defined in class Geos (*intersection(r)@Geos*) to be applied to the object of class Regions which is also an object of class Geos, (ii) casting[5] the result to be of class Regions (*return(Regions)*). In the body of method *intersection* (class Geos), appears method *intersect of class Geo*. This method is redefined in classes Region and Line. At run time, when method *intersect* is sent to an object (of class Geo), the system chooses the code to be executed according to the type (Region or Line) of that object (overloading).

Thanks to this well known mechanism [Mey86], we obtained a generic intersection operation.

Unfortunately, when the intersection of two sets of lines is interpreted as the set of intersection points between two sets of lines (Query 5), a new method has to be written (*pintersection* of class Lines). Its code is almost identical to the code of the method for the other interpretation. The only change is the call of method *pintersect* of class Line.

In summary, this implementation of geometric functions showed the power of overloading through inheritance. Note also with the *Arcs* hierarchy, that inheritance allows to elegantly check the following integrity constraints: (i) with a line

[5]Casting and explicitly choosing which method code should be executed in a superclass are extensions of the current version of $CO_2$.

is associated a connected graph and (ii) the graph associated with a simple polygon boundary is a simple cycle. We conclude this section with a discussion on the implementation of variant types in our application.

### 5.3.3 Variant types

The two hierarchies with roots Geos and Geo allow, to a certain extent, (i) to model geometric objects with various types, (ii) to manipulate an heterogeneous set of objects (e.g., display on a single map a mixture of regions, roads and cities).

If classes Geos, Regions, Lines and Points are visible at the map level, then the user might decide between heterogeneous data (use of class Geos) or homogeneous data (use of classes Regions, Lines or Points).

If only class Geos is visible, then the user has no way of enforcing homogeneity since objects of class Geos represent homogeneous as well as heterogeneous data.

# 6 User Interface

User interfaces represent an important part of applications code. Appropriate tools for constructing powerful interfaces are not yet available for applications which manipulate graphic data (CAD-CAM, geography, cartography,...).

In this prototype, it was not our objective to design and implement a complete interface for geographic applications for two reasons:

- Functionalities required at the interface level are not well understood yet. [EF88, Voi91] are among the first attempts to describe the functionalities of a user interface for GIS.
- The *Looks* system designed to support the construction of user interfaces in the $O_2$ environment was not intended to handle geographic applications, which require powerful display tools for geometric types (vectors).

We chose to extend *Looks* with basic functionalities for map display and querying. Our objective was a better understanding of the minimal functionalities that should be provided by future versions of *Looks* in order to support the construction of interfaces for applications manipulating graphic data.

In this section, we first describe a few requirements for GIS interfaces. Then we present the current extension of *Looks* for displaying and querying maps.

## 6.1 Requirements for GIS user interfaces

Two basic functions of database interfaces are: (i) the display of values of objects and schema and (ii) interactive queries. The specificities of geographic database user interfaces are:

1. Most of the time the geometry is not stored in raster format but in vectorized format. Atomic objects are (i) points, (ii) sets of end points for representing a line and (iii) border lines for representing 2-dimensional objects. A geographic user interface must be able to display the vectorized geometry of the objects as well as the alphanumeric description of these objects.

   There exists a large number of parameters for customizing map display, specified independently of the database. The value of these parameters may be changed during the display and chosen interactively by the user. Customization of map display includes features such as:

   - mapping colors, texture, legends to geometric objects,
   - highlighting a given object,
   - zooming and change of scale,
   - using bitmaps as a background of displayed maps,
   - displaying several maps in the same window,...

2. When querying the database, while some parameters are entered with the keyboard (or dragged with the mouse when identified as icons), other parameters correspond to the currently displayed geometry and are usually identified with the mouse. For example, one might ask for the name of the states adjacent to a state selected by clicking inside the corresponding geometric polygon. Some common ways of geometrically selecting objects for querying, already mentioned in Section 3.1 are:

   - Point selection: select a 1-dimensional or a 2-dimensional object with the mouse. This implies the execution of a geometric test, e.g., Region (Line) contains point?
   - Clipping: select all objects intersecting with a given window.
   - Windowing: select all objects inside a given window.

## 6.2 Current implementation

### 6.2.1 Looks

*Looks* [PCL*90] allows the interactive manipulation of objects and values directly or through methods called by menus. It also allows the specification of the edition of objects stored in the database according to their structure and according to the existence of predefined editors. To each atomic type, corresponds an editor (string editor, integer editor,...). To each constructor (set, list, tuple), corresponds an editor as well. One displays a complex object by building complex editors which recursively call predefined editors. Currently, the presentation style of objects is unique. However, the capability of refining generic presentations in customized ones is under study [Mam90].

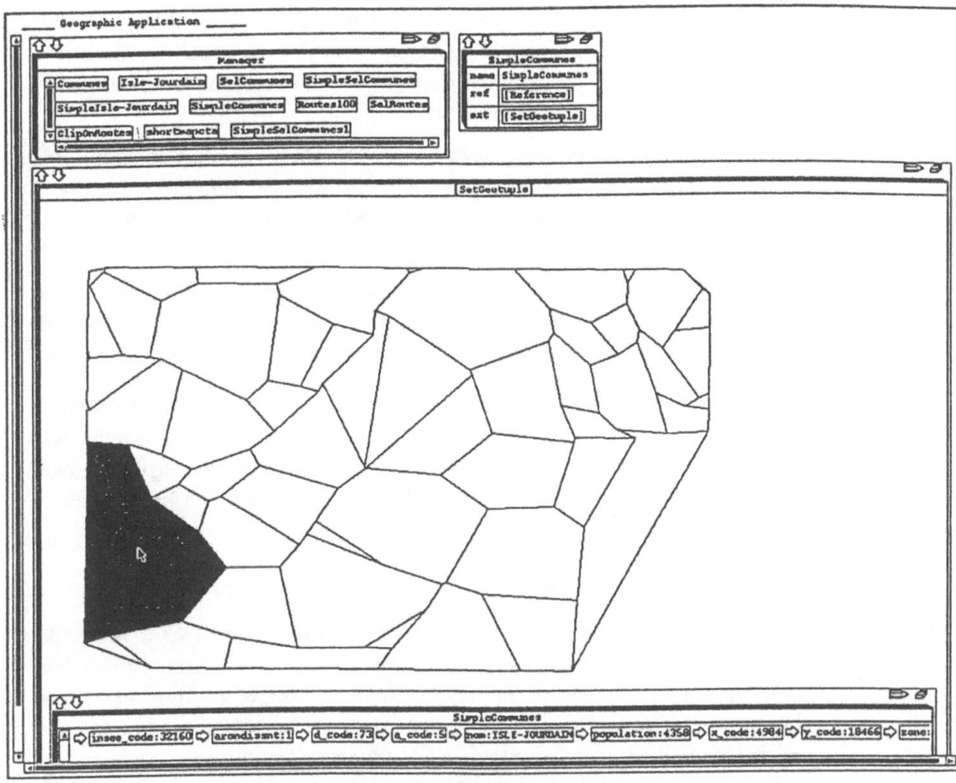


Fig. 9: Map display

### 6.2.2 Looks extension

1. *Display*
   Since one interesting feature of *Looks* is its extensibility through the capability of building new specific editors, we enriched *Looks* with *ad hoc* editors able to deal with our geometric types (Regions, Lines, Points); i.e., we developed editors displaying points, segments and polygons. The geometric part of a map is displayed in a window whose size is pre-defined. To obtain the description of a particular geometric object, the user clicks inside the corresponding region (or onto the line). A point selection is executed: a method is called which finds the object, and an editor containing the list of its attributes is displayed.

   An example of map display is shown in Figure 9. It corresponds to the following user actions on a database of districts in the southwest of France.

   (a) Display the database: maps are represented by icons in the standard *Looks* window with header "Manager".

(b) Select map "SimpleCommunes"[6]: the field "Display-level-one" of the menu associated with icon "SimpleCommunes" is selected by the user. The system answers by displaying an object in the standard *Looks* window with header "SimpleCommunes".

(c) Display "SimpleCommunes": this object has three attributes: name, ref (coordinates system) and ext (map extension). By selecting "Display-level-one" associated with field "ext" of "SimpleCommunes", the map is displayed in the central window.

(d) Display the description of a given district: by clicking inside the black polygon, the latter is highlighted and the list of descriptive attributes (name and value) is displayed in the window at the bottom of the screen.

Currently, it is not possible to superimpose two maps onto the same window, because of the bijection in *Looks* between an object and the window which contains it. If the user wants to overlay two maps, he/she can explicitly call method *map overlay*: a new map is created whose geometry can be displayed.

2. *Queries*
There are two ways of running queries on maps:

   - By *Looks* menus offered to the end user in order to call $CO_2$ methods on a class. The methods may only have alphanumeric arguments. For example, if one asks for an overlay between two maps, method *overlay* is chosen on the screen and its argument is given by the user through the arguments collector of the interface. Methods may also have a geometric argument. This is the case for methods *windowing* and *clipping* (see 6.1.2) for which the user selects with the mouse a particular area on the screen.
   - By using the query language $O_2$Query. At the time of implementation, this was not possible (see Section 4). In the future, this will be done by opening a window dedicated to queries in which the user can edit his/her queries.

# 7 Conclusion

In this paper, we studied the implementation of spatial query languages for geographic applications, by means of an object-oriented database system. To demonstrate the feasibility of such an approach, a prototype has been implemented with version V1 of the $O_2$ system developed by Altaïr [BDK91].

An interface for displaying and querying spatial objects (maps) has been developed. It is an extension of the $O_2$ interface tool *Looks* which, as *Looks* itself, is implemented in Aïda [DD88] and is connected to $O_2$ (Section 6). Simple queries to the geographic database are methods sent from the interface to a class "Maps". Such a class and its methods written in the programming language of $O_2$, $CO_2$ implement

[6] A "commune" in French administration is a county.

a subset of a spatial query language. A simple spatial data model was chosen for representing geometric objects. An $O_2$ schema for implementing a few geometric operations was proposed (Section 5) and is partially implemented. A quadtree-like index [Sam89] is under development for accelerating access to geometric objects.

At the current state of development, several conclusions may be drawn:

1. The object-oriented approach for developing an application turned out to be extremely efficient: three modules were developed in parallel, independently of each other. The integration only took a few days.

2. The object-oriented data model of $O_2$ is extremely well adapted to the representation of application dependent objects and to the implementation of extensible application dependent functions (Section 5).

3. With the current tools available for generating interfaces, designing a customizable interface is still a complex task for standard applications. For geographic applications which deal with spatial objects, it is an even more challenging task.

4. The query language $O_2$Query presents the advantage of modeling complex structures, which is forbidden by any "extended SQL" query language. But, in spite of its power, it was not used for answering geographical queries because it was not fully integrated in $CO_2$ at the implementation time. The designer who would like, instead of using $O_2$Query, to develop relational-like query capabilities in $CO_2$ would need some extensions to $CO_2$. Extensions for relational operations are suggested in Appendix 2.

5. It is not clear whether the encapsulation of application dependent objects should be totally respected or partially violated. How much of the semantics of the geometric objects should be visible at the user level? This complicates the task of designing a geographical query language and perhaps explains why there is no standard emerging for such applications.

**Acknowledgements**
The authors acknowledge the GIP Altaïr members who made available the $O_2$ system and provided assistance in the development of the prototype. The authors are strongly indebted to J.C. Mamou whose expertise in *Looks* and Aïda was extremely useful and who suggested with S. Cluet the map schema which was actually implemented. The clarity and the quality of the paper was significantly improved thanks to the careful reading of the first draft by S. Abiteboul, C. Delobel and particularly by F. Bancilhon. Thanks also go to G. Ferran, R. Güting and D. Plateau for helpful discussions and comments on the paper. The implementation was done by B. Amann, D. Charlet, A. Souf, A. Voisard and J.H. Yapi. A sample of geographic data was provided by IGN. This work also benefited from the design and implementation with R. Jeansoulin of a first prototype using the graphical and geographical software GOODIES [Jea89]. The help of F. Chauveau and P. Razac for installing $O_2$ at the Cedric lab was extremely appreciated.

## References

[AB88] S. Abiteboul and C. Beeri. *On the Power of Languages for the Manipulation of Complex Objects.* Technical Report 846, INRIA, May 1988.

[Abe89] D. J. Abel. SIRO-DBMS: a Database Toolkit for Geographical Informations Systems. *Int. J. Geographical Information Systems*, 3(2):103–116, 1989.

[AFS88] S. Abiteboul, P.C. Fisher, and H.J. Schek. *Nested Relations and Complex Objects.* LNCS 361, Springer-Verlag, 1988.

[AK89] S. Abiteboul and P. Kanellakis. Object Identity as a Query Language Primitive. In *Proceedings of the ACM-SIGMOD Conference on the Management of Data*, Portland, Oregon, June 1989.

[BB81] R. Barrera and A. Buchmann. Schema definition and query language for a geographical database system. In *IEEE Computer Architecture for Pattern Analysis and Image Database Management*, 1981.

[BBKV87] F. Bancilhon, T. Briggs, S. Khoshafian, and P. Valduriez. FAD, a powerful and simple database language. In *Proc. of the Conference on Very Large Data Bases (VLDB)*, Brighton, 1987.

[BCD89] F. Bancilhon, S. Cluet, and C. Delobel. A Query Language for the $O_2$ Object-oriented Database System. In *Proc. of the 2nd Worshop on Database Programming Languages*, Salishan, Oregon, 1989.

[BDK91] *The $O_2$ BOOK*, Morgan Kaufmann, 1991, F. Bancilhon, C. Delobel, and P. Kanellakis, eds

[BDQV90] K. Bennis, B. David, I. Quilio, and Y. Viémont. Géotropics: database support alternatives for geographic applications. In *4th International Symposium on Spatial Data Handling*, Zurich, 1990.

[CJ88] A.F Cardenas and T. Joseph. Picquery: a high level query language for pictorial database management. *IEEE Transactions on Software Engineering*, 14(5): pp 630–638, May 1988.

[Dav88] B. David. Le Modèle Spatiarel: Une extension du modèle relationnel pour gérer des données géographiques. In *4es Journées BD avancées BD3*, Bénodet, France, May 1988.

[DD88] M. Devin and P. Duquesnoy. Manuel de référence AIDA version 1.2, Ilog, 1988.

[EF88] M. Egenhofer and A. Frank. Towards a spatial query language: user interface considerations. In *VLDB*, Los Angeles, 1988.

[Ege89a] M. Egenhofer. A formal definition of binary topological relationships. In *Foundations on Database Organization and Algorithms (FODO)*, LNCS 367, Springer-Verlag, 1989.

[Ege89b] M. Egenhofer. *Spatial SQL: A Spatial Query Language.* Technical Report, Department of Surveying Engineering, University of Maine, 1989.

[Ege89c] M. Egenhofer. A topological data model for spatial databases. In *Design and Implementation of Large Spatial Databases*, LNCS 409, pp 271–286, Springer-Verlag, 1989.

[Fra82] A. Frank. Mapquery: data base query language for retrieval of geometric data and their graphical representation. *Computer Graphics*, 16, 1982.

[Fra84] A. Frank. Requirements for database systems suitable to manage large spatial databases. In *First International Symposium on Spatial Data Handling*, pp 38–60, Zurich, 1984.

[GCK*89] G. Gardarin, J.P. Cheiney, G. Kiernan, D. Pastre, and H. Stora. Managing complex objects in an extensible relational DBMS. In *VLDB*, Amsterdam, 1989.

[GNT91] M. Gargano, E. Nardelli, and M. Talamo. A Model for Complex data: Completeness and Soundness Theorem. In [GSS91].

[GSS91] *Proc. of Internal Worshop on DBMS's for Geographical Applications* (Capri, May 15-17, 1991), BRA series Springer-Verlag, G. Gambosi, M. Scholl, H.-W Six Eds.(this volume).

[Gut88] R.H. Güting. Geo-relational algebra : a model and query language for geometric database systems. In *Conference on Extending Database Technology (EDBT '88)*, LNCS 303, pp 506–527, Springer-Verlag, 1988.

[Gut89] R.H. Güting. Gral: an extensible relational database system for geometric applications. In *VLDB*, Amsterdam, August 1989.

[Jea89] R. Jeansoulin. Le système d'information géographique orienté-objet GOODIES. In *7ième congrès AFCET-RFIA*, Paris, 1989.

[LR89] C. Lécluse and P. Richard. The $O_2$ Database Programming Language. In *VLDB*, Amsterdam, August 1989.

[Mam90] J.C. Mamou. Toonmaker: an approach to create presentations for object-oriented DBMS. 1990, submitted for publication.

[Mey86] B. Meyer. Genericity versus inheritance. *Proc. of the First ACM Conference on Object-Oriented Programming Systems, languages and Applications, SIGPLAN Notices*, 21(11):391–405, 1986.

[Mey88] B. Meyer. *Object-Oriented Software Construction.* Prentice Hall, 1988.

[OBB89] A. Ohori, P. Buneman, and V. Breazu-Tannen. Database programming in machiavelli - a polymorphic language with static type inference. In *Proc. of the ACM-SIGMOD Conference on Management of Data*, pp 46–57, Portland, Oregon, 1989.

[ODD89] B. C. Ooi, Ron Sack Davis, and K. J. Mc Donell. Extending a DBMS for geographic applications. In *5th Int. Conf. on Data Engineering*, Los Angeles, 1989.

[O290] O. Deux et al. The Story of O2. *IEEE transactions on Knowledge and Data Engineering*, 2(1), 1990.

[PCL*90] D. Plateau, R. Cazalens, D. Lévêque, J.C. Mamou, and B. Poyet. Building user interfaces with the looks hyper-objects system. In *Proc. Eurographics Workshop on Object-Oriented Graphics*, Koenigswinter, 1990.

[QDDA89] I. Quilio, P. Dupuis, B. David, and J.M. Ambrosini. Report on design of architecture and interface of Geotropics. ESPRIT-TR 2427-0036, Tropics Project, July 1989.

[Sam89] H. Samet. *The Design and Analysis of Spatial Data Structures*. Addison-Wesley, 1989.

[SFSS90] D. Stemple, L. Fegaras, T. Sheard, and A. Socorro. Exceedings the limits of polymorphism in database programming languages. In *Conference on Extending Database Technology (EDBT '90)*, LNCS 416, pp 269–285, Springer-Verlag, 1990.

[SMO87] R. Sack-Davis, K.J. McDonell, and B.C. Ooi. GEOQL - a query language for geographic information system. In *Australian and New Zeland Association for the Advancement of Science Congress*, Townsville, Australia, August 1987.

[SMSE87] T. R. Smith, S. Menon, J.L. Star, and J.E. Estes. Requirements and principles for the implementation and construction of large-scale geographic information systems. *International Journal of Geographical Information Systems*, 1(1): pp 13–31, 1987.

[Smy90] C. S. Smyth. A reference data model for spatial and geographic applications. In *4th International Symposium on Spatial Data Handling*, pp 869–878, Zurich, July 1990.

[Sou90] A. Souf. *Implantation d'un langage de requêtes géographiques avec le SGBD O2*. Technical Report, Paris 6, rapport de DEA, September 1990.

[SR86] M. Stonebraker and L. A. Rowe. The design of POSTGRES. In *Proc. ACM SIGACT-SIGMOD*, pp 340–355, 1986.

[SV89] M. Scholl and A. Voisard. Thematic map modeling. In *Design and Implementation of Large Spatial Databases*, LNCS 409, pp 167–190, Springer-Verlag, 1989.

[SW88] H. Six and P. Widmayer. Spatial searching in geometric databases. In *Proceedings of the 4th International Conference on Data Engineering*, pp 496–503, Los Angeles, 1988.

[Ull88] J.D. Ullman. *Principles of Database and knowledge-base Systems.* Computer Science Press, 1988.

[Voi91] A. Voisard. Towards a Toolbox for Geographic User Interfaces. submitted for publication.

[WH87] T. C. Waugh and R.G. Healey. The GEO-VIEW design, a relational database approach to geographical data handling. *Int. J. of Geographical Information Systems*, 1(2):101–118, 1987.

# *Appendix 1: $O_2$ schema for the geometry of a map*

class *Spatial*
  */information common to all geometric classes

class *Arc* inherits Spatial
  tuple[begin:Point
    end:Point]
  method *colinear*(a:Arc):boolean
  */checks whether the two arcs are colinear

  method *pintersect*(a:Arc):Point
  */intersection of two linear segments

class *Arcs* inherits Spatial
  list (a:Arc)
  method *colinear*(a:Arcs):boolean
  */two lists of arcs are colinear iff at least one arc of one list is
  */colinear with one arc of the other list: calls method colinear of class Arc

  method *pintersect*(a:Arcs):boolean
  */two lists of arcs are intersecting iff at least one arc of one list intersects
  */one arc of the other list: calls method pintersect of class Arc

class *ConnGraph* inherits Arcs
  method *connected*:boolean
  */checks whether the list of arcs is a connected graph

class *Cycle* inherits ConnGraph
  method *cyclic*:boolean
  */checks whether the connected graph is a cycle

class *Geo* inherits Spatial
  tuple[any]
  method *adj*(g:Geo):boolean

```
   body(error)

  method intersect(g:Geo):Geos
   body(error)


class Point inherits Geo
  tuple[x:real
       y:real]


class Region inherits Geo
  tuple[...
       c:Cycle]
  method border:Cycle
   body(return *self.c)

  method adj(r:Region):boolean
   body(return(self→ border→colinear(r→border))

  method intersect(r:Region):Regions
  */polygon intersection


class Line inherits Geo
  tuple[...
       c:ConnGraph]
  method adj(l:Line):boolean
   body(return(*self.c→colinear(*l.c)))

  method intersect(l:Line):Lines
  */the intersection of the two lines is a set of lines colinear with both argument
  */lines: calls method colinear of class ConnGraph inherited from class Arcs

  method pintersect(l:Line):Points
  */the pintersection of two lines is the set of intersection points between arcs of each
  */line: calls method pintersect of class ConnGraph inherited from class Arcs
```

```
class Geos inherits Spatial
   set(g:Geo)
   method adjacent(g:Geos):boolean
      body(...Geo t,u...
         if t→adj(u)...)

   method intersection(g:Geos):Geos
      body(
         Geos s,v
         Geo t,u
         s=new(Geos)
         for each t in *self
           for each u in *r
              v=new(Geos)
              v=t→intersect(u)
              s+=v
         return(s))


class Regions inherits Geos
   set(r:Region)
   method intersection(r:Regions):Regions
      body(return(Regions)(self→intersection(r)@Geos))

   method border:Lines


class Lines inherits Geos
   set(l:Line)
   method intersection(l:Lines):Lines
      body(return(Lines)(self→intersection(l)@Geos))

   method pintersection(l:Lines):Points
      body(
         Points v,s
         Line t,u
         s=new(Points)
         for each t in *self
           for each u in *l
              v=new(Points)
              v=t→pinter(u)
              s+=v
         return(s))
class Points inherits Geos
   set(p:Point)
```

# *Appendix 2:* *Extension of $CO_2$ for implementing relational operations*

We show below that generic operations of the relational algebra can be implemented with $CO_2$ given the following extensions:

1. casting,

2. given a hierarchy of classes and a method m overloaded in each class of the hierarchy, the possibility of sending method m@j from a method of subclass i, where m@j refers to the code of method m of superclass j,

3. a few built-in functions to be defined below.

Without loss of generality, we illustrate the mechanism on projection. Two hierarchies are defined with roots: classes *Relation* and *Tuple*. Relation is of type: set(Tuple).Tuple has no structure defined. Any relation with schema R(*A1: D1,..., An: Dn*) is an object of class R with structure set(TupleR) where TupleR is a class with structure:

*[A1: D1,...,An: Dn].*

Class R is a subclass of class Relation. TupleR is a subclass of class Tuple.

Let us look at the implementation of projection. Two methods are needed:

1. method *Projection* of class Relation, overloaded in each subclass representing a relation; the body of this method is given in Figure 1. It consists in defining a new object c, in applying to each tuple, method *Proj* and in adding the resulting tuple to c.

2. method *Proj* is a *built-in* function which given as arguments a list of attribute names (a list of strings) and a source tuple, returns the projected tuple.

In each subclass *RelationS* (structure: set(TupleS)) of Relation, method Projection is redefined as follows (Figure 1): the body consists in applying to the object (self) method Projection as defined in class Relation (Projection@Relation) and to cast the result, i.e., to force the resulting set of tuples to belong to subclass RelationS (return(RelationS)). In subclass *TupleS* method *Proj* is redefined similarly: apply proj@Tuple and enforce the resulting tuple to belong to class TupleS.

```
class Relation
    type set(Tuple)
    method Projection(listattr:list(string)):Relation
        body {
        Relation c
        Tuple t,t'
        c=new(Relation)
        for each t in *self
            t'=t→Proj(listattr)
            c+={t'}
        return(c)}

class RelationS inherits Relation
```

```
    type set(TupleS)
    method Projection(listattr:list(string)):RelationS
        body {
        return(RelationS)(self→Projection(listattr)@Relation)}

class Tuple
    type any
    method Proj(listattr:list(string)):Tuple
    body { built-in }

class TupleS inherits Tuple
    type tuple(A1:D1,...,An:Dn)
    method Proj(listattr:list(string)):TupleS
      body {return(TupleS)(self→Proj(listattr)@Tuple)}
```

Fig. 1: Projection

A projection is performed as follows:
RelationS s
r=new(RelationT)
r=(RelationT)s→projection(listattr)

A target relation is created, i.e., an object r of class RelationT and r is assigned the result from sending method Projection to the source relation (object s of class RelationS) casted to class RelationT. Observe, however that it is *assumed that class RelationT already exists.*

Similarly, one can program Selection using as built-in method (of class Tuple): *select*(Attributename:string,Comparator:string,Value:string):Tuple (or *select*(Attributename:string,Comparator:string,Attributename:string):Tuple). Sent to a tuple, it gives the tuple itself if it satisfies the selection criteria. Cartesian product and Join would use as built-in method *concat*(t:Tuple):Tuple. It concatenates the tuple to which the method is sent with the tuple in argument.

# Loading Spatial Features into the Incomplete Pyramid Data Structure

Walid G. Aref
Hanan Samet

Spatial data usually is stored in images where spatial features are classified using image processing techniques. However, for efficient processing, spatial features must be loaded into appropriate in-core spatial data structures. The pyramid data structure is of particular interest because it can handle both overlapping and non-overlapping features. The incomplete pyramid, a variant of the pyramid, has several advantages over the pyramid in the context of spatial processing and answering queries on spatial features. Algorithms are presented for loading overlapping and non-overlapping features into the incomplete pyramid from a number of disk-based pointerless quadtree feature representations. The running time for these algorithms is linearly proportional to the perimeter of the spatial feature being loaded and is independent of the pyramid depth.

## 1 Introduction

The large volume of spatial data imposes the need to store spatial features in disk files. Many disk-based data representations exist for storing spatial data. However, for efficient processing and query answering spatial data has to be, partially or if possible totally, loaded into suitable in-core spatial data structures.

The pyramid is proposed as one in-core representation for spatial features because its hierarchical and multi-resolution nature facilitates the performance of a number of useful operations. Nevertheless, spatial data is often represented using non-pyramidal methods. One reason is that most images (a common form of spatial data) are traditionally represented by use of methods such as binary arrays, runlength codes, boundary codes, or polygons (i.e., vectors). Other reasons include the nature of the application (e.g., runlength codes are particularly useful for raster-like display devices). Also, from an implementation perspective, some of these representations are pointer-based and hence are more suitable as main memory structures while other representations are pointerless and hence can be useful as disk-based representations. Therefore, techniques that can switch efficiently between these representations are needed. This paper discusses a number of such techniques and presents algorithms for some of them.

Our presentation is primarily in the context of loading variants of pointerless quadtree representations (a disk-based representation) into the pyramid data structure (a main memory representation). The quadtree is a refinement of the array representation of an image that attempts to save storage by taking advantage of some regularity in the image by decomposing the array into homogeneous disjoint two-dimensional blocks centered at predetermined positions. Image elements are now two-dimensional blocks instead of pixels. The quadtree can be implemented as a list or as a tree. The list representation is a pointerless representation that facilitates sequential access but is inefficient for random access to specific image elements. It is more suitable for disk-based storage. The tree representation is an attempt to balance the costs of sequential and random access.

In this paper, we provide a family of efficient algorithms for loading spatial features into a variant of the pyramid data structure, termed the *incomplete* pyramid, from several pointerless quadtree feature representations. In practice, pointerless quadtree representations are widely used for storing and manipulating spatial data (e.g., [1, 6, 12, 15]). The pointerless quadtree representation (in the form of lists) can further be grouped into two categories. The first treats the image as a collection of leaf nodes (e.g., the linear quadtree [3]). The second represents the image in the form of a traversal of the nodes of its quadtree (e.g., DF-expressions [4]). By virtue of being pointerless, these representations are suitable for disk-based storage of spatial data. In addition, they are relatively compact methods of aggregating regions with the same features or attributes. This is especially true when they are compared with the array representation. Finally, mapping feature raster data into one of these representations is well-known [7, 9, 11].

In previous work [2], we introduced the incomplete pyramid data structure as an in-core candidate for representing spatial features. The incomplete pyramid is a multiresolution data structure, in contrast to the region quadtree which is a variable resolution data structure. The incomplete pyramid is of particular interest because it can handle both overlapping and non-overlapping features. It also permits direct access of nodes at all levels. We make extensive use of this property. The incomplete pyramid is an efficient query answering tool [2]. We summarize these results in this paper as well.

The rest of the paper is organized as follows. In Section 2 we classify spatial features into two categories according to how they interfere in space (i.e., whether they overlap or not). In Section 3 we give a brief description of the incomplete pyramid data structure along with some of its important characteristics that we utilize in this paper. In Section 4 we review two appropriate pointerless quadtree representations for storing spatial data. The first is a DF-expression [4] (described in Section 4.1), and the second is a linear quadtree [3] (described in Section 4.2). Sections 5 and 6 present and analyze, respectively, algorithms for loading the incomplete pyramid by spatial features represented using these representations. The novel aspect of this work is that both overlapping and non-overlapping features are permitted. In addition, these algorithms make use of the fact that the incomplete pyramid is an in-core structure and hence parts of it (the pyramid nodes) can be randomly accessed. Thereby avoiding, whenever possible, the descent of the pyramid nodes in tree order.

## 2 Overlapping vs. Non-overlapping Features

We classify spatial features into two groups: overlapping and non-overlapping. Examples of overlapping features are cities, roads, lakes, and counties that comprise a given region. Examples of non-overlapping features are landuse maps where every land region has only one usage and is assigned a feature name to distinguish it from regions with different usages. Figures 1 and 2 show a collection of non-overlapping and overlapping features in a 16 × 16 feature space.

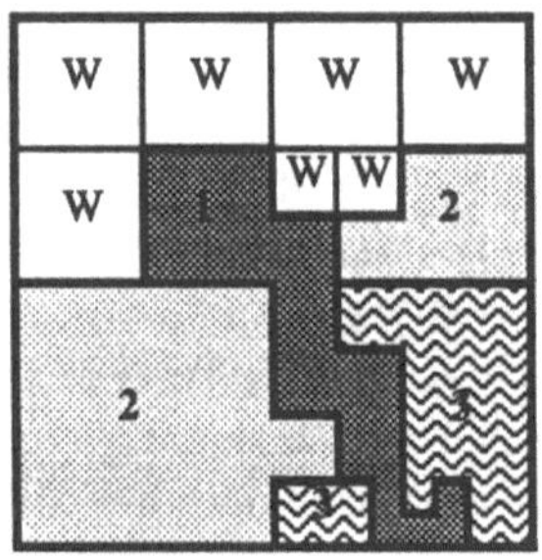


Figure 1: Three non-overlapping spatial features 1, 2, and 3 in a 16 × 16 feature space.

## 3 The Incomplete Pyramid Data Structure

The incomplete pyramid data structure was first introduced in [2]. It is an adaptation of the pyramid data structure [13]. This section is organized as follows. We first review the pyramid and briefly cover its use for spatial query processing. Next we show how to modify the pyramid to form the incomplete pyramid variant.

The pyramid is a multiresolution data structure. This is in contrast to variable-resolution hierarchical data structures such as the region quadtree [5, 8]. Also, the pyramid is used to store a collection of related layers each representing one overlay of the feature space. In this sense, the pyramid can be thought of as an aggregation of complete quadtrees where each quadtree represents one overlay. Features in one quadtree do not overlap in space. This is due to the definition of the region quadtree and its decomposition rules. On the other hand, features in the pyramid from the different spatial overlays may or may not overlap. We use Figures 3 and Figure 4 for illustration.

Figure 3 shows the region quadtree representations of two overlays. Each overlay contains one of the features given in Figures 2b and 2d. This is for illustration only. However, an overlay may contain more than one non-overlapping feature. Nodes in a region quadtree are either leaf or gray. Leaf nodes correspond to those blocks that are either entirely inside the feature region in the image (labeled *black*) or are entirely outside (labeled *white*). Each non-leaf node is said to be *gray* (i.e., its corresponding

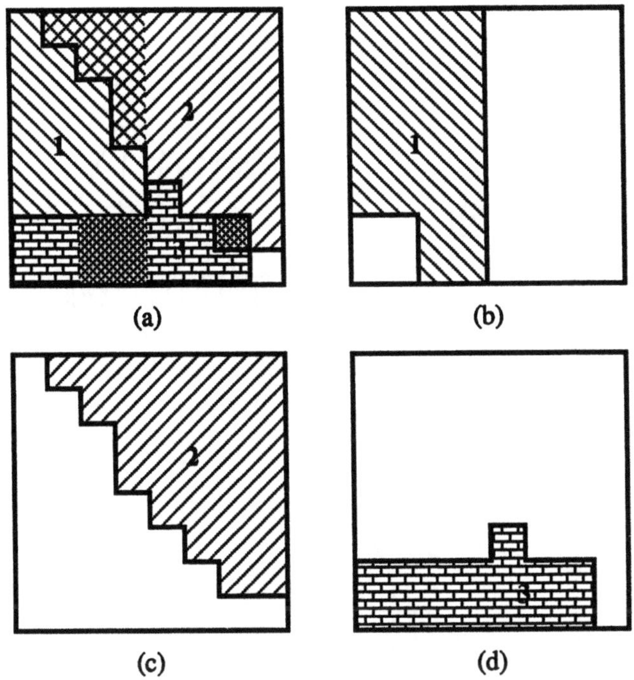


Figure 2: (a) the result of overlaying the three overlapping spatial features 1, 2, and 3 shown in (b), (c), and (d). The features lie in a $16 \times 16$ feature space. W denotes an empty region (i.e., white).

block is neither entirely inside nor entirely outside the feature region). The parent of a leaf node is a gray node (i.e., leaf nodes are maximal).

Figure 4 shows the pyramidal representation of the two overlays of the same region. Notice that, for example, feature 1 of overlay 1 (Figure 2b) and feature 3 of overlay 3 (Figure 2d) overlap and both are loaded into the pyramid. More formally, given a $2^d \times 2^d$ image, the pyramid is defined as a tree where the root node represents the whole image space, and then a recursive decomposition into quadrants is performed just as in quadtree construction, except that we keep subdividing until we reach the individual pixels. The leaf nodes of the resulting tree represent the pixels while the nodes immediately above the leaf nodes correspond to image blocks of size $2^{d-1} \times 2^{d-1}$. The non-leaf nodes are assigned a value that is a function of the nodes immediately below them (i.e., their four sons) such as the average gray level. For our purposes, this function is the Boolean OR set operation. In other words, we say that a certain feature is present in the parent node when this feature is present in at least one of its four sons.

Nodes in the pyramid have the following internal structure: each node has a fixed number $m$ of feature bit-slots and summarizing variables that contain information about the descendants of the node. For now, we are interested in the feature bits.

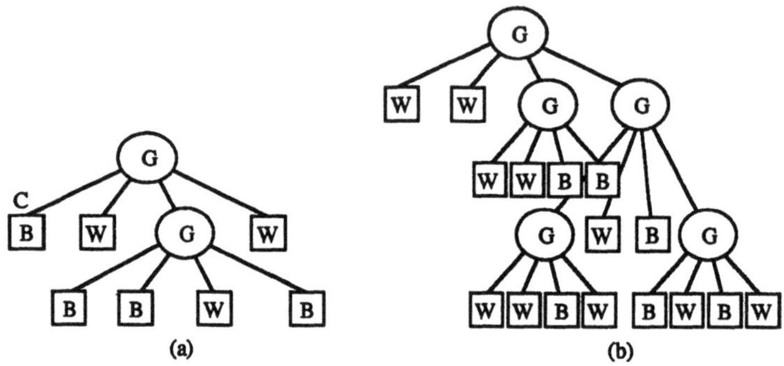


Figure 3: The region quadtree representation of two overlays: (a) the overlay containing the feature given in Figure 2b, (b) the overlay containing the feature given in Figure 2d.

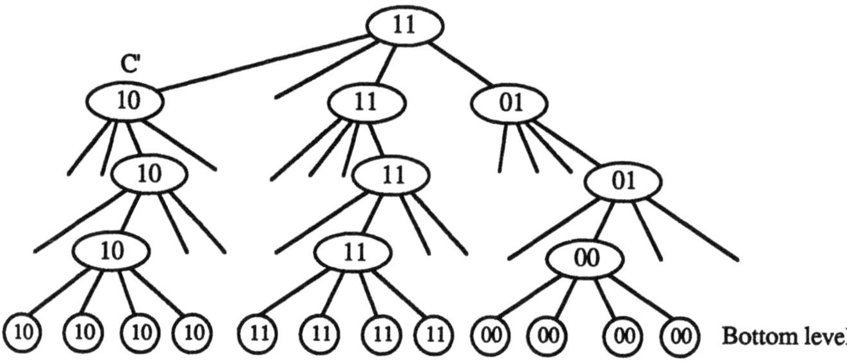


Figure 4: The pyramid representation of spatial features.

Each bit-slot corresponds to a feature stored in the pyramid. Each feature is encoded into its bit-slot through the whole pyramid independent of the features in the other bit-slots. In the following, we show how one feature is encoded into the pyramid. The same technique is applied for the remaining features. In this discussion, we use the term *node* to denote a one bit-slot.

As can be seen from Figures 3 and 4, the way spatial features are encoded into the nodes of the pyramid hides the information with respect to which nodes correspond to maximal blocks in the recursive decomposition. As an example, consider block C of Figure 3. C is a maximal quadtree block. Its feature is represented in the pyramid by the first bit-slot of pyramid node C' of Figure 4. In the pyramid, there is no specific information to indicate whether C' is a maximal block or not. This information, although not required, can speed-up processing time significantly. In many algorithms, when operating on the feature corresponding to the first bit-slot, processing can stop at node C' without a need to access the nodes of higher resolution below C'.

In an attempt to solve this problem, Shaffer and Samet [10] modify slightly the way spatial features are coded into the pyramid so that the information about maximal quadtree blocks of a given overlay is preserved inside the pyramid. Their suggested method is called the GL coding method. Figure 5 illustrates this coding method where the overlay of Figure 3a is represented in the pyramid data structure using the GL coding method. The GL method works as follows. If an intermediate

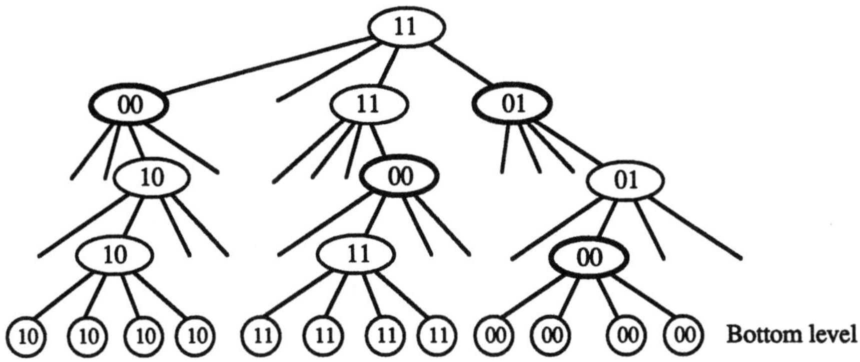


Figure 5: Example of GL coding. Nodes with thick border are leaf nodes for either the first feature or the second feature or both.

node is gray, then its bit slot value is 1. If an intermediate node is a leaf node, then its bit slot value is 0. Since it is necessary to determine the actual value of a leaf node (i.e., black or white), this value is stored in all of the descendants of the leaf node down to the pixel level (i.e., 1 and 0, respectively). If a bottom-level node is a leaf node, then its value is stored in the node itself since it has no descendants. Clearly, a gray node cannot appear in the bottom level.

Although it succeeds in locating maximal quadtree blocks inside the pyramid, the GL encoding method has two deficiencies. First, if pyramid nodes were to be accessed directly without descending from the root of the pyramid, it would be a difficult task (logarithmic) to know whether a given node is gray or black since both are coded with the value 1. Notice that some algorithms may make use of this capability if it exists (e.g., the selection operation presented in [2]). In order to distinguish between gray and black nodes using the GL coding method, one has to either descend or ascend the pyramid nodes and test whether a value 0 is encountered or not. The number of nodes to be accessed in this process of descent or ascent to distinguish between gray and black nodes is of the order of the depth of the pyramid. In addition to the fact that this is dependent on the pyramid depth and hence a log factor is to be introduced as a computational overhead, it also violates the essence of direct accessing the pyramid nodes and gaining the desired information.

The second deficiency in the GL method is that when loading one overlay containing some spatial features (e.g., lakes) into the pyramid data structure, eventually, all the pyramid nodes have to be accessed. This slows down the loading of spatial features significantly. This is especially true if we consider the fact that the pyramid contains $m$ overlays (one overlay per bit-slot in a pyramid node) and each one has to be loaded independently. Therefore, saving the loading time of overlays is significant.

One way to overcome the two deficiencies of the GL method is by using the modified GL coding method suggested by Aref and Samet in [2]. The difference between the normal GL coding scheme and the modified scheme is that in the modified scheme when a leaf node (a 0 value) appears at an intermediate level, then all its descendants have a 0 value except for those nodes at the bottom level which are assigned the values 1 or 0 depending on whether they are black or white, respectively. If a leaf node appears at the bottom level then it is assigned the value 1 or 0 depending on whether it is black or white, respectively. Notice that using this coding scheme, we are able to overcome the deficiency that arises with normal GL coding. In particular, when nodes are accessed directly, we can distinguish between gray and leaf nodes. As an example, consider Figure 6. Assume that we want to access the second bit-slot of the pyramid only. When node A is accessed directly, we know it is a gray node since its value is 1, while node B is a leaf node since its value is 0. To determine if node B is black or white, we access node C directly at the bottom level of the pyramid which indicates that B's value is black (since C has a value of 1). In addition, in contrast to the GL method, in the process of loading

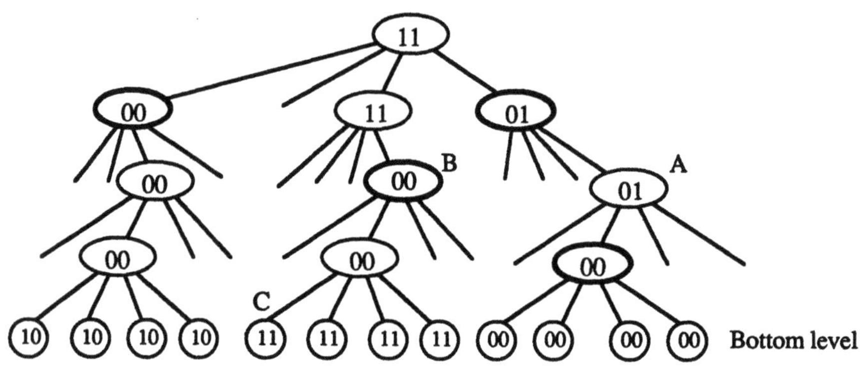


Figure 6: Example of modified GL coding. Nodes with thick border are leaf nodes for either the first feature or the second feature or both.

spatial overlays using the modified GL method, descending down the pyramid nodes can stop at maximal quadtree blocks without accessing the nodes below them, i.e., all the other nodes are left unchanged (reset to 0) except for the bottom level. The bottom level has to be loaded for both the GL and the modified GL methods. The bottom level of the pyramid contains the whole overlay at full resolution and can be loaded as a single entity by a method other than pyramid traversal. This is possible since, as we will see below, the pyramid is implemented as a stack of arrays and hence the bottom level of the pyramid has contiguous storage.

Combining the modified GL encoding of spatial features with the pyramid data structure yields what we term the *incomplete* pyramid. Since the encoding of a feature starts at the root of the pyramid and stops at a leaf node which can be at any level, the pattern of ones inside the pyramid looks like a partially filled pyramid as shown in Figure 7. Another illustration of the difference between the regular

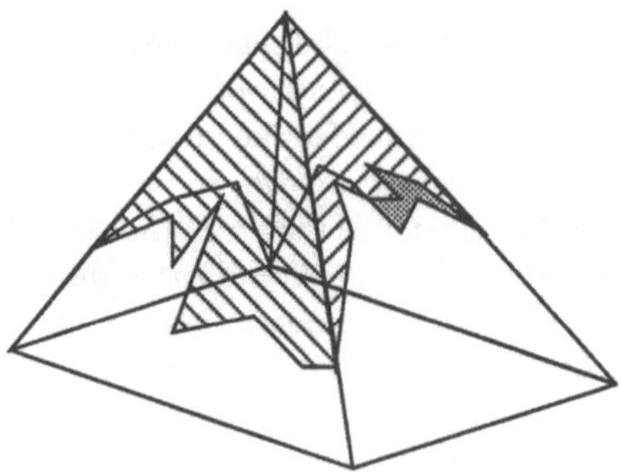

Figure 7: The incomplete pyramid where the shaded areas correspond to features represented by patterns of ones descending from the root down to leaf nodes.

pyramid and the incomplete pyramid is that the regular pyramid is an aggregation of complete quadtrees each representing a feature in the image space, while the incomplete pyramid is an aggregation of region quadtrees with some enhancements that improve the run-time performance.

The pyramid is implemented by a linear array in a heap-like fashion [16]. Each pyramid node is represented by an array entry. Given a parent node $p$ with relative index $j$ at level $i$, we can access the $k^{th}$ son of $p$ (where $k = 0, 1, 2, \text{or } 3$) by the following formula:

$$son(i, j, k) = 4^i + 4 \times j + k$$

The number of array elements at levels 0 (the root level) through $i-1$ is $(4^i - 1)/3$.

Also, there is a correspondence between the pyramid nodes and the spatial database blocks. Assume a coordinate system such that the origin is at the upper-left corner, $x$ increases to the right, and $y$ increases downwards. A block of size $2^i \times 2^i$ $(0 \leq i \leq d)$, having $(r, c)$ as the coordinates of its upper-left corner $(0 \leq r, c \leq T-1)$ such that $r \bmod 2^i = c \bmod 2^i = 0$ corresponds to the pyramid node at level $d - i$ whose relative index within this level is equal to

$$(bit\_interleave(r, c)) \gg (2 \times i),$$

where $\gg$ is the shift-right operation. Thus $y \gg x$ means that $y$ is shifted to the right by $x$ bit positions.

The pyramid representation has an overhead of one third extra storage because of the need to store intermediate nodes. In other words, for a $T \times T$ image where $T = 2^d$, the entire pyramid requires $4 \times (T \times T - 1)/3$ nodes. This is the same amount of storage necessary for a complete quadtree [8] where every node of the quadtree has been expanded down to the pixel level.

Even though it is a space-redundant data structure, the incomplete pyramid has major benefits in its multiresolution nature as well as the run-time improvements that it permits. Many queries can be answered with the same accuracy at lower levels of resolution. Examples of these queries are window-based queries that include the exist query (i.e., determining whether or not a spatial feature exists inside a window), and the report query (i.e., reporting the identity of all the features that exist inside a window). In [2], we showed that these queries can be answered in $O(n \log \log T)$ time for a window of size $n \times n$ in a feature space (e.g., an image) of size $T \times T$ (e.g., pixel elements). The significance of this result is that even though the window contains $n^2$ pixel elements, the worst-case time complexity of the algorithms is almost linearly proportional (and not quadratic) to the window diameter, and does not depend on other factors. The techniques used to answer the report and exist queries can be used to answer other queries as well (e.g., the containment query).

*Some Properties of the Incomplete-Pyramid*

- **Property 1**: Direct Access
  The physical location of any node in the incomplete pyramid can be directly accessed in constant time [14]. The incomplete pyramid is implemented as a linear array with the root at location 0, nodes at depth 1 in locations 1 – 4, followed by the nodes at depths 2 through $d$ (the maximum level). Given the values of the $x$ and $y$ coordinates of the upper-left corner of a block in the image and its size (i.e., $2^i \times 2^i$), we can compute the pyramid address of the 'corresponding' node, say $Q$, in constant time as follows:

$$pyr\_address(i, x, y) = (4^i - 1)/3 + (bit\_interleave(x, y) \gg 2i)$$

  The same formula applies for computing the incomplete pyramid address of a given node in constant time.

- **Property 2**: Parent and Son Addresses
  Given the address (or the index inside the incomplete pyramid array) of a node, we can generate the address of the node's parent, or any of its sons, in constant time.

- **Property 3**: Contents of a Node
  Given the address (or the index inside the incomplete pyramid array) of a node, we can retrieve the features contained or overlapping with this node's block in at most two direct accesses to the incomplete pyramid.

  Procedure *features*, given in Figure 8, shows how features stored in $Q$ can be retrieved from the incomplete pyramid structure in constant time. $Q$ can be at an intermediate or bottom level in the pyramid.

```
set_of_features procedure features(i, x, y);
begin
  value integer i, x, y;
  set_of_features s;
  global integer d;
  if(i = d) then /* bottom level*/
    s ← contents(pyr_address(d, x, y))
  else  /* features in the node or in one of its sons */
    s ← contents(pyr_address(i, x, y))
        ∪ contents(pyr_address(d, x, y));
  return (s);
end;
```

Figure 8: Retrieving the features stored in a node of the incomplete pyramid.

These properties along with other interesting properties are explained further in [2].

# 4 Pointerless Quadtree Representations

In this section, we review two pointerless quadtree feature representations: the DF-expression and the linear quadtree. The former is a linear list resulting from a preorder traversal of the active nodes in the incomplete pyramid data structure, while the latter is a collection of leaf nodes of a quadtree representing the spatial features. Other pointerless quadtree representations exist, but they are of less practical use. The reader is referred to [7] for further details.

## 4.1 DF-expressions

A DF-expression [4] is a pointerless quadtree representation of a feature. It can also be used with an incomplete pyramid data structure. It is in the form of a preorder tree traversal of the bit-slot(s) corresponding to the feature(s) in the active nodes of the incomplete pyramid. The result is a string consisting of the symbols 'G', 'B', 'W', corresponding to gray, black, and white, respectively. Other symbols are used when *multiple* features are stored in one DF-expression, as presented below. In the following, we show how a DF-expression is used to store multiple features regardless of whether or not they overlap in space. We first describe the simple one-feature case. More details about operations based on DF-expressions can be found in [7].

*One Feature*
Using the symbols ('G', 'B', and 'W') and applying a preorder traversal to the bit-slots of the active nodes in the incomplete pyramid corresponding to some spatial feature, we get the DF-expression string representing this feature. For example, the DF-expression of the feature of Figure 9 is given by:

GGWWWB
GWWGWWBWW
W
GGBWBBWGWBWGWBWBGWWGWBBBW.

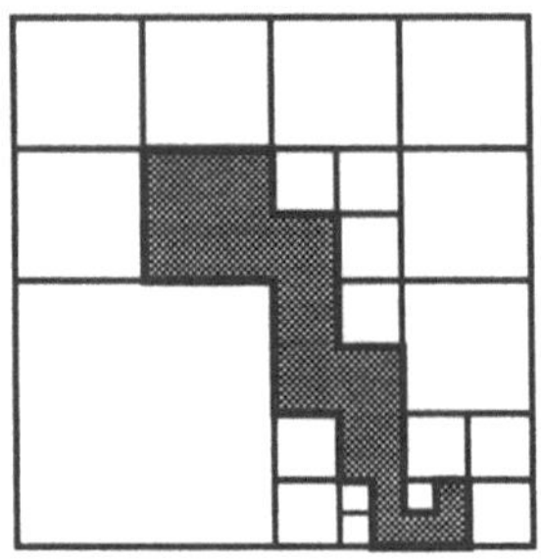

Figure 9: A spatial feature in a 16 × 16 feature space.

*Multiple Features*
Although multiple features can be represented by multiple DF-expressions (i.e., one DF-expression for each feature), there are several advantages to storing multiple features in a single DF-expression. It is economical from the standpoint of saving execution time when the same operation must be applied to all the features. An example of such an important operation is down loading and storing features into and out of the on-line incomplete pyramid data structure. In Section 5 we show that by storing multiple features together in one DF-expression, one traversal of the the incomplete pyramid is sufficient. This is in contrast to traversing the incomplete pyramid several times, once for each feature.

To support multiple features, we add more symbols to the DF-expression alphabet. In particular, in addition to 'G', 'B', and 'W', we also use $i$, $\{i_1, i_2, ..., i_n\}$, to correspond to an index for feature $i$, and a set of feature indexes, respectively.

*Non-Overlapping Features*
Representing multiple non-overlapping features by a single DF-expression is straightforward. It is the same as for the one-feature case with one slight modification. Whenever a black symbol 'B' appears in the DF-expression string, the index of the corresponding feature is concatenated to it (for example, 'B3' refers to a black block covered by feature 3). This is the same as coding colored images by a DF-expression. Notice that since features do not overlap in space, only one feature index is stored with each black symbol. As an example, the following string is the DF-expression for the image in Figure 1 which consists of 3 different features.

GGWWWB1
GWWGWWB1B2B2
B2
GGB1B3B1B1B3GB2B1B3GB3B1B3B1GB3B3GB3B1B1B1B3.

*Overlapping Features*
The representation of overlapping features requires the attachment of more information to gray and black symbols. The problem is that (1) a node may be gray for some features and black for several other features; and (2) a black node may correspond to several overlapping features. For example, for the three overlapping features in Figure 2, we have the following DF-expression:

GG1GWB2WWB2WGWB2WB2
B2
GB1B1B3B1,3
GGB2B2B3B2B2B3GB2,3B2B3W.

Note the use of commas to separate multiple overlapping features.

### 4.2 Linear Quadtrees

A linear quadtree only supports non-overlapping features. It represents a quadtree as a collection of the leaf nodes comprising it. This collection of leaf nodes is stored as a sorted linear list. Each node consists of three fields:

1. The address field: contains the two-dimensional locational code of the node's corresponding block. The locational code is composed by interleaving the bits of the binary representations of the value of the $x$ and $y$ coordinates of the block's upper left corner.
2. The depth field: contains the size of each block (i.e., the depth of the corresponding node in the quadtree).
3. The value field: contains the index of the feature associated with this node.

For example, the first twelve nodes of the linear quadtree for the non-overlapping features shown in Figure 1 are

(0000,2,W),(0100,2,W),(0200,2,W),(0300,2,1),(1000,2,W),(1100,2,W),(1200,3,W),
(1210,3,W),(1220,3,1),(1230,3,2),(1300,2,2),(2000,1,2),(3000,3,1),(3010,3,3), ... .

where (*address*, *depth*, *value*) correspond to the address field (coded in base 4), the depth field, and the value fields of a node, respectively.

## 5 Algorithms

We present several algorithms for loading non-overlapping and overlapping features into the incomplete pyramid data structure. For non-overlapping features we consider two algorithms based on how the input features are represented, either as a DF-expression or as a linear quadtree. For overlapping features we consider only the case where features are represented as a DF-expression. Recall that a linear quadtree only supports non-overlapping features.

One major benefit of the incomplete pyramid is that the subtrees descending from an active leaf node (recall that this is also the leaf node of the corresponding linear quadtree) are never accessed while loading the features (except for the bottom part which contains a raster representation of the feature space). This saves time when loading the features without sacrificing the spatial processing power of the regular pyramid as shown in [2]. Our three algorithms take advantage of this property.

Our presentation makes use of the following predefined functions:

- all_sons_reported($p$) : returns *true* if all sons of $p$ reported to it, and returns *false* otherwise. This routine is described in greater detail in Section 5.1.2.
- bit_slot($f$) : returns the bit-slot in a node of the incomplete pyramid corresponding to the feature index $f$.
- data($l$) : reads the current data element from the list $l$.
- depth($q$) : returns the value stored in the depth field of node $q$ of the linear quadtree.
- eol($l$) : returns *true* when there are no more data elements in $l$; otherwise, it returns *false*.
- father($p$) : computes the address of $p$'s father in the incomplete pyramid. $p$ is an address of a node in the incomplete pyramid.
- incr_intermed($p$) : increments the 2-bit report field associated with node $p$ in the incomplete pyramid. This 2-bit field is described in greater detail in Section 5.1.2.
- load_pyr_base($x, y, s, v$) : fills a block in the bottom level of the incomplete pyramid with the value $v$. $v$ is usually a feature index. This block is specified by its size $s$ and the location of its upper-left corner ($x$ and $y$).
- next($l$) : advances to the next data element in the list $l$.
- number($s$) : returns *true* when the data element $s$ contains a number; otherwise, it returns *false*.
- or_bits($p, q$) : is equivalent to contents($p$) $\leftarrow$ contents($p$) OR contents($q$).
- pyr_node_address($d, x, y$) : returns the address of the node at depth $d$ in the incomplete pyramid that corresponds to a block whose upper-left corner coordinate values are $x$ and $y$.
- rc_to_pi($r, c$) : bit-interleaves $r$ and $c$ to form an index into the incomplete pyramid or a quadtree address field.

- set_node($i, p$) : sets the $i$th bit-slot of the node in the incomplete pyramid pointed at by $p$ to 1 and sets the other bit-slots to 0.

- set_pixel($i, v$) : sets the node in the incomplete pyramid corresponding to a bottom-level pixel whose index is $i$ to the value $v$.

- size($q$) : returns the size of the block corresponding to the quadtree node $q$.

- son($p$,$i$) : computes the address of the $i^{th}$ son of the node pointed at by $p$ in the incomplete pyramid.

- son_x($x$,$i$) : computes the $x$-coordinate value of the upper left corner of the block corresponding to the $i^{th}$ son given that the upper left corner of his parent's block has an $x$-coordinate value of $x$.

- son_y($y$,$i$) : computes the $y$-coordinate value of the upper left corner of the block corresponding to the $i^{th}$ son given that the upper left corner of his parent's block has a $y$-coordinate value of $y$.

- store_features($l, p$) : stores the set of feature indexes currently at the front of the list $l$ into the node of the incomplete pyramid pointed at by $p$ and, as a side effect, advances to the next data element in the list $l$. This operation is used for loading multiple overlapping features in the same node $p$. Refer to Section 4.1 for the exact format for storing multiple features in DF-expressions.

- value($q$) : returns the value stored in the value field of node $q$ of the linear quadtree.

- x($q$) : returns the value of the $x$-coordinate stored in the address field of the quadtree node $q$. This involves accessing the depth field as well so that we can determine the number of significant bits in the address field.

- y($q$) : returns the value of the $y$-coordinate stored in the address field of the quadtree node $q$.

## 5.1 Non-Overlapping Features

We can speed-up the loading process by taking advantage of the fact that features do not overlap in space. We scan the image space once, one block at a time, and load the intersecting feature with this block (if any). Since features are non-overlapping, we know that at most one feature intersects this block.

We give two new algorithms for loading the incomplete pyramid with non overlapping features. The first algorithm assumes that the input is represented using a DF-expression while the second assumes that it is represented using a linear region quadtree. Since the intermediate nodes of the incomplete pyramid include summary information about the existence of features in the subtrees below them, both algorithms access each intermediate node a constant number of times independent of the number of features being loaded (which is non-trivial to achieve).

### 5.1.1 The DF-Expression Algorithm

In this section, the input is assumed to be a DF-expression. The basic idea behind the algorithm is that the DF-expression is scanned at the same time that the incomplete pyramid is traversed in postorder. Actions in the incomplete pyramid are performed according to the current DF-expression symbol.

```
1  Boolean recursive procedure load_nonoverlap(l, p, x, y, size)
   begin
     reference pointer list l;
     value pointer pyr_node p;
     value integer x, y, size;
     quadrant i;
2    if data(l) = 'G' then
       begin /* Gray node */
3        for i in {'NW', 'NE', 'SW', 'SE'} do
           begin
4            l ← next(l);
5            load_nonoverlap(l, son(p, i), son_x(x, i), son_y(y, i), size/2);
6            or_bits(p, son(p, i)); /* Store features in the parent. */
           end;
       end
7    else if data(l) = 'B' then
         begin /* Black node */
8          l ← next(l);
9          which_slot ← bit_slot(data(l));
10         set_node(which_slot, p);
11         load_pyr_base(x, y, size, which_slot);
12         l ← next(l);
         end;
     /* No action is necessary for a white node. */
   end;
```

Figure 10: Loading multiple non-overlapping features represented by a single DF-expression into the incomplete pyramid.

The algorithm performs a depth-first search in the incomplete pyramid guided by the symbols in the DF-expression. The code for the algorithm is given in Figure 10. The 'G' symbol instructs the algorithm to descend the incomplete pyramid by one level (line 5). Once a 'B' symbol is encountered, the corresponding node in the incomplete pyramid is set to include this feature in it (lines 8 and 9). The 'B' symbol is followed by the feature index which is used to set the corresponding bit slot in the leaf node. After processing a gray node's son, the features it represents are propagated into the gray node by the OR Boolean operation (line 6). Note that each non-leaf (i.e., gray) node has to report to its parent as soon as all its sons report back to it. This way each node reports to its parent only once. In other words, each non-leaf node is accessed a constant number of times. The code for loading blocks in the bottom level of the incomplete pyramid is given in Figure 11. Procedure

```
Boolean procedure load_pyr_base(x, y, size, val)
begin
  value integer x, y, size, val;
  integer size_sqr, base_xy, index;
  size_sqr ← size ↑ 2;
  if(size_sqr > 1) then
    begin /* The corresponding node is above the pixel level. */
      base_xy ← rc_to_pi(x, y);
      for index ← base_xy to base_xy + size_sqr - 1 do
        set_pixel(index, which_slot);
    end;
end;
```

Figure 11: Loading the base of the incomplete pyramid with a given feature index.

load_nonoverlap is initially invoked by the call load_nonoverlap(df_list, pyr_root, 0, 0, pyr_base_size).

### 5.1.2 The Linear Quadtree Algorithm

In this section, the input is assumed to be a linear list of the leaf nodes. Each node contains the following information:

- The locational code of the node's corresponding block.
- The block's size (or equivalently, the node's depth in the quadtree).
- The value of this node: white (i.e., the node is empty) or black (i.e., a feature covers the node's corresponding block). In the latter case, the value of the node is a feature index identifying the feature covering the block.

The algorithm performs the following four steps for each leaf node $q$:

1. Directly access the corresponding node in the incomplete pyramid, say $p$.
2. Mark $p$ with the feature stored in the value field of $q$,
3. Report $p$'s features to $p$'s father, say $P$. We assume that each intermediate node of the incomplete pyramid has an extra 2-bit field to count the number of direct sons that report to it. Node $p$ is said to *report* to its father $P$ by adding its features to $P$, and incrementing $P$'s counter. Once all of $P$'s sons have reported to $P$, $P$ also has to report to its father, say $G$. The same may apply for $G$ and so on up to the root node. The root node does not report to any other node.

4. If $q$ is BLACK (i.e., the feature covers $q$), then fill the base-level nodes in the incomplete pyramid descending from $p$ with the same feature index. The base-level nodes below $p$ do not report to their father. Note that if the bottom level of the incomplete pyramid is initialized with the image raster representation, then this step is superfluous. In other words, it is important to note that performing the filling step for all input leaf nodes of the linear quadtree is equivalent to loading the raster representation of the image in the bottom level of the incomplete pyramid.

```
Boolean procedure HelpLoadMap(quad_node)
begin
  value quadtree_node quad_node;
  integer quad_depth, quad_size; /* Quadtree node depth, size */
  integer quad_x, quad_y; /* The (x,y) coordinates of the upper left corner */
  integer which_slot; /* Bit-slot index */
  pyr_node p_addr, f_addr; /* Temporary incomplete pyramid pointers */
  quad_depth ← depth(quad_node);
  quad_size ← size(quad_node);
  quad_x ← x(quad_node);
  quad_y ← y(quad_node);
  p_addr ← pyr_node_address(quad_depth, quad_x, quad_y);
  which_slot ← bit_slot(value(quad_node));
  set_node(which_slot, p_addr);
  load_pyr_base(quad_x, quad_y, quad_size, which_slot);
  do
    begin
      f_addr ← father(p_addr);
      or_bits(f_addr, p_addr);
      incr_intermed(f_addr);
      p_addr ← f_addr;
    end
  until not (all_sons_reported(p_addr));
end;
```

Figure 12: Loading a linear region quadtree into the incomplete pyramid.

The code for the algorithm is given in Figure 12. Note that it is applied to each leaf node (whether black or white) in the linear quadtree. Also, it involves use of procedure load_pyr_base given in Section 5.1.1.

## 5.2 Overlapping Features

Overlapping features are stored in one DF-expression in the format described in Section 4.1. Our algorithm is similar to the one for loading non-overlapping features presented in Section 5.1.1. The difference is that the algorithm of this section takes into account the case that (1) a node can be gray for some features as well as black for other features, and (2) that a black node can correspond to several overlapping features.

The algorithm performs a depth-first search guided by the symbols in the DF-expression. The code for the algorithm is given in Figure 13. For each 'G' symbol the algorithm descends the incomplete pyramid by one level (line 8). In such a case, the algorithm checks if there are features to be stored along with this gray node (line 4). If so, their feature indexes are read from the DF-expression and the corresponding bit-slots are set in the node of the incomplete pyramid (lines 5 and 6). The node will not report to its father until all its sons report to it (lines 8 and 9). Once a 'B' symbol is encountered, the corresponding node in the incomplete pyramid is set to include the feature(s) in it (lines 11 and 12). The 'B' symbol is followed by the set of feature indexes to be used to set the corresponding bit slots in the leaf node. After processing a gray node's son, the features gathered in this son are propagated into the node by the OR Boolean operation (line 9). Procedure load_overlap is initially involved by the call load_overlap(df_list, pyr_root, 0, 0, pyr_base_size).

# 6 Analysis

In this section, we analyze the three algorithms for loading features into the incomplete pyramid data structure. Our discussion is simplified by the following notation:

- NDF: algorithm to load non-overlapping features from a DF-expression.
- NLQ: algorithm to load non-overlapping features from a linear quadtree.
- ODF: algorithm to load overlapping features from a DF-expression.

The algorithms are compared on the basis of the length of the input (as dictated by the specific coding method), the space complexity of each algorithm, and the worst-case execution time.

## 6.1 Length of the Input

Assume that $m$ features are loaded into the incomplete pyramid. Let $T \times T$ be the size of the feature space, $L$ be the sum of the number of their non-white leaf nodes ( which is linearly proportional to the total perimeter of the features), $S_f$ be the number of bits necessary to encode the feature index, $d$ be the depth of the incomplete pyramid (and the linear quadtree as well), $W$ be the number of white nodes in the linear quadtree encoding of the features. Notice that $T = 2^d$, and that $G$, the number of gray nodes, is $(L + W - 1)/3$.

For the NLQ algorithm, the length of each node is the sum of the lengths of the three fields that comprise it. The address field (locational code) takes $2 \cdot d$ bits.

```
1 Boolean recursive procedure load_overlap(l, p, x, y, size)
  begin
     reference pointer list l;
     value pointer pyr_node p;
     value integer x, y, size;
     quadrant i;
2    if data(l) = 'G' then
        begin /* Gray node */
3          l ← next(l);
4          if number(data(l)) then
              begin
5                store_features(l, p); /* Load features into the gray node. */
6                load_pyr_base(x, y, size, contents(p));
              end;
7          for i in {'NW', 'NE', 'SW', 'SE'} do
              begin
8                load_overlap(l, son(p, i), son_x(x, i), son_y(y, i), size/2);
9                or_bits(p, son(p, i)); /* Store features in the parent. */
              end;
        end
     else
10      if data(l) = 'B' then
           begin /* Black node */
11            l ← next(l);
12            store_features(l, p);
13            load_pyr_base(x, y, size, contents(p));
           end;
     /* No action is necessary for a white node. */
  end;
```

Figure 13: Loading multiple overlapping features represented as a single DF-expression into the incomplete pyramid.

The depth field takes $\log d$ bits. And the value field takes $S_f$ bits. Finally, the total number of nodes in a linear quadtree is $L + W$.

Interestingly, NDF and ODF have the same input length (when $L$ is the same for both of them). The extra commas in ODF (in the worst-case) are equal in number to the extra 'B' symbols in NDF.

Therefore, the lengths of the inputs are:

- NDF: $(L + W - 1)/3 + L + W + L \cdot S_f$
- NLQ: $(2 \cdot d + \log d + S_f) \cdot (L + W)$
- ODF: $(L + W - 1)/3 + L + W + L \cdot S_f$

## 6.2 Space Complexity of the Loading Algorithms

The NLQ algorithm accesses a leaf node directly, and then moves upwards (i.e., from the addressed leaf node towards the root). The NLQ algorithm uses some extra space (2 bits for each intermediate node) to guarantee that each non-leaf node waits until all its children report their features to it before reporting once to its parent.

The NDF and the ODF algorithms need no explicit extra space to keep track of how many nodes report to their parents since this information is inherent in the postorder processing of the nodes. Recall that both algorithms perform either a postorder traversal or a combination of a preorder and a postorder traversal in order to propagate feature indexes through the incomplete pyramid. Therefore, the only storage required is the stack area to maintain the recursive calls and the actual argument values at each call. These sum up to 3 integers and 2 pointers per call, and the maximum number of calls is the same as the maximum depth ($d$) of the incomplete pyramid.

Thus the space complexity (in addition to the space occupied by the incomplete pyramid) of the three algorithms is summarized as follows:

- NDF: $d \cdot (3 \cdot int\_size + 2 \cdot pointer\_size)$
- NLQ: $2 \cdot (4^{d-1} - 1)/3) \cong T^2/6$
- ODF: $d \cdot (3 \cdot int\_size + 2 \cdot pointer\_size)$

## 6.3 Worst-case Execution Time Complexity

First, let us examine the case that the features do not overlap. Here we consider only the NDF and NLQ algorithms. Our analysis is based on the cost of calculating the address of a node's son and parent, or of accessing a leaf node directly (as in the case of the NLQ algorithm). Recall from Section 3 that explicit pointers to sons or parents are not stored in the incomplete pyramid.

The NDF algorithm descends through the incomplete pyramid via a combination of a preorder and a postorder traversal. This implies that for each non-leaf (gray) node we must perform four son-address calculations while descending, one per child.

For each leaf node the NLQ algorithm performs an address calculation to directly access the corresponding leaf node in the incomplete pyramid. On the other hand, each non-leaf node is accessed four times. Each child node has to compute its parent's address once the child wants to report to it. Hence four parent-address calculations are performed per non-leaf node, once from each of its children.

When features are permitted to overlap, we use the ODF algorithm. It has the same run-time complexity as the NDF algorithm. The only difference is that more input symbols may have to be read when encountering a gray or a leaf node. This is because a gray or leaf node may contain more than one overlapping feature with its corresponding block. However, this is compensated by the fact that less 'B' symbols are input (in contrast to the NDF algorithm). Note that for comparison purposes, $L$ is the same for each of the three representations.

The following formulas summarize the execution times for all three algorithms. $A_S$ is the time required to compute the address in the incomplete pyramid of a node's son, given the node's address in the incomplete pyramid, and to access the son's location. $A_P$ is the time required to compute the address in the incomplete pyramid of a node's parent given the node's address in the incomplete pyramid and to access the parent's location. $A_L$ is the time required to compute and access a leaf node's address in the incomplete pyramid given its corresponding block's parameters (size and the location of its upper left corner). Notice that the definition of the incomplete pyramid means that $A_S, A_P,$ and $A_L$ are constants (i.e., they have constant execution times).

- NDF: $4 \cdot A_S \cdot G$
- NLQ: $4 \cdot A_P \cdot G + A_L \cdot L$
- ODF: $4 \cdot A_S \cdot G$

Recall that $G$, the number of gray nodes, is $(L + W - 1)/3$.

# 7 Conclusion

Algorithms for loading spatial features into the incomplete pyramid data structure from two pointerless quadtree representations have been presented. The algorithms differ on the basis of the representation of the input feature(s). Two of these algorithms only work for non-overlapping features, while the third algorithm only works for overlapping features. The execution time for all three algorithms is linear in the length of the perimeter of the features loaded, and is independent of the depth of the incomplete pyramid.

Analysis of the space and execution time requirements of the algorithms revealed that the DF-expression is superior to the linear quadtree. This is not surprising because the principal drawback of the DF-expression is its sequential nature (i.e., it does not support random access) which the linear quadtree does. However, the loading algorithms only require that the whole tree be processed in sequence and hence the DF-expression is adequate.

At the moment, the conversion algorithm has only been implemented for the linear quadtree. It takes approximately 9 seconds on a VAX 11/785 to load a $512 \times 512$ image. We are presently implementing the algorithms for overlapping and non-overlapping DF-expressions.

Our work shows the usefulness of the incomplete pyramid data structure. The algorithms for loading features into the incomplete pyramid are much faster than that for the regular complete pyramid. This speed-up arises from the fact that for the incomplete pyramid there is no need to fill all the intermediate nodes descending from a leaf node.

## 8 Acknowledgements

The support of the National Science Foundation under Grant IRI-9017393 is gratefully acknowledged.

## References

[1] D. J. Abel. SIRO-DBMS: a database tool-kit for geographical information systems. *International Journal of Geographical Information Systems*, 3(2):103–116, April–June 1989.

[2] W. G. Aref and H. Samet. Efficient processing of window queries in the pyramid data structure. In *Proceedings of the 9th ACM-SIGACT-SIGMOD-SIGART Symposium on Principles of Database Systems*, pages 265–272, Nashville, TN, April 1990.

[3] I. Gargantini. An effective way to represent quadtrees. *Communications of the ACM*, 25(12):905–910, December 1982.

[4] E. Kawaguchi, T. Endo, and M. Yokota. Depth-first expression viewed from digital picture processing. *IEEE Transactions on Pattern Analysis and Machine Intelligence*, 5(4):373–384, July 1983.

[5] A. Klinger. Patterns and search statistics. In J. S. Rustagi, editor, *Optimizing Methods in Statistics*, pages 303–337. Academic Press, New York, 1971.

[6] J. A. Orenstein. Spatial query processing in an object-oriented database system. In *Proceedings of the 1986 ACM SIGMOD International Conference on Management of Data*, pages 326–336, Washington, DC, May 1986.

[7] H. Samet. *Applications of Spatial Data Structures: Computer Graphics, Image Processing, and GIS.* Addison-Wesley, Reading, MA, 1990.

[8] H. Samet. *The Design and Analysis of Spatial Data Structures.* Addison-Wesley, Reading, MA, 1990.

[9] C. A. Shaffer. *Application of Alternative Quadtree Representations.* PhD thesis, University of Maryland, College Park, MD, June 1986. Technical Report Computer Science TR 1672.

[10] C. A. Shaffer and H. Samet. An in-core hierarchical data structure organization for a geographic database. Technical Report Computer Science TR 1886, University of Maryland, College Park, MD, July 1987.

[11] C. A. Shaffer and H. Samet. Optimal quadtree conversion algorithms. *Computer Graphics, Vision, and Image Processing,* 37(3):402–419, March 1987.

[12] C. A. Shaffer, H. Samet, and R. C. Nelson. QUILT: A geographic information system based on quadtrees. *International Journal of Geographical Information Systems,* 4(2):103–131, April–June 1990.

[13] S. Tanimoto and T. Pavlidis. A hierarchical data structure for picture processing. *Computer Graphics and Image Processing,* 4(2):104–119, June 1975.

[14] L. Tucker. *Computer Vision Using Quadtree Refinement.* PhD thesis, Polytechnic Institute of New York, Brooklyn, May 1984.

[15] T. C. Waugh and R. G. Healey. The GEOVIEW design: A relational data base approach to geographical data handling. *International Journal of Geographical Information Systems,* 1(2):101–118, April–June 1987.

[16] J. W. J. Williams. Algorithm 232: Heapsort. *Communications of the ACM,* 7(6):347–348, June 1964.

# Global Order Makes Spatial Access Faster

Andreas Hutflesz Peter Widmayer Cora Zimmermann

## Abstract

We present a secondary storage access scheme for geometric objects in a dynamic environment that aims at minimizing disk seek operations for spatial proximity queries. This goal is achieved by preserving the natural spatial order of objects on the secondary storage medium not only within blocks, but also beyond block boundaries. It turns out that window queries can be answered extremely fast as compared with schemes that do not take disk seeks into account.

## 1 Introduction

Database management systems for geographical applications should support the retrieval of spatially close objects — so-called *proximity queries* — efficiently. For instance, a *window query* is a special proximity query that asks for all objects geometrically intersecting a query window. The shape and position of complex geometric objects, like polygons, is usually approximated by an orthogonal smallest *bounding box*, serving as a simple geometric key. In this paper, we study the problem of supporting window queries on large, dynamic sets of multidimensional (orthogonal) rectangles stored on secondary memory.

In contrast to previous investigations of the subject [1,2,3,4,5,6,7,9,16,17,18,19], we do not make the simplifying assumption that the access cost is the same for each storage block. Instead, we consider *seek* operations to determine the access cost: after a disk seek operation, a sequence of consecutive blocks can be transferred for free in our model. This is motivated by the fact that for most disks, seek plus latency times dominate block transmission time by a large factor. We propose an access structure that aims at minimizing the number of disk seeks for window queries, as opposed to minimizing the number of accessed blocks.

For simplicity, we assume disk space to consist of blocks in a linear arrangement; block addresses run from 0 to some maximum integer. That is, an access to the sequence $a, a+1, \ldots, a+k$ of blocks costs only one disk seek [21,22] (for ease of description, we identify blocks with their addresses). Certainly, the simplicity of this cost measure leaves room for abuse: for instance, accessing all blocks on the entire disk takes only one disk seek. We refrain from abusing the cost measure by insisting that for each operation, only the minimum necessary set of blocks be accessed.

Spatial access structures that aim at minimizing the number of accessed blocks gain their efficiencies from order preserving mappings of rectangles to blocks in the sense that all rectangles stored in a common block are spatially close to each other. Hence, an accessed block will hopefully contribute several rectangles to the answer of a window query. Since no attempt is made to preserve order beyond block boundaries, we call these structures *locally order preserving*.

For minimizing disk seeks, local order preservation does not suffice. In addition, *global order* should be preserved as well: rectangles stored in consecutive blocks should be spatially close together. As a consequence, a disk seek will hopefully contribute several consecutive blocks to the answer of a window query. For point objects, a globally order preserving multidimensional linear hashing scheme, dynamic $z$-hashing, is known [8]. It is based on linear hashing [12] (see also [10]) and is, hence, especially useful for uniformly distributed points; it does not support rectangles.

It seems not to make sense to use dynamic $z$-hashing for rectangles transformed into points (the transformation is described, e.g., in [6,17]), because the point distribution will not be uniform. Also, at least for high map data density, clipping will often lead to high redundancy and inefficiency. We suggest an access structure that covers the data space with overlapping subspaces, such that global order is preserved on the basis of dynamic $z$-hashing.

Let us briefly recall the main idea of dynamic $z$-hashing; for details, consult [8]. The data space is partitioned into rectangular regions. Block addresses are associated with regions along a (space-filling) $z$-curve through the data space; along the $z$-curve, numbers created by bit interleaving, also known as Morton encoding [14,15,20], occur in sorted order. For four different partitions, $z$-curves are shown in Fig. 1.1.

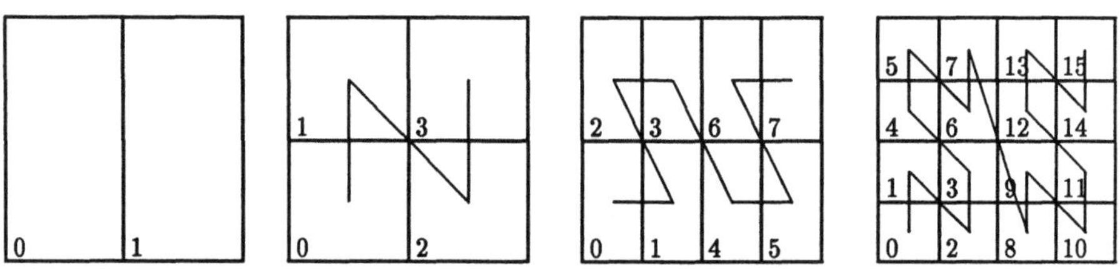


**Fig. 1.1:** Four $z$-curves and their partitions

At any stage, the structure addresses blocks 0 to $m$, for the appropriate value of $m$. Upon a block split, triggered by some control function, the address space is extended to include blocks $m+1$ and $m+2$. The block $k$ to be split is selected such that its contents can be stored in the new blocks according to a refined $z$-curve address. For a sequence of block splits, see Fig. 1.2 (a) through (c). Like in linear hashing, at any stage there are (at most) two different hash functions ($z$-curves), one of them being a refinement of the other by considering one more bit of one dimension. Therefore, $k = (m+1)/2$, and block $k$ is split into blocks $2k$ and $2k+1$. Unlike in linear hashing, block $k$ becomes empty, i.e., is not used currently. Whenever the set of blocks 0 through $m$ contains two adjacent unused blocks $j$ and $j+1$, the next block to be split is chosen such that it splits into $j$ and $j+1$ (the split taking Fig. 1.2 (c) to Fig. 1.2 (d) is of this type, with $j = 4$). Analytic and experimental performance results for dynamic $z$-hashing can be found in [8]; especially, there are at most $\log m$ unused blocks at any stage.

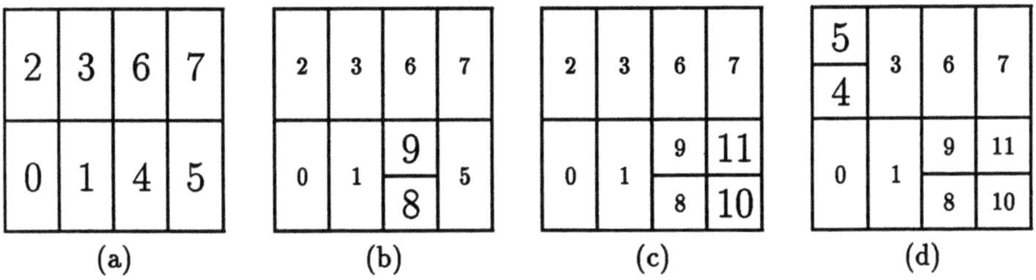


**Fig. 1.2:** Splitting blocks in dynamic $z$-hashing

## 2 Dynamic lz-hashing

Let us illustrate by a two-dimensional example the basic idea and operations of *lz*-hashing, the layered $z$-hashing structure we propose for maintaining a set of multidimensional rectangles. A more formal description of the *lz*-hashing operations can be found in the appendix.

Let us try to use dynamic $z$-hashing as far as possible to store rectangles. Assume that block capacity is three rectangles, and that the six rectangles shown in Figure 2.1 are to be stored in a dynamic $z$-hashing structure $Z_1$. Figure 2.1 shows a possible partition of the data space into two grid cells, forced by the control function (at the end of this section we will discuss control functions).

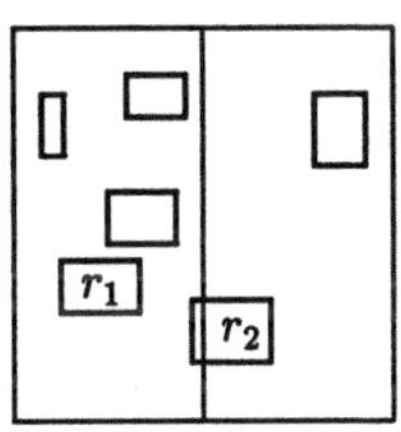


**Fig. 2.1:** Rectangles to be stored using dynamic $z$-hashing

Clearly, not all six rectangles can be associated with the two blocks corresponding to the two grid cells: more than three rectangles lie in the left cell, and one rectangle intersects both cells. For the left cell, an overflow block storing, e.g., rectangle $r_1$, solves part of the problem. Overflow blocks, however, are undesirable, since their addresses do not fit nicely into the $z$-curve sequence and therefore weaken the global order. The problem that $r_2$ intersects two cells is even more severe (see [19] for a discussion of several alternative strategies).

In this situation, we propose to store the two rectangles $r_1$ (chosen arbitrarily from the left cell) and $r_2$ in an additional dynamic $z$-hashing structure $Z_2$ (see Fig. 2.2 with the block addresses in the lower left corner of the cells).

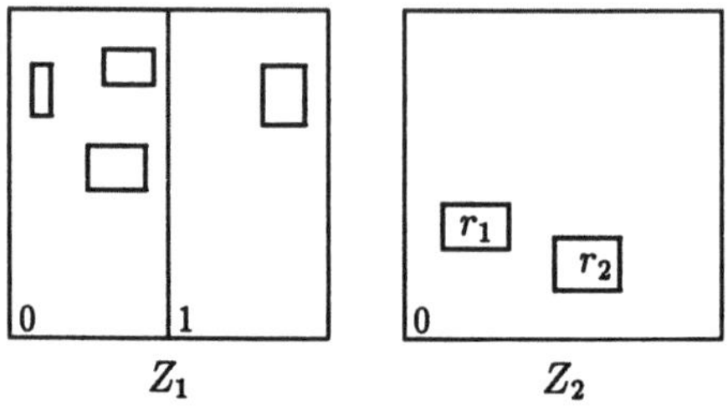


**Fig. 2.2:** Six rectangles stored with *lz*-hashing

For more rectangles, more *layers* of dynamic *z*-hashing structures, each of them covering the entire data space, are used; we call the resulting layered dynamic *z*-hashing structure *lz*-hashing. However, it is not enough to simply repeat dynamic *z*-hashing for each layer. To see why, let us try to insert rectangle $r$ (whose location and shape are shown in Fig. 2.3), starting with the situation depicted in Fig. 2.2. As a rule, an additional rectangle is stored in $Z_1$ if it fits entirely into a grid cell of $Z_1$ and the corresponding block is not full. Otherwise, the rectangle is stored in $Z_2$ (or higher index layers). In our example, $r$ fits into the grid cell of $Z_1$ corresponding to block 0 (cell 0 for short), but block 0 is already full. Therefore, we try to insert $r$ into $Z_2$; here, $r$ lies in cell 0, and block 0 is not full, so the insertion is successful. Now assume that the control function forces a split in $Z_1$. As explained in Section 1, dynamic *z*-hashing determines the block to be split without taking individual block load into account. In the example, block 1 is split, resulting in the situation shown in Fig. 2.3.

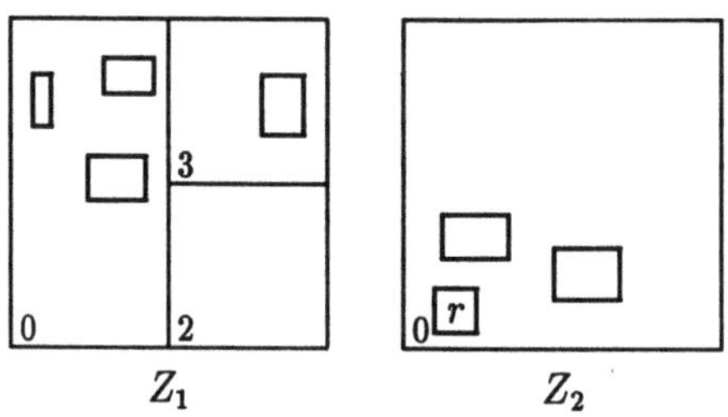


**Fig. 2.3:** Seven rectangles stored in four blocks with *lz*-hashing

Assume that for $Z_2$ the control function forces a split, too. Dynamic *z*-hashing splits block 0 in $Z_2$ vertically into a left and a right grid cell. If this split goes through the center of the cell as in $Z_1$, all rectangles intersecting this center split line cannot be stored in $Z_2$. This is particularly undesirable, because a cell partition of $Z_2$ that is identical to parts of the cell partition of $Z_1$ is not better than $Z_1$ for the rectangles on these partition lines. Instead, a different cell partition should be used in $Z_2$ in a way that makes it less likely for rectangles to intersect cell boundaries in $Z_1$ *and* in $Z_2$. Hence, split positions in $Z_2$ and in $Z_1$ should be chosen fairly far apart from each other (see also [19] for a discussion of this issue). On the other hand, a cell split far out of its center leads to low block load, because future block splits occur for each block with the same frequency, and hence cell sizes tend to differ considerably. The two conflicting goals of having split positions throughout all layers with maximal distance from each other and of having split positions at or at least close to the cell centers is resolved heuristically. In Section 3, we describe in detail a split strategy we used in our performance tests. For the moment, let us choose a split line slightly to the left of the center (see Fig. 2.4).

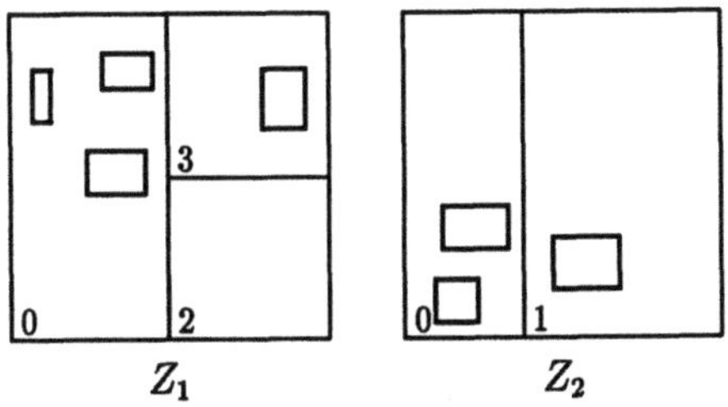


**Fig. 2.4:** Seven rectangles stored in five blocks with *lz*-hashing

To see how flexibly the *lz*-hashing cells accomodate rectangles, note that rectangles can not only migrate from lower (index) to higher (index) layers, but also in the opposite direction. Before illustrating this effect, let us assume that one more rectangle has been inserted (see Fig. 2.5).

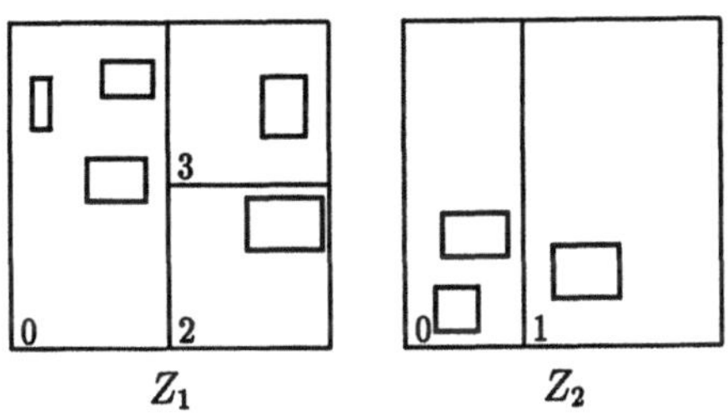


**Fig. 2.5:** Eight rectangles stored with *lz*-hashing

After the insertion, assume that block 0 of $Z_1$ splits horizontally at its center. The rectangle intersected by the new split line migrates to $Z_2$, but in $Z_2$ it also intersects a split line. Hence, an extra layer $Z_3$ is needed to store this rectangle (see Fig. 2.6).

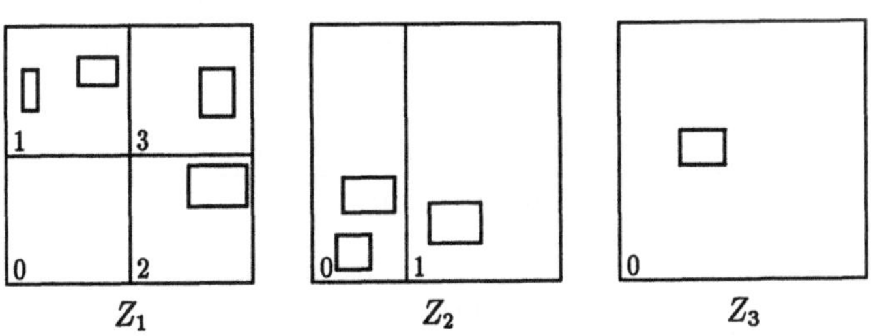


**Fig. 2.6:** Eight rectangles before downwards migrations

As a consequence, the two blocks 0 and 1 in $Z_1$ generated by the split, taken together, do not fill one block, even though before the split, block 0 (see Fig. 2.5) was full and also had overflow rectangles migrated to $Z_2$. In our example, these overflow rectangles can now easily be stored in block 0 of $Z_1$ (see Fig. 2.7), and they indeed should.

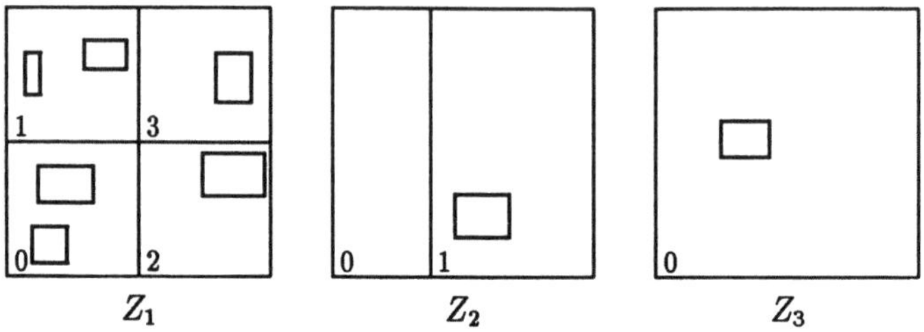


**Fig. 2.7: Eight rectangles after downwards migrations**

In general, after the split of a block for which overflow rectangles exist, these rectangles migrate from higher layers if they can be stored in one of the two blocks created by the split, unless this would cause a block to become overfull. If the migration of rectangles causes blocks (on higher layers) to become less than full, the very same downwards migration is applied, so as to propagate the effect through higher layers.

In the description of operations, we so far did not restrict the growth of more and more layers at all. However, because a high number of layers entails a certain overhead for all operations, it may be beneficial to restrict the number of layers, e.g., to four (see [19] for experiences with a layered grid file), and to allow for clipping and for overflow blocks on the highest layer.

Let us now summarize the structural *lz*-hashing invariant motivated by means of example situations, for an unrestricted number of layers:

> If a rectangle $r$ is stored in block $b$ on layer $l$, then the cell of $b$ encloses $r$, and no layer lower than $l$ has a cell enclosing $r$ whose block is not full.

The examples have been used to sketch the operations we propose to preserve this *lz*-hashing invariant; a precise definition can be found in the appendix.

Finally, let us suggest a suitable control function. For each layer, a block split is triggered whenever the average load exceeds a desired load factor, specified by the user of *lz*-hashing; the desired load factor need not be the same for all layers. This simple mechanism is sufficient for a growing structure; shrinking is symmetric in the sense that block merge operations are triggered by desired load factors. To properly take into account the overflow rectangles stored on other layers, the average load for a layer is defined as

$$\frac{\text{number of stored rectangles, assuming that each layer has its own overflow blocks}}{\text{number of blocks (without overflow blocks)} \cdot \text{block capacity}}$$

This implies that load can be more than 100%. For any given layer, the number of stored rectangles cannot be determined from that layer alone: on the one hand, overflow rectangles from lower layers are stored here, too, and on the other hand, the overflow rectangles of this layer are stored on higher layers. We propose to calculate the number of stored rectangles with a special accounting mechanism that disregards the layer where a rectangle is stored, and instead counts a rectangle for the lowest layer where it is enclosed in one grid cell. For instance, rectangle $r_1$ in Fig. 2.2 counts for $Z_1$, even though it is stored in $Z_2$, whereas rectangle $r_2$ counts for $Z_2$, because it overlaps two cells of $Z_1$.

## 3 Performance Evaluation

To compare the performances of *lz*-hashing and the R-tree [4], we have implemented these two structures in Logitech Modula–2 on an IBM–AT under MS–DOS; both support insertions and window queries (for a detailed description of the implementation, including extensive arguments on design decisions and performance results, see [23]).

For *lz*-hashing, we restricted the number of layers to four, a good compromise for query windows with several different areas. Since split positions do not necessarily coincide with the center of the interval to be split, we need to explicity keep track of split positions. Relative to split positions, bucket addresses are based on a binary trie in each dimension, with e.g. 0 for left and 1 for right. Therefore, we store a binary tree of split positions for each layer and each dimension, and we use this tree to read off the bit sequence (pseudokey) corresponding to a given coordinate value. The internal storage space requirement for these trees turned out not to be higher than that for the root directory of the R-tree. We compared the *lz*-hashing performance with that of the R-tree for various choices of desired load factors. In case of identical load factors for all layers, 70% resulted in the best average performance. We feel that in general, identical load factors for all layers produce a fairly large number of overflow blocks and result in poor query performance and space utilization, especially for layer $Z_1$. Therefore, we evaluated the performance of *lz*-hashing for load factors varying from 100% to 40% in steps of 20% for layers $Z_1$ to $Z_4$.

In all our experiments, we used the following simple split strategy. For split layer $Z_i$, let $I$ be the projection of the grid cell to be split onto the split dimension. Let $c$ be the center position of interval $I$, and let $l$ be its length. The existing split positions of all layers divide $I$ into subintervals. Then the split position is taken as follows:

1. If there exists a split position of layer $Z_i$ inside $I$, then take it.

2. Otherwise: if $c$ is not a split position of another layer, then take $c$.

3. Otherwise: select a split position inside the larger of the two subintervals neighboring $c$. If the two subintervals have equal lengths, we choose the left subinterval. Let $c'$ be the center of the chosen subinterval. Let $f$ be a parameter that can be chosen between 0 and $\frac{1}{4}$, indicating a desired relative distance to the center. If $|c' - c| \leq l \cdot f$, then take $c'$. Otherwise, take $c' - l \cdot f$ if the subinterval is to the left of $c$ (symmetrically, take $c' + l \cdot f$ if it is to the right).

In all tests, we chose $f = \frac{1}{32}$ to get split positions close to the center.

For the R-tree, we have implemented two of the originally published split algorithms [4], the linear and the quadratic algorithm. It turned out that the quadratic algorithm outperformed the linear in each single test case. Therefore, only the former is compared to *lz*-hashing. The minimum node utilization (a parameter of the R-tree split algorithms) was chosen to be 40%, the directory block size was 2048 bytes. For all measurements, the root directory was assumed to be in main memory, thus causing no accesses to secondary storage.

The test data were two-dimensional bounding boxes (rectangles) of 48400 polygons from a German land-use survey. The sum of the rectangle areas is 2.4 times the area of the data space (the *cover quotient* is 2.4). Nevertheless, there are several clusters with more than 10 rectangles covering the same point of the data space.

In our experiments, we kept track of the following performance indicators. First, we calculated the data block utilization; it is defined as the number of stored rectangles, divided by the number of used data blocks times the block capacity. Most importantly, of course, we measured the number of disk seek operations necessary to answer window queries for square windows of different areas, each measurement averaged over 20 window queries at random positions.

In the first series of experiments, we inserted all 48400 rectangles in random order into the initially empty structures, for all block capacities ranging from 8 to 32 in steps of 2.

It turns out that for small block capacities, the R-tree has higher space utilization, and for block capacities above 16, *lz*-hashing is superior (see Fig. 3.1).

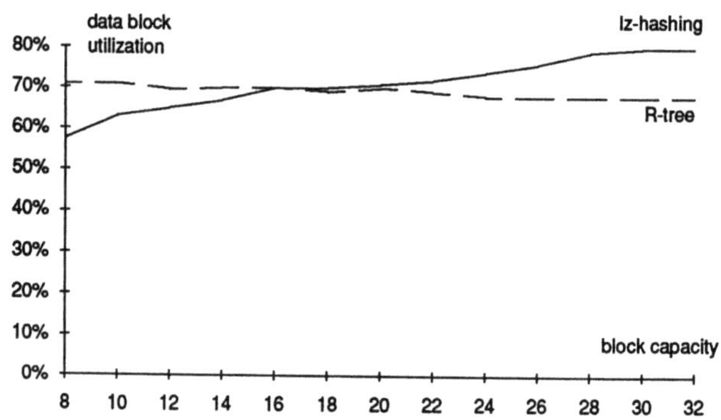


**Fig. 3.1:** Space utilization of data blocks

Window queries with window area 1% of the area of the data space retrieve roughly 550 rectangles, a realistic number to be displayed on a screen. The number of blocks to be read is slightly lower for the R-tree, but the number of disk seek operations is between 60% and 90% higher for the R-tree, as compared with *lz*-hashing (see Fig. 3.2).

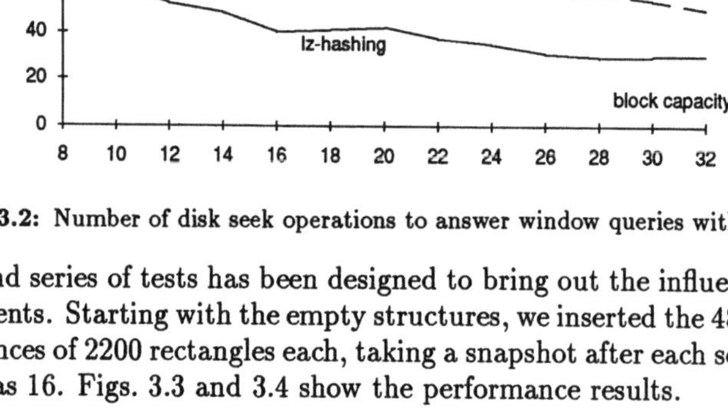


**Fig. 3.2:** Number of disk seek operations to answer window queries with area 1%

A second series of tests has been designed to bring out the influence of different cover quotients. Starting with the empty structures, we inserted the 48400 rectangles in 22 sequences of 2200 rectangles each, taking a snapshot after each sequence. Block capacity was 16. Figs. 3.3 and 3.4 show the performance results.

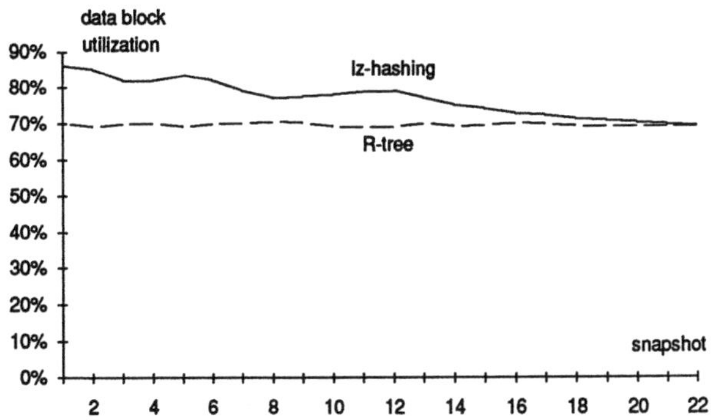


**Fig. 3.3:** Space utilization of data blocks

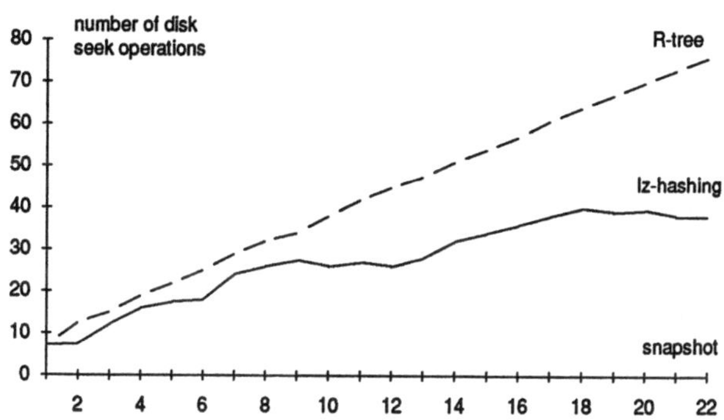


**Fig. 3.4:** Number of disk seek operations to answer window queries with area 1%

For unrealistically small cover quotients, say below 1, window query performance of *lz*-hashing is only slightly better than that of the R-tree. As soon as larger and more realistic cover quotients, from 1 to our maximum value of 2.4, are considered, window query performance improves substantially for *lz*-hashing.

Fig. 3.5 shows the behavior of both access structures for window queries of 20 different areas, varying from 0.25% to 5% of the area of the data space. These values were measured after insertion of all 48400 rectangles into structures with block capacity 16. The number of block accesses is almost identical for the two structures, but the number of disk seek operations is up to three times higher for the R-tree. This clear superiority of *lz*-hashing over the R-tree in window query efficiency is achieved by a higher number of external accesses during insertions. Here, *lz*-hashing needs four times as many external accesses as the R-tree, on the average over all experiments. Hence, *lz*-hashing should be preferred over R-trees only in applications where proximity queries are clearly more frequent or important than insertions.

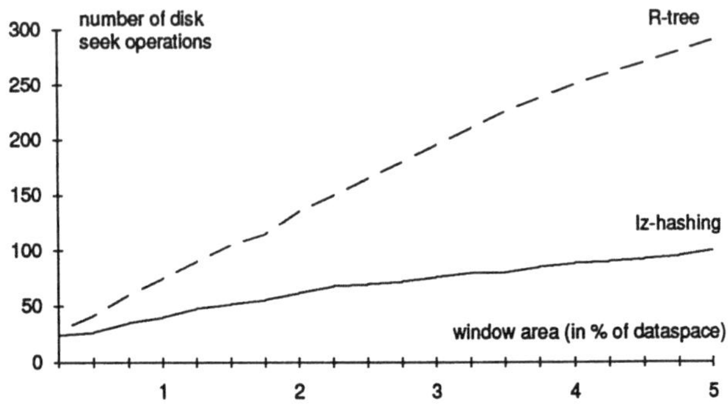


**Fig. 3.5:** Number of disk seek operations to answer window queries of different areas

We also evaluated the performance of the two structures for two other data sets, taken from the same database, one with a smaller cover quotient (2.04) and without clusters, and another with a cover quotient of 5.78. As expected, *lz*-hashing performance improves for the former and becomes worse for the latter. Nevertheless, window queries can be answered much faster with *lz*-hashing than with the R-tree, in both cases.

In comparison with the load factors we used for the above experiments, lower load factors cause the following changes in performance:

- the data block utilization becomes lower;
- the number of external accesses for insertions diminishes;
- the number of block accesses to answer window queries grows;
- the number of disk seek operations to answer window queries shrinks, due to the reduced number of overflow blocks.

In the light of these correlations, together with the freedom to choose the load factors (and also the number of layers), *lz*-hashing can be tailored specifically to many different data sets and applications.

# 4 Conclusion

In this paper, we have presented *lz*-hashing, an access scheme for geometric objects of non-zero size that aims at minimizing disk seek operations for window queries. To achieve this goal, *lz*-hashing is designed to preserve the natural geometric order of objects not only within, but also beyond block boundaries. We have evaluated the performance of *lz*-hashing in comparison with the R-tree [4]. In all our experiments, *lz*-hashing outperformed the R-tree in terms of disk seeks necessary to answer window queries. It seems promising to exploit the inherent flexibility of *lz*-hashing for tuning it to specific applications, so as to improve performance further.

# Acknowledgement

We thank an anonymous referee for helpful suggestions, and W. Halboth, G. Pfefferle, and G. Reich for their typesetting assistance.

## References


1. Beckmann, N., Kriegel, H.-P., Schneider, R., Seeger, B.: The R*-tree: an efficient and robust access method for points and rectangles, Proc. ACM SIGMOD International Conference on Management of Data, 1990, 322–331.
2. Faloutsos, C., Sellis, T., Roussopoulos, N.: Analysis of Object Oriented Spatial Access Methods, Proc. ACM SIGMOD International Conference on Management of Data, 1987, 426–439.
3. Günther, O.: Efficient Structures for Geometric Data Management, Springer Lecture Notes in Comp. Sci., Vol. 337, 1988.
4. Guttman, A.: R-Trees: A Dynamic Index Structure for Spatial Searching, Proc. ACM SIGMOD International Conference on Management of Data, 1984, 47–57.
5. Henrich, A., Six, H.-W., Widmayer, P.: The LSD-tree: spatial access to multidimensional point and non-point objects, Proc. 15th International Conference on Very Large Data Bases, 1989, 45–53.
6. Henrich, A.: Der LSD-Baum: eine mehrdimensionale Zugriffsstruktur und ihre Einsatzmöglichkeiten in Datenbanksystemen, Doctoral thesis, Fernuniversität -GHS-Hagen, Germany, 1990.
7. Hinrichs, K.H.: The grid file system: implementation and case studies of applications, Doctoral Thesis No. 7734, ETH Zürich, 1985.
8. Hutflesz, A., Six, H.-W., Widmayer, P.: Globally Order Preserving Multidimensional Linear Hashing, Proc. IEEE 4th International Conference on Data Engineering, 1988, 572–579.
9. Hutflesz, A., Six, H.-W., Widmayer, P.: The R-file: an efficient access structure for proximity queries, Proc. IEEE 6th International Conference on Data Engineering, 1990, 372–379.
10. Kriegel, H.-P., Seeger, B.: Multidimensional Order Preserving Linear Hashing with Partial Expansions, Proc. International Conference on Database Theory, 1986, 203–220.
11. Kriegel, H.-P., Seeger, B.: PLOP-Hashing: A Grid File without Directory, Proc. IEEE 4th International Conference on Data Engineering, 1988, 369–376.
12. Litwin, W.: Linear Hashing: A New Tool for File and Table Addressing, Proc. 6th International Conference on Very Large Data Bases, 1980, 212–223.
13. Nievergelt, J., Hinterberger, H., Sevcik, K.C.: The Grid File: An Adaptable, Symmetric Multikey File Structure, ACM Transactions on Database Systems, Vol. 9, 1, 1984, 38–71.
14. Orenstein, J.A., Merrett, T.H.: A Class of Data Structures for Associative Searching, Proc. 3rd ACM SIGACT/SIGMOD Symposium on Principles of Database Systems, 1984, 181–190.
15. Orenstein, J.A.: Spatial Query Processing in an Object-Oriented Database System, Proc. ACM SIGMOD International Conference on Management of Data, 1986, 326–336.
16. Seeger, B., Kriegel, H.-P.: Techniques for Design and Implementation of Efficient Spatial Access Methods, Proc. 14th International Conference on Very Large Data Bases, 1988, 360–371.
17. Seeger, B.: Entwurf und Implementierung mehrdimensionaler Zugriffsstrukturen, Doctoral thesis, Universität Bremen, Germany, 1989.

18. Sellis, T., Roussopoulos, N., Faloutsos, C.: The $R^+$-Tree: A Dynamic Index for Multi-Dimensional Objects, Proc. 13th International Conference on Very Large Data Bases, 1987, 507–518.

19. Six, H.-W., Widmayer, P.: Spatial Searching in Geometric Databases, Proc. IEEE 4th International Conference on Data Engineering, 1988, 496–503.

20. Tropf, H., Herzog, H.: Multidimensional Range Search in Dynamically Balanced Trees, Angewandte Informatik 2, 1981, 71–77.

21. Wang, J.-H., Yuen, T.-S., Du, D.H.-C.: On Multiple Random Accesses and Physical Data Placement in Dynamic Files, IEEE Transactions on Software Engineering, Vol. 13, 8, 1987, 977–987.

22. Weikum, G., Neumann, B., Paul, H.-B.: Konzeption und Realisierung einer mengenorientierten Seitenschnittstelle zum effizienten Zugriff auf komplexe Objekte, Proc. GI-Fachtagung Datenbanksysteme in Büro, Technik und Wissenschaft, Darmstadt, 1987, 212–230.

23. Zimmermann, C.: Multilayer dynamic $z$-hashing, Diploma thesis, Universität Karlsruhe, Germany, 1989.

# Appendix

This appendix serves to precisely define the operations window query, exact match, insertion, deletion, split, and merge in full generality. Let $N$ be the number of layers $Z_1, \ldots, Z_N$ organized by *lz*-hashing. For each layer, we keep track of the expansion level $L$ and the expansion status $E$, as defined for dynamic $z$-hashing [8]. For clarity of the presentation, the description of the algorithms does not include the accounting mechanism explained at the end of Section 2. For the highest layer $Z_N$, the notion *block* always includes potential overflow blocks.
A call of $WindowQuery(w, i, s)$ with $i = 1$ returns in $s$ all rectangles (of layers $Z_i, \ldots, Z_N$) overlapping window $w$:

```
procedure  WindowQuery (w, i, var s);
begin
    if ( i = N ) then
        s := ClippingLayerWindowQuery(w, Z_N)
    else
        WindowQuery(w, i + 1, s);
        s := s ∪ LayerWindowQuery(w, Z_i)
    endif
end  WindowQuery.
```

*LayerWindowQuery* performs a window query on one dynamic $z$-hashing structure without overflow blocks; *ClippingLayerWindowQuery* performs a window query on one dynamic $z$-hashing structure with overflow blocks and clipping.
An exact match with rectangle $r$ is carried out by calling $ExactMatch(r, i, found, a)$ with $i = 1$. *ExactMatch* starts searching for $r$ on layer $Z_i$. Variable *found* returns *true* if rectangle $r$ is stored in $Z_j$, $j \geq i$, *false* otherwise. Variable $i$ returns the index of the layer where the search ends (successfully or unsuccessfully), and $a$ returns the corresponding block address(es). That is, $a$ is one block address if the search ends on a layer $Z_j$, $j < N$, and a set of block addresses if the search ends on layer $Z_N$. If the search ends unsuccessfully, block $a$ is not full; this fact is used for insertion (see *Insert* below).

```
procedure  ExactMatch (r, var i, var found, var a);
begin
    if ( i = N ) then
        compute all addresses a of blocks whose cells overlap r;
        if ( r is stored in block a ) then
            found := true
        else
            found := false
        endif
    else  { i < N }
        if ( r is enclosed in a cell of Z_i ) then
            compute address a of this cell;
            if ( r is stored in block a ) then
                found := true
```

```
            elsif ( block a is not full ) then
                found := false
            else
                i := i + 1;
                ExactMatch(r, i, found, a)
            endif
        else { r overlaps more than one cell of Z_i }
            i := i + 1;
            ExactMatch(r, i, found, a)
        endif
    endif
end ExactMatch.
```

Rectangle $r$ is inserted by calling $Insert(r, i)$ with $i = 1$; this inserts $r$ on layer $Z_i$ or on a higher layer.

```
procedure Insert (r, i);
begin
    ExactMatch(r, i, found, a);   { this may change i }
    if ( found = true ) then
        return { r is already stored in Z_i }
    endif;
    if ( i = N ) then
        store r in all blocks a
    else
        store r in block a
    endif
end Insert.
```

Rectangle $r$ is deleted by calling $Delete(r, i)$ with $i = 1$. If $r$ is deleted from a full block of layer $Z_j$, $j \geq i$, an overflow rectangle of this block, i.e., a rectangle enclosed in the block's cell that is stored on a higher layer, migrates downwards from that layer, to preserve the structural invariant (see Section 2).

```
procedure Delete (r, i);
begin
    ExactMatch(r, i, found, a);   { this may change i }
    if ( found = false ) then
        return { r could not be found }
    endif;
    if ( i = N ) then
        remove r from all blocks a
    else
        remove r from block a;
        if ( block a was full before the deletion ) then
            w := window corresponding to block a;
            WindowQuery(w, i + 1, s);
            let r be an arbitrary element of s;
```

```
            store r in block a;
            Delete(r, i + 1)    { downwards migration }
        endif
    endif
end Delete.
```

After an insertion, a split operation on layer $Z_i$ forced by the control function is carried out by *Split*($i$).

```
procedure Split (i);
begin
    compute block a of Z_i to be split, as in dynamic z-hashing;
    compute split position p according to the specified split strategy;
    if ( i = N ) then
        foreach r ∈ block a do
            if ( r lies totally below p ) then
                store r in block 2a
            elsif ( r lies totally above p ) then
                store r in block 2a + 1
            else
                store r in both blocks 2a and 2a + 1
            endif
        endforeach
    else { i < N }
        foreach r ∈ block a do
            if ( r lies totally below p ) then
                store r in block 2a
            elsif ( r lies totally above p ) then
                store r in block 2a + 1
            else
                Insert(r, i + 1)    { upwards migration to avoid clipping }
            endif
        endforeach;
        w := the window corresponding to block a;
        WindowQuery(w, i + 1, s);
        s := all rectangles of s enclosed in w;
        foreach r ∈ s do    { downwards migration, if possible }
            if ( r lies totally below p and block 2a is not full ) then
                store r in block 2a;
                Delete(r, i + 1)
            elsif ( r lies totally above p and block 2a + 1 is not full ) then
                store r in block 2a + 1;
                Delete(r, i + 1)
            endif
        endforeach
    endif
end Split.
```

After a deletion, a merge operation on layer $Z_i$ forced by the control function is carried out by calling $Merge(i)$.

```
procedure  Merge (i);
begin
    compute block a of Z_i created by merging blocks 2a and 2a + 1;
    if  ( i = N )  then
        foreach   r ∈ blocks 2a or 2a + 1   do
            store r in block a, removing clipping duplicates
        endforeach
    else  { i < N }
        foreach   r ∈ blocks 2a or 2a + 1   do
            if  ( block a is not full )  then
                store r in block a
            else
                Insert(r, i + 1)    { upwards migration }
            endif
        endforeach;
        if  ( block a is not full )  then    { downwards migration, if possible }
            w := the window corresponding to the center line of block a;
            WindowQuery(w, i + 1, s);
            s := all rectangles of s enclosed in the window of block a;
            foreach   r ∈ s   do
                if  ( block a is not full )  then
                    store r in block a;
                    Delete(r, i + 1)
                endif
            endforeach
        endif
    endif
end  Merge.
```

# Evaluation of Spatial Access Methods with Oversize Shelves

Oliver Günther

## Abstract

In this paper we evaluate a new technique to improve the performance of spatial access methods by minimizing redundancy: the *oversize shelf*. Oversize shelves are additional disk pages that are attached to the interior nodes of a tree–based spatial access method (such as the $R^+$–tree or the cell tree). These pages are used to accomodate very large data objects in order to avoid their excessive fragmentation. Whenever inserting a new object into the tree, one now has to decide whether to store it on an oversize shelf or to insert it into the corresponding subtrees. For this purpose, we developed an analytic model for the behavior of dynamic spatial access methods under insertions and deletions. The model yields a threshold value for the size of an object, such that it is more favorable to put it on the oversize shelf if and only if its size is greater than the threshold value. Otherwise the insertion into the corresponding subtrees is preferable. Practical experiments indicate that this policy results in a more efficient computation of search queries.

## 1 Introduction

Non–standard database applications such as cartography or environmental information systems are becoming increasingly important, and geometric or spatial data objects (such as points, polygons, or polyhedra) play a crucial role in many of these areas. An important class of geometric operators that needs special support is the class of spatial search operators. Retrieval and update of spatial data is usually based not only on the value of certain alphanumeric attributes but also on the spatial location of a data object. A typical spatial database query may ask for all objects that contain a given point *(point search)* or that overlap a given search space *(range search)*. In all of these cases, it is necessary to retrieve those objects from the database that occupy a given location in space.

Given a large number of objects in the spatial database, it is not efficient to check every single object for intersection with the search space. To support such search operations, one needs to use *spatial access methods* (or *index structures*) that enable the user to access the objects in a certain spatial neighbourhood very efficiently.

Research in spatial data management has produced a wide range of access methods to provide efficient support for these spatial search operators. Most of these methods have originally been designed to manage large sets of multi–dimensional points and to support exact match or partial match queries. Usually, the points in the database are organized in a number of buckets, each of which corresponds to some partition of the data space. The buckets are then accessed by means of a search tree or some hashing scheme.

The grid file (Nievergelt et al. 1984), for example, uses a directory and a grid–like partitioning of the data space to answer an exact match query with exactly two disk accesses. Furthermore, there are multi–dimensional hashing schemes without directory (Kriegel and Seeger 1986, Kriegel and Seeger 1988), multilayer grid files (Whang 1985, Widmayer et al. 1988) and so–called hash trees (Ouksel 1985, Otoo 1986, Freeston 1987), where the directory is organized as a tree structure. Tree–based access methods are usually a generalization of the B–tree (Bayer and McCreight 1972) to higher dimensions, such as the K–D–B–tree (Robinson 1981).

The generalization of these *point access methods* to manage *extended* spatial objects (such as polygons) rather than just points is not trivial. As discussed by Seeger and Kriegel (1988), there are three basic methods: *clipping*, where objects are divided along the partitioning hyperplanes of the underlying access structure; *overlapping regions*, where different buckets may correspond to mutually overlapping partitions, such that each object can be assigned as a whole to one partition; and *transformation*, where extended spatial objects are mapped into higher–dimensional points.

Of course, transformation only works for relatively simple objects, such as rectangles. More complicated entities, such as arbitrary polygons (as they typically occur in geographic applications), can only be represented by means of their bounding boxes (Nievergelt and Hinrichs 1987). In this case, the access method cannot solve a given search problem completely. As it is always the case when using bounding boxes, the index only yields a set of boxes whose enclosed objects *may* intersect the search space. One is left with the problem of testing the *objects* for intersection with the search space.

Overlapping region schemes, on the other hand, tend to experience major performance problems when the spatial database contains objects whose size is large relative to the total size of the data space. This is especially typical for geographic applications where one has to represent objects of widely varying size (such as buildings and states) in the same spatial database. Each insertion of a new data object may increase the overlap. Eventually, the overlap between regions may become large enough to render the index inefficient: one ends up searching a significant portion of the whole index for a single point query. A well–known example where this behavior has been observed is the R–tree (Guttman 1984, Greene 1988).

In the case of clipping schemes, the problems with large objects are of a somewhat different nature. During insertion, each new data object is divided along the partitioning hyperplanes of the index structure, and usually several leaf node entries

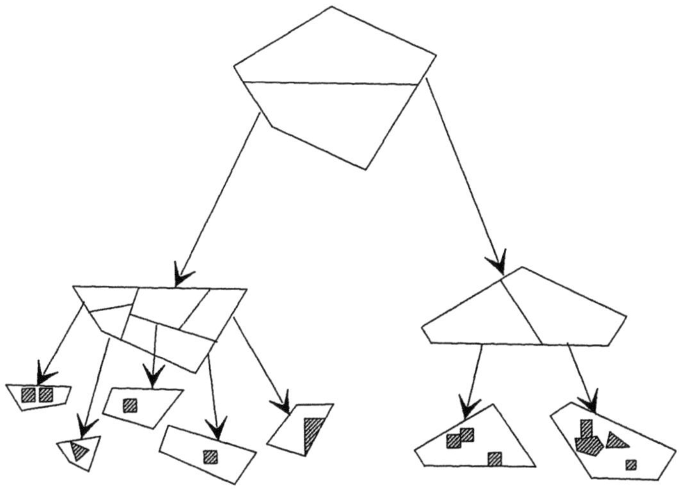

Figure 1: A cell tree with cells (dark shading)

are created to represent just one object. The net result of this effect is not only an increase in the average search time but also an increase of the frequency of node overflows, which makes updates much more complicated. Examples for clipping-based schemes where these problems may occur include the R$^+$-tree (Sellis et al. 1987) and the cell tree (Günther 1988, Günther 1989).

In the following section, we will give a brief review of the cell tree design. Section 3 describes the performance problems of clipping-based access methods in some more detail, and presents the idea of *oversize shelves* as a possible solution. Oversize shelves are additional disk pages that are attached to the interior nodes of a tree-based spatial index structure. These pages are used to accomodate very large data objects in order to avoid their excessive fragmentation. In section 4, this concept is evaluated for the case of the cell tree by means of an analytic model for the tree behavior under insertions and deletions. From the model it becomes obvious that, once the database becomes very populated, most objects will be stored on oversize shelves. In section 5, these considerations are generalized to other spatial access methods, such as the R$^+$-tree. In section 6 we present the results of a practical performance analysis that shows that oversize shelves result in a better average performance in all test cases, and section 7 contains our conclusions and plans for future research.

## 2 The Cell Tree

The cell tree uses clipping to manage large sets of polyhedra in arbitrary dimensions. It corresponds to a decomposition of the data space into disjoint convex partitions, such that the interior nodes correspond to a hierarchy of nested convex polyhedra and each leaf node corresponds to one of the partitions (Fig. 1).

Objects to be inserted into the cell tree are first partitioned into a small number of convex *components* whose union is the original object. All of these components are assigned the same object identifier and inserted into the cell tree one by one. Due to clipping, each component may have to be subdivided into several *cells* during insertion because it overlaps more than one partition. This problem is discussed in more detail in section 3. Each cell is stored in one leaf node of the cell tree.

A point search can be conducted as follows. The search point is inserted into the root of the tree and tested against each partition in the root for inclusion. Once the containing partition has been found, the point is inserted into the corresponding subtree. This step is repeated until a leaf of the tree has been reached. Then all cells in the corresponding leaf partition are examined whether they contain the search point. The solution consists of those objects that contain at least one of the cells that qualify. A similar algorithm exists for range searches (Günther 1988).

In analogy to the R– and $R^+$–tree, a cell tree is a height–balanced tree and each tree node corresponds to exactly one disk page. Therefore the computation of a search operator requires only a small number of disk accesses. The cell tree is a fully dynamic index structure; insertions and deletions can be interleaved with searches and no periodic reorganization is required.

If an insertion causes a disk page to overflow, the corresponding partition and cell tree node have to be split and its descendants are distributed among the two resulting nodes. Each split may propagate up the tree, i.e., it may be necessary to split the ancestor node as well, and so on.

# 3 Redundancy and Oversize Shelves

As mentioned above, one major disadvantage of all spatial access methods that are based on clipping is that each new data object may have to be subdivided in order to be stored because it overlaps several partitions (Fig. 2). Each fragment is stored in one leaf node of the cell tree; another possibility would be to store a copy of the whole object in each of these leaf nodes. This *fragmentation* or *redundancy* effect is getting worse as the spatial database continues to become more populated. Each split of a partition leads to a decrease of the average partition size and therefore to an increase in the number of fragments per object. The net result of this fragmentation effect is not only an increase in the average search time but also an increase of the frequency of node overflows, which in turn leads to an increase in fragmentation, and so on. It is important to minimize the probability of such an endless loop.

Note that the fragmentation effect is essentially the *redundancy induced by region boundaries* as described by Orenstein (Orenstein 1989). Other than in the case of Orenstein's z–order (Orenstein 1986), however, there is no inherent advantage of having redundancy in a cell tree; redundancy is an undesirable side effect.

In order to minimize the negative implications of redundancy in the case of tree–based access methods we recently proposed a new concept called *oversize shelves* (Günther 1990). Each interior node $N$ is enhanced by a short linked list of extra disk pages that contain those (sub-)objects that would have been fragmented too much by the insertion into the subtree under $N$ (Fig. 3).

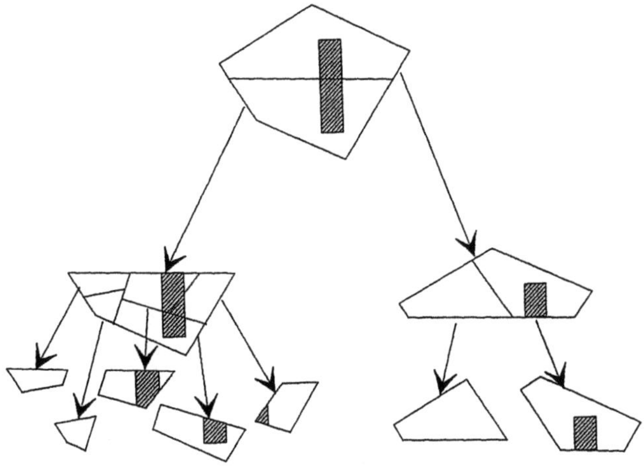

Figure 2: Fragmentation in a cell tree

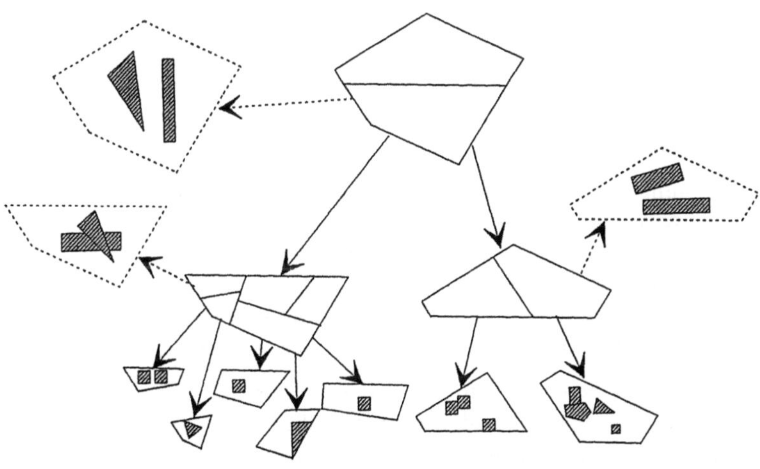

Figure 3: A cell tree with oversize shelves

This concept is very similar to the idea of oversize shelves in a library. All those books that are too tall for the main shelf (i.e. the shelf where the rest of the books with the same call number are located) are put on a special oversize shelf. This saves space because the library can use shelves with a lower height for most of the books. On the other hand, library users may have to accept a small increase in search time because they have to search the oversize shelf as well if they do not find a book on the main shelf.

In the case of spatial index structures, the tradeoff is of a somewhat different nature. The storage of a (sub-)object on an oversize shelf decreases the number of leaf node entries and therefore in the long run the height of the tree. On the other hand, each oversize shelf has to be searched whenever the corresponding tree node is involved in a search operation. Additional oversize shelves therefore tend to increase the average search time per node.

# 4 The Analytic Model

The decision if a new component should be inserted into the subtree or stored on the oversize shelf is an optimization problem and depends on various parameters. Because most search operations are I/O–bound, the objective of this optimization is to minimize the expected number of page faults per search. For the point search this number can be computed as the depth of the tree plus the average number of oversize shelf pages encountered on a path from the root to a leaf.

We will first prove a theorem that solves the optimization problem for the point search in a cell tree, then generalize our results to the range search case and to other spatial access methods. The following parameters are used to characterize the spatial database and the index.

| | | |
|---|---|---|
| $N$ | ... | number of cells stored in the leaves of the cell tree |
| $a$ | ... | average space utilization of the cell tree pages |
| $k$ | ... | average number of cells that fit on a disk page |
| $n$ | ... | maximum number of children per interior node |
| $s$ | ... | average number of cells that is split when a leaf node is split |
| $v$ | ... | average number of disk pages per interior node (incl. oversize shelf) |

The theorem is based on a model that simulates the behavior of a cell tree as cells are dynamically inserted and deleted from the tree. Obviously, the parameter $n$ is constant whereas the remaining parameters $N, a, k, s$, and $v$ change over time. In particular, $s$ is typically increasing as the database becomes more populated because in that case there will usually be more overlap between cells and it will become increasingly difficult to find splitting hyperplanes that intersect only a small number (possibly zero) cells. To make optimal decisions in accordance with the following theorem, it is therefore necessary to keep track of those parameters throughout all tree updates.

*Theorem* 1.
Let $C$ denote a component that is to be inserted into the subtree under the node $L$, which is located at height $i$ (leaves are at height 0). Furthermore, let $q$ denote the

number of partitions in $L$ that overlap $C$. Under certain simplifying assumptions, the expected number of page faults per *point search* is minimized by storing $C$ on $L$'s oversize shelf if and only if $q > q_i$ where

$$q_1 = \min\{n, \frac{a^2 n(k-s)\ln(na)}{vk}\}$$

$$q_i = \min\{n, \frac{an^2 + an}{n - q_{i-1} + an}\} \qquad (i \geq 2) \tag{1}$$

*Proof.*
In order to make the model tractable, the following simplifying assumptions have been made.

S1. The depth of the cell tree is represented by the real number $\log_{na} N/ak$ rather than by an integer. This real number is a very close approximation of the actual depth of the tree. However, it allows us to circumvent the difficulties that arise from the stepwise character of the depth function.

S2. All parameters are considered constant for the duration of one insertion.

Under these assumptions, the insertion of a component $C$ into an oversize shelf corresponding to a node $L$ at height 1 causes an increase in the expected number of page faults per search by

$$O_1 = \frac{a^2 n}{N}$$

due to the following considerations. Because there are $N$ cells, the expected number of leaves is $N/(ak)$ and the expected number of nodes at height 1 is $N/(a^2kn)$. Hence, the probability that node $L$ is encountered during a given point search is $a^2kn/N$. The insertion of $C$ into $L$'s oversize shelf increases the shelf's expected size (in disk pages) by $1/k$. Therefore the expected number of page faults per search increases by $a^2kn/N \cdot 1/k$ or $a^2n/N$.

As a corollary we obtain that the storage of a component on an oversize shelf at tree height $i$ leads to an increase in the expected number of page faults per search of

$$O_i = \frac{a^{i+1}n^i}{N} \tag{2}$$

On the other hand, the insertion of a cell into a leaf node increases the depth of the cell tree from

$$\log_{na} \frac{N}{ak}$$

to

$$\log_{na} \frac{N + \frac{k}{k-s}}{ak}$$

The factor of $k/(k-s)$ is due to the fact that each split of a full leaf node splits $s$ cells in the average. I.e., it creates $s$ new cells, which leaves $k-s$ more cells to be inserted in the average before the next leaf node split will occur. Hence, the insertion of one new cell actually creates $1+s/(k-s)$ or $k/(k-s)$ new cells in the average. Because each interior node occupies $v$ disk pages (including oversize shelf), the insertion of a cell into a leaf node therefore increases the expected number of page faults per search by

$$\begin{aligned} I_0 &= v\log_{na}\frac{N+k/(k-s)}{ak} - v\log_{na}\frac{N}{ak} \\ &= v\log_{na}(1+\frac{k}{N(k-s)}) \\ &= \frac{v\ln(1+\frac{k}{N(k-s)})}{\ln(na)} \\ &\approx \frac{vk}{(k-s)N\ln(na)} \end{aligned}$$

for $N$ sufficiently large.

The insertion of a component $C$ into a node $L$ at height 1 causes $q$ cells to be forwarded to the leaves under $L$, where $q$ is the number of partitions of $C$ that overlap $L$. The insertion therefore increases the expected number of page faults per search by

$$I_1(q) = qI_0 = \frac{qvk}{(k-s)N\ln(na)}$$

Hence, it is more advantageous to store $C$ on $L$'s oversize shelf if and only if

$$I_1(q) > O_1$$

or because of $q \leq n$

$$q > \min\{n, \frac{a^2n(k-s)\ln(na)}{vk}\} =: q_1 \tag{3}$$

For $i \geq 2$, the insertion of a component $C$ into a node $L$ at height $i$ causes $q$ subcomponents to be forwarded to the descendant nodes of $L$ at height $i-1$. Some of these subcomponents will be stored on the corresponding oversize shelves, some others will be inserted into the subtrees. Assuming that the probability for insertion is $p_{i-1}$, this leads to an increase in the expected number of page faults per search of

$$I_i(q) = (1-p_{i-1})qO_{i-1} + p_{i-1}qI_{i-1}(\overline{q_{i-1}}) \tag{4}$$

where $\overline{q_{i-1}}$ is the expected number of partitions at height $i-1$ overlapping a given subcomponent.

If the component $C$ is inserted into the node $L$ at height $i$ it is therefore more favorable to put $C$ on $L$'s oversize shelf if and only if $O_i < I_i(q)$, i.e., if

$$\frac{a^{i+1}n^i}{N} < (1-p_{i-1})q\frac{a^in^{i-1}}{N} + p_{i-1}qI_{i-1}(\overline{q_{i-1}})$$

or

$$q > \frac{a^{i+1}n^i}{(1-p_{i-1})a^in^{i-1} + p_{i-1}I_{i-1}(\overline{q_{i-1}})N}$$

Hence, we obtain the following threshold value $q_i$ for $q$ $(i \geq 2)$.

$$q_i^* := \frac{a^{i+1}n^i}{(1-p_{i-1})a^in^{i-1} + p_{i-1}I_{i-1}(\overline{q_{i-1}})N} \tag{5}$$

From equations (2) and (4) we get

$$\begin{aligned}
q_i^* &= \frac{a^{i+1}n^i}{(1-p_{i-1})a^in^{i-1} + p_{i-1}N(1-p_{i-2})\overline{q_{i-1}}O_{i-2} + p_{i-1}Np_{i-2}\overline{q_{i-1}}I_{i-2}(\overline{q_{i-2}})} \\
&= \frac{a^{i+1}n^i}{(1-p_{i-1})a^in^{i-1} + p_{i-1}(1-p_{i-2})\overline{q_{i-1}}a^{i-1}n^{i-2} + p_{i-1}Np_{i-2}\overline{q_{i-1}}I_{i-2}(\overline{q_{i-2}})} \\
&= \frac{an^2}{n - np_{i-1} + p_{i-1}n\overline{q_{i-1}}\frac{(1-p_{i-2})a^{i-1}n^{i-2}+p_{i-2}I_{i-2}(\overline{q_{i-2}})N}{a^in^{i-1}}} \\
&= \frac{an^2}{n - np_{i-1} + p_{i-1}n\overline{q_{i-1}}\frac{1}{q_{i-1}^*}} \qquad \text{(because of (5))}
\end{aligned}$$

If $q$ is less than $q_i^*$ it is more efficient to insert $C$ into the subtrees under $L$, otherwise it is better to put $C$ on $L$'s oversize shelf. The insertion probability $p_i$ can therefore be approximated closely as follows.

S3. $$p_i \approx \min\{q_i^*/n, 1\}$$

Hence, one obtains

$$q_i^* = \begin{cases} \dfrac{an}{\overline{q_{i-1}}} q_{i-1}^* & q_{i-1}^* \geq n \\[2ex] \dfrac{an^2}{n - q_{i-1}^* + \overline{q_{i-1}}} & \text{otherwise} \end{cases}$$

From there and from $\overline{q_{i-1}} \leq an$, it is obvious that

$$q_i^* > n \Rightarrow \forall j \geq i : q_j^* > n$$

On the other hand, $q$ can be no greater than $n$. Hence, one may use the following threshold value $q_i$ instead of $q_i^*$:

$$q_i := \min\{n, \frac{an^2}{n - q_{i-1} + \overline{q_{i-1}}}\}$$

The main question remaining is how to approximate $\overline{q_{i-1}}$, the expected number of partitions at height $i-1$ that overlap a given subcomponent. Certainly $\overline{q_{i-1}}$ is no greater than $an$ because each node on level $i-1$ has $an$ partitions in the average and only a certain subset of those will overlap a given subcomponent. Moreover, there has to be a positive correlation between $q_i$ and $\overline{q_{i-1}}$ for the following reason. Any component handed down from level $i$ to level $i-1$ is a fragment of some component $C$ that has been inserted into the node $L$ on level $i$. If the number of fragments is large then most of the fragments will cover the corresponding partition *completely* because they belong to the interior of $C$. In this case, $\overline{q_{i-1}}$ would be close to $an$. On the other hand, if the number of fragments is small then most of the fragments will belong to the boundary of $C$ and therefore cover only parts of the corresponding partition. $\overline{q_{i-1}}$ may then be significantly less than $an$.

Based on these considerations, we approximate $\overline{q_{i-1}}$ as follows.

S4. $$\overline{q_{i-1}} \approx an(1-\frac{1}{q_i})$$

Inserted into equation (6), this together with equation (3) proves Theorem 1. □

An interesting observation is how $q_i$ depends on $s$, the average number of cells that is split when a leaf node is split. As the database becomes more populated, there usually will be more overlap between cells in the database, which corresponds to an increase of $s$. $a, n$, and $k$, on the other hand, remain relatively constant. This results in a smaller $q_1$ and correspondingly in smaller $q_i$ for $i \geq 2$. Hence, it becomes more profitable to put even smaller components on the oversize shelves when the database (and correspondingly the overlap) grows. This makes sense because each component that is stored on an oversize shelf will not contribute to a further fragmentation of the data space.

Other properties of the function $q_i$ are investigated in (Günther 1990). That report also contains a generalization of Theorem 1 to range queries. The resulting theorem is given here without proof for brevity reasons. The search space size $r$ is measured in percent of the size of the data space ($r = 0$: point query, $0 < r \leq 1$: range query). Note that for $r = 0$ this theorem coincides with Theorem 1.

*Theorem* 2.

Let $C$ denote a component that is to be inserted into the subtree under the node $L$, which is located at height $i$ (leaves are at height 0). Furthermore, let $q$ denote the number of partitions in $L$ that overlap $C$. Under certain simplifying assumptions, the expected number of page faults per *range search* with search space size $r$ is minimized by storing $C$ on $L$'s oversize shelf if and only if $q > q_i^R$ where

$$q_1^R = \min\{n, \frac{(an)^{1-r}a(k-s)\ln(na)}{vk}\}$$

$$q_i^R = \min\{n, \frac{(an)^{1-r}n + an}{n - q_{i-1}^R + an}\} \qquad (i \geq 2)$$

## 5 Generalization

The concept of oversize shelves and the analytic model described above can be applied to other spatial access methods as well. This generalization is fairly straightfor-

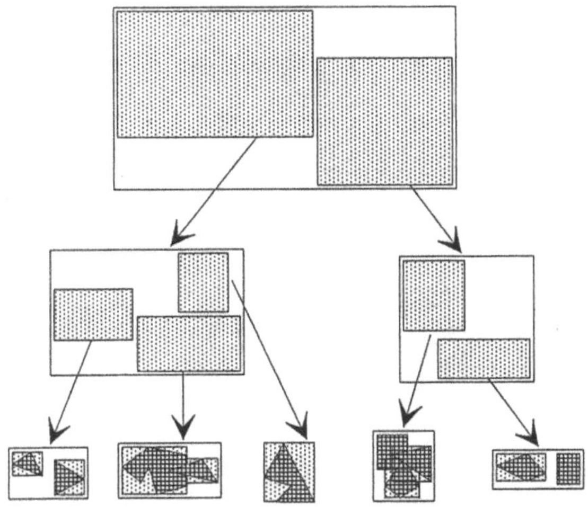

Figure 4: An R$^+$–tree with data objects (dark shading)

ward in the case of access methods that also use the *clipping* approach. An example for such a structure is the R$^+$-tree, a close relative of the cell tree.

R$^+$–trees (Fig. 4) are balanced trees that correspond to a nesting of $d$–dimensional intervals. Each node $N$ corresponds to a disk page $D(N)$ and an interval $I(N)$. If $N$ is an interior node then all intervals corresponding to the immediate descendants of $N$ are *disjoint* subsets of $I(N)$ and stored on the disk page $D(N)$. If $N$ is a leaf node then $D(N)$ also contains a number of intervals that are subsets of $I(N)$ (but not necessarily disjoint). Each of these data intervals is the bounding box of some fragment of the data object.

The concept of oversize shelves can be applied immediately to this structure. The analytic model presented in section 4 also holds for this case except that some of the parameters now have a slightly different meaning:

$N$ ... number of intervals stored in the leaves of the R$^+$-tree
$a$ ... average space utilization of the R$^+$-tree pages
$k$ ... average number of intervals that fit on a disk page
$s$ ... average number of intervals that is split when a leaf node is split

The parameter $n$ in the cell tree model is identical to $k$ in this case because the maximal number of children of an interior R$^+$-tree node is equal to the number of intervals that can be stored there.

Moreover, oversize shelves can be applied to spatial access methods that are based on the overlapping regions approach, such as the R–tree. As discussed in the introduction, these methods may suffer from very large objects as well because a large amount to overlap may render the whole index inefficient. Oversize shelves can help to prevent this type of breakdown. The changes that have to be made to the analytic model in order to capture the overlapping regions technique are subject to our current work.

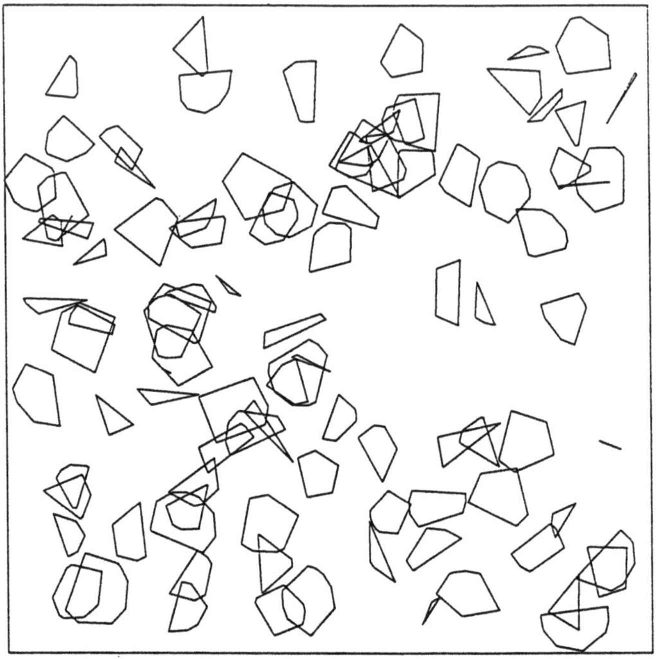

Figure 5: An example database

# 6 Performance Evaluation

The two–dimensional cell tree with oversize shelves has been implemented at FAW Ulm on a SUN 3/260 workstation with a 68020 processor and a 68881 floating point coprocessor.

All databases used in this performance evaluation are randomly generated sets of convex polygons. For each polygon, an ellipse with random major and minor axis and random orientation was generated, and a random number of points on that curve (at least three) were selected as vertices of the polygon. Then each polygon was displaced at random within the given two–dimensional data space; see Figure 5 for an example database. For simplicity, no concave data objects are included as concave objects can always be represented as unions of convex cells.

There are two major parameters that characterize such a geometric database; the number $N$ of data objects (the *database size*) and their *average size* $S$, measured in percent of the size of the data space. From there, various other parameters can be computed: the product $N$ x $S/100$, for example, yields the *coverage* of the data space, which often is a good indicator for the overlap between data objects. We took measurements for four different databases containing 500, 1000, 5000, and

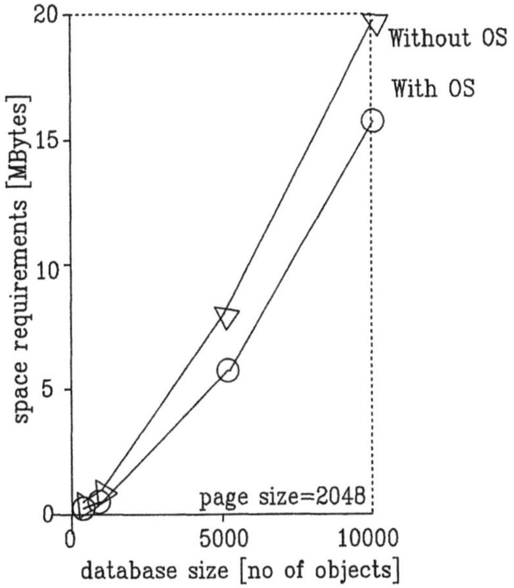


Figure 6: Space Requirements vs. Database Size

10000 convex polygons of average size 0.05 (i.e., 0.05 percent of the size of the data space).

The size of the objects in the database is distributed similar to an Erlang distribution: most objects are comparatively small (size about 0.01), but some may have a size of up to 10, i.e., they cover about 10% of the size of the data space. This type of distribution is very typical for geographic applications.

All cell trees were constructed by means of consecutive insertion of the data objects into an initially empty tree. For each tree level $i$ we used the corresponding value $q_i$, as defined in Theorem 1, for the construction of the cell tree. The average search space size was assumed to be 0.1 percent of the data space size (see below). The page size was 2048 bytes.

The first thing measured were the space requirements of the cell tree vs. database size. Our results are presented in Figure 6. The results show that oversize shelves lead to some space savings (up to 10%).

In order to measure the search performance, we defined a benchmark of search operations using three different search space sizes $r$, each measured in percent of the size of the data space: 0 (point query), 0.1, and 10 (range queries). For each search space size, 100 points or random convex polygons were generated and used as search spaces. We obtained average CPU times per search operation, as shown in Figure 7. It can be seen that performance gains of up to 60% were obtained from using oversize shelves.

At least as important as the average *CPU time* per search, however, is the number of *disk accesses* performed during a search operation. Search operations

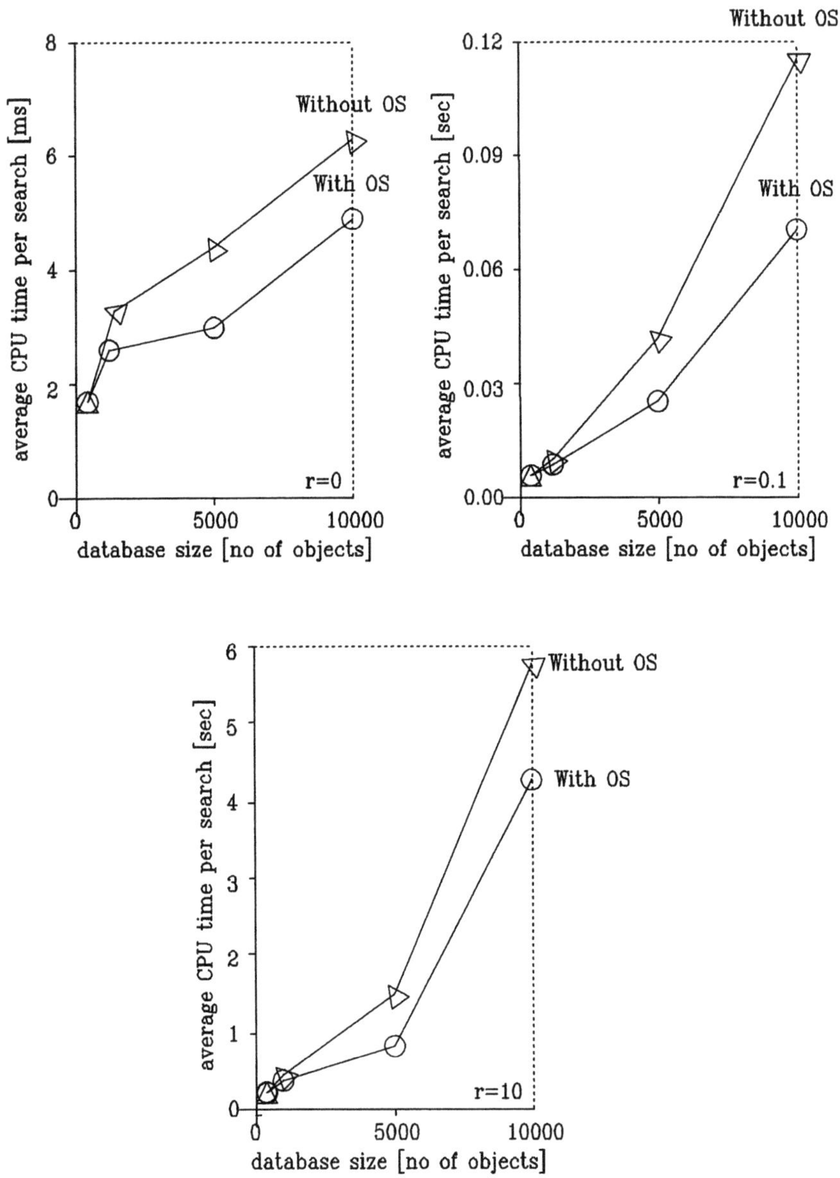


Figure 7: CPU Time vs. Database Size

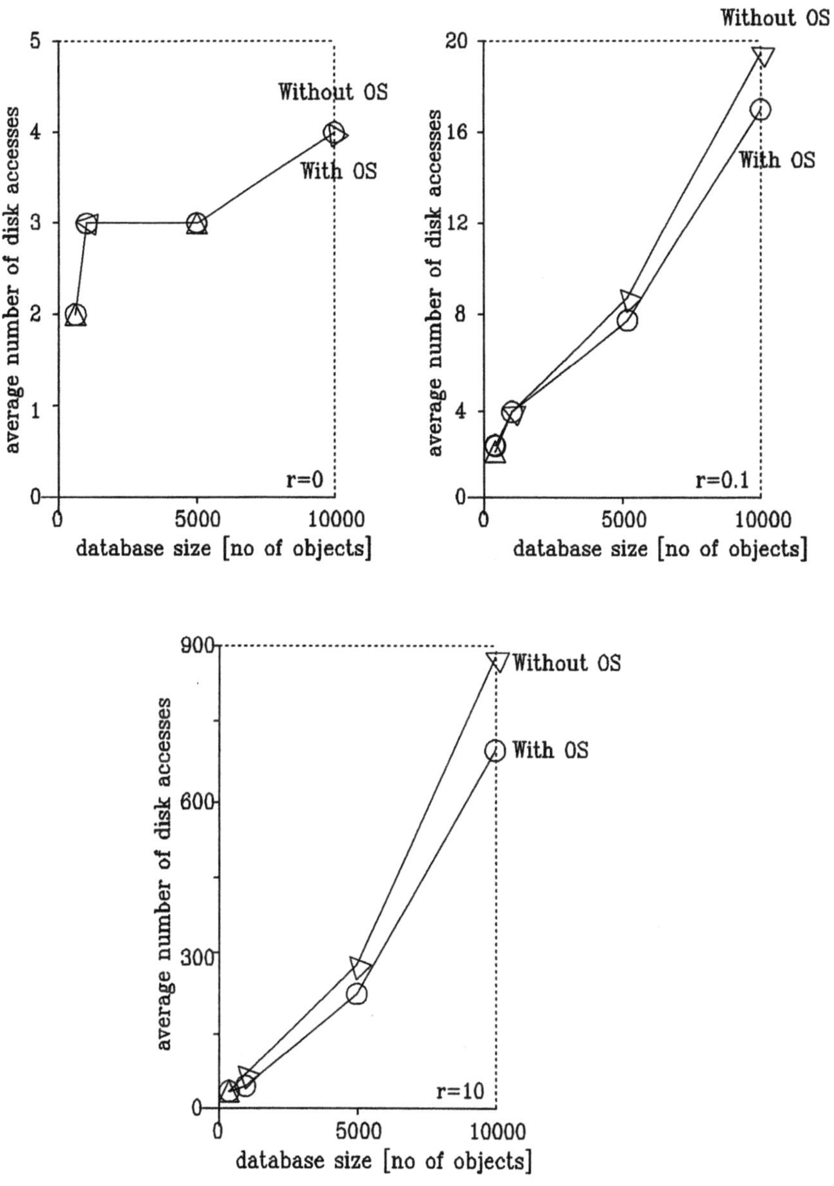


Figure 8: Disk Accesses vs. Database Size

may be I/O-bound or CPU-bound, depending on the size of the application and the main memory available.

As shown in Figure 8, oversize shelves result in a more efficient computation of range queries in terms of disk accesses as well, although the savings are not quite as large (up to 20%). For point searches, oversize shelves have no impact on performance.

## 7 Conclusions

In conclusion, the idea of oversize shelves has been proven to be an efficient means to improve the search performance of spatial access methods. When a database is not very populated, most components are stored in the leaves of the chosen access method; only very large components which cover a large fraction of the data space will be put on oversize shelves. As the database becomes more populated, the threshold size at which a cell is stored on an oversize shelf is decreasing. More components will be stored on oversize shelves, which means that they will not contribute to a further fragmentation of the data space.

Initial experiments confirm our conjecture that oversize shelves in fact improve the search performance. So far, up to 60% savings have been reported.

The performance of oversize shelves has been optimized using a new analytic approach to model the behavior of dynamic spatial index structures under insertions and deletions. While the model has been presented in the context of cell trees, it can be generalized to other tree-based access methods such as R– or the $R^+$–tree. We also intend to extend the model towards a more general mechanism to evaluate and compare the performance of tree-based spatial access methods analytically rather than just experimentally.

## Acknowledgments

Thanks to Volker Gaede, Thomas Kämpke (both FAW Ulm), and Hartmut Noltemeier (Universität Würzburg) for valuable discussions, and to Rudi Rapp (FAW Ulm) for his careful typesetting. The implementation of the oversize shelves is also by Volker Gaede.

## References

Bayer, R., McCreight, E.M.: Organisation and maintenance of large ordered indices. Acta Informatica 1 (3) 1–21 (1972)

Freeston, M.: The BANG file: A new kind of grid file. Proc. ACM SIGMOD, 1988

Greene, D.: An implementation and performance analysis of spatial data access methods. Proc. IEEE Fifth International Conference on Data Engineering, 1988

Günther, O.: Efficient structures for geometric data management, Lecture Notes in Computer Science No.337. Springer–Verlag, Berlin 1988

Günther, O.: The design of the cell tree: An object–oriented index structure for geometric databases. Proc.IEEE Fifth International Conference on Data Engineering, 1989

Günther, O.: Analysis of spatial access methods. FAW Technical Report No. 90017, 1990

Günther, O., Bilmes, J.: Tree–based access methods for spatial databases: Implementation and performance evaluation. IEEE Trans. Knowledge and Data Engineering 3 (3) 342-356 (1991)

Guttman, A.: R–trees: A dynamic index structure for spatial searching. Proc. ACM SIGMOD, 1984

Hutflesz, A., Six, H.–W., Widmayer, P.: Twin grid files: Space optimizing access schemes. Proc. ACM SIGMOD, 1988

Kriegel, H.P., Seeger, B.: Multidimensional order preserving linear hashing with partial expansions. Proc. International Conference on Database Theory. Lecture Notes in Computer Science No. 243. Springer–Verlag Berlin 1986

Kriegel, H.P., Seeger, B.: PLOP-Hashing: A grid file without directory. Proc. IEEE Fourth International Conference on Data Engineering, 1988

Nievergelt, J., Hinterberger, H., Sevcik, K.C.: The grid file: An adaptable, symmetric multi–key file structure. ACM TODS 9 (1) 38–71 (1984)

Nievergelt, J., Hinrichs, K.: Storage and access structures for geometric databases, Proc. Int. Conf. on Foundations of Data Organization. Plenum Press, New York 1987, pp. 335–345

Orenstein, J.A.: Query processing in an object–oriented database system. Proc. ACM SIGMOD, 1986

Orenstein, J.A.: Redundancy in spatial databases, Proc. ACM SIGMOD, 1989

Otoo, E.J.: Balanced multidimensional extendible hash tree, Proc. 5th ACM SIGACT/ SIGMOD Symp. on Principles of Database Systems, 1986

Ouksel, M.: The interpolation based grid file, Proc. 4th ACM SIGACT/SIGMOD Symp. on Principles of Database Systems, 1985

Robinson, J.T.: The K–D–B–tree: A search structure for large multidimensional dynamic indexes, Proc. ACM SIGMOD, 1981

Sellis, T., Roussopoulos, N., Faloutsos, C.: The $R^+$–tree: A dynamic index for multi–dimensional objects. Proc. 13st International Conference on Very Large Databases, 1987

Seeger, B., Kriegel, H.P.: Techniques for design and implementation of efficient spatial access methods, Proc. 14th International Conference on Very Large Databases, 1988

Whang, K.–Y., Krishnamurthy, R.: Multilevel grid files, IBM Research Lab., Yorktown Heights, 1985

# An Efficient Map Overlay Algorithm Based on Spatial Access Methods and Computational Geometry*

Hans-Peter Kriegel, Thomas Brinkhoff, Ralf Schneider

## Abstract

Geographic database systems, known as geographic information systems (GISs) particularly among non-computer scientists, are one of the most important applications of the very active research area named spatial database systems. Additionally to other spatial database systems where query and manipulation processing are emphasized, the most important purpose of a GIS is to analyze geographical data. The basic building block for analysis operations in GIS is the operation map overlay. However, available map overlay algorithms suffer from poor performance. In this paper, we present an efficient map overlay algorithm using as ingredients spatial access methods and state-of-the-art computational geometry concepts. An experimental performance analysis of our implemented map overlay algorithm demonstrates the fruitfulness of the marriage of spatial access methods with computational geometry methods for improving the efficiency of GISs. The possibility of coupling these two areas is based on the fact that both use spatial order relations in order to increase the performance of query processing.

**Keywords:** map overlay, computational geometry, spatial access methods

## 1 Introduction

Geographic information systems are one of the most important applications of spatial database systems. Basically, they consist of two parts: First, components to query and manipulate geographical data and second, components to manage and store the data. However, the main purpose of a GIS is to analyze geographical data. Map overlay is one of the most important analysis operations in a GIS, e.g. C.D. Tomlin's map analysis package (MAP) [Tom90] is completely based on the map overlay operation. The map overlay combines two or more maps of different topics into a single new map. The goals are to derive new maps, to find correlations between the information encoded in maps, and to process complex queries.

The algorithms presented in the past (e.g. in [Whi 78], [Fra 83], [FW 87]) assume that the input maps of the map overlay are kept in main memory or in sequential files on secondary storage. The following two important requirements of future GISs demand for new approaches: First, the database system of a GIS must be able to manage very large volumes of data. The large amount of data (in the order of Giga- and Terabytes) is additionally increased by pursuing the goal to manage scaleless and seamless databases [Oos 90]. Second, the database system has to support spatial access to parts of the database, such as maps, and to the objects of a map. Such access is a necessary condition for efficient query and manipulation processing.

Pursuing these goals we want to take advantage of spatial access methods (SAMs). In the past few years many access methods were developed which allow to organize large sets of spatial objects on secondary storage. There are three basic techniques which extend multidimensional point access methods (PAMs) to multidimensional spatial access methods [SK 88]: clipping, overlapping regions, and transformation. Point access

* This work was supported by grant no. Kr 670/4-3 from the Deutsche Forschungsgemeinschaft (German Research Society) and by the Ministry of Environmental and Urban Planning of Bremen

methods such as the grid file [NHS 84], PLOP-hashing [KS 88], the BANG file [Fre 87], and the buddy tree [SK 90] can be extended by these techniques. Additionally, there are access methods which are designed for managing simple spatial objects directly. They use one of the above techniques inherently, e.g. the R-tree [Gut 84] and the R*-tree [BKSS 90] use overlapping regions, or the cell tree [Gün 89] uses clipping.

The use of SAMs as an ingredient in GISs is absolutly necessary to guarantee good retrieval and manipulation performance, in particular for large maps. The use of SAMs enables us to overlay only relevant parts of the seamless database. Additionally, we have the possibility to partition the maps and to combine the results of these partitions in order to increase the performance of the overlay. In addition to using access methods, we increase the performance of the map overlay operation in our approach applying a plane-sweep algorithm. Thus we combine spatial access methods and computational geometry in order to improve the efficiency of GISs. The combination of these two areas is based on the fact that both use spatial order relations. To our knowledge this combination, although self-suggesting, has never been performed before.

The next section explains and formally defines the terms 'thematic map' and 'map overlay'. Our plane-sweep overlay algorithm is presented in section 3. A analytical and experimental performance analysis of the overlay operation is presented in section 4. The combination of the algorithm and spatial access methods follows in section 5. The paper concludes with a summary and an outlook to future work.

# 2 Definitions

## 2.1 Thematic maps

In this paper the term map is used for *thematic maps*. Those emphasize one or more selected topics, e.g. land utilization, population density etc. Thematic maps are generally represented by *choropleth maps* which separate areas of different properties by boundaries [Bur 86], e.g. forests, lakes, roads, or agriculturally used areas (see figure 1).

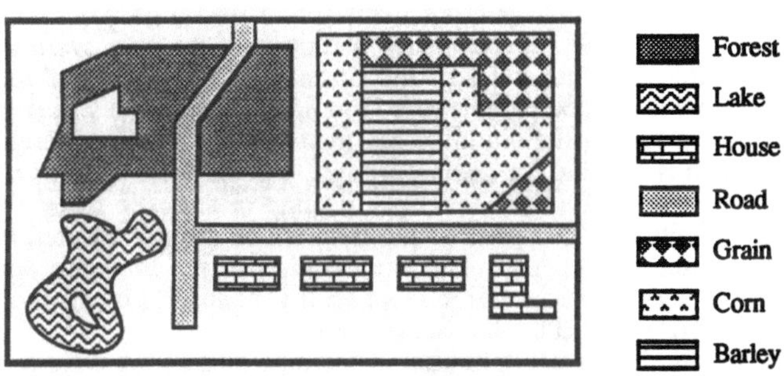


**Fig. 1:** Example of a thematic map

We consider in this paper only maps modelled by a *vector representation* because there are two main disadvantages of raster representations [Oos 90]: (1) Raster data depends on a specific projection. Therefore, there are problems when combining raster maps from different sources. A scaleless database can not be realized using a raster representation. (2) Objects in raster maps generally are not handled individually. Thus, a support by access methods is more difficult. Additionally, raster data are more voluminous.

We assume that the connected areas with the same property are described by *simple polygons with holes*, and that the used data structures are able to handle such polygons explicitly [KHHSS 91b]. A polygon is simple if there is no pair of nonconsecutive edges sharing a point. A simple polygon with holes is a simple polygon where simple polygonal holes may be cut out (see figure 2). There may be other areas in such a hole. The areas of a map are disjoint but they do not need to cover the map completely. Each area refers to exactly one *thematic attribute*. In figure 1 these characteristics of a thematic map are depicted by an example which visualizes the land utilization of a part of a map.

**Fig. 2:** Different polygons

In the following a formal definition of thematic maps is presented where $\cap_r$ denotes the regularized intersection [Til 80] and T is the set of values of the thematic attributes of the map M:

$$M := \{ t = (t.P, t.A) \mid t.P \text{ is a simple polygon with holes}, t.A \in T \},$$
$$\text{where } t_1, t_2 \in M, t_1 \neq t_2 \Rightarrow t_1.P \cap_r t_2.P = \emptyset$$

Maps of different topics describing the same part of the world are called *map layers*. The map overlay operation combines such layers.

## 2.2 Map overlay

The *map overlay* combines two or more maps (input maps) of different topics into a single new map (output map). An established definition of the map overlay operation on vector representation does not exist in the literature. The power of the algorithms presented in literature is not comparable (see [Bur 86]). Therefore it was necessary in our opinion to develop a formal definition of the map overlay which generalizes all these proposals and which is suitable for many purposes, e.g. all those claimed in MAP [Tom 90]. Our definition is based on the above specification of thematic maps. One important property of our definition is that parts of the map which are not covered by each input map may occur in the output map. The combination of the thematic attributes or of geometric or topological properties of the input areas is controlled by an *overlay function* f, where f is defined or selected by the user of the GIS.

Before we present the formal definition of the map overlay we want to illustrate this operation by an example. Figure 3 depicts two input maps: the map 'land utilization' from figure 1 and another map 'soil pollution'. In the output map all areas should be reported, which are forests or agriculturally used land and where the degree of soil pollution is greater than 2.

In the following we present the formal definition of map overlay where $M^*$ denotes the set of all thematic maps. Furthermore, we introduce for each input map $M_i$ a tuple $r_i$ to process the empty parts of the map correctly. $r_i$ consists of $r_i.P$ which represents the non-covered area of the map and where $r_i.A \notin T_i$. By not combining different $r_i$'s, we prevent generating areas in the output map which are not covered by any area. For this we use the intersection operation $\cap'$, which is the same as the regularized intersection except for $r_i.P \cap' r_j.P = \emptyset$. To eliminate areas which should not appear in the output map, we

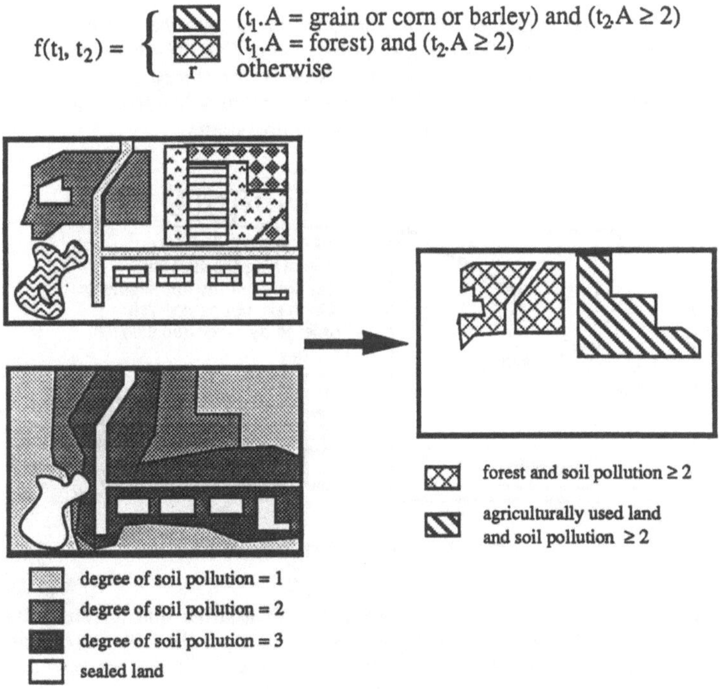


**Fig. 3:** Example of a map overlay and an overlay function

introduce an attribute r indicating an unwanted tuple. The set of connected but not expandable parts of A is denoted sep (A).

$$\text{overlay: } M^* \times M^* \times \{f \mid f: (M_1 \cup \{r_1\}) \times (M_2 \cup \{r_2\}) \rightarrow (T_3 \cup \{r\})\} \rightarrow M^*$$

$$(M_1, M_2, f) \rightarrow M_3 \ ,$$

$$\text{where } M_3 = \{\ t = (t.P, t.A) \mid t.P \in \text{sep}\,(t_1.P \cap' t_2.P)\ ,\ t.P \neq \varnothing\ ,$$
$$t.A = f\,(t_1, t_2),\ t.A \neq r,\ t_1 \in M_1 \cup \{r_1\}\ ,\ t_2 \in M_2 \cup \{r_2\}\ \}$$

Due to clarity we restricted the definition to an overlay with two maps and an overlay function which processes only tuples of the maps. The generalization of the definition for overlaying more than two maps is straightforward. For an integration of topological properties, e.g. neighborhood, only the parameter list of the overlay function f must be extended.

## 3 The plane-sweep overlay

A desirable overlay algorithm should define and utilize an order relation on the objects in the plane to enable a spatial partition of the input maps. Plane sweep is a technique of computational geometry which fulfills this demand [PS 88]: the coordinates of objects *(event points)* are projected onto the x-axis and are processed according to the order relation on this axis. Event points are stored in a queue called *event point scheduler*. If event points are computed during processing, the event point scheduler must be able to insert event points after initialization. A vertical line sweeps the plane according to the event points from left to right. This line is called *sweep line*. The state of the plane at the sweep line position is recorded in a table called *sweep line status*. Event points which are passed by the sweep line are deleted from the event point scheduler. Figure 4 depicts an example of the event point scheduler and the sweep line status.

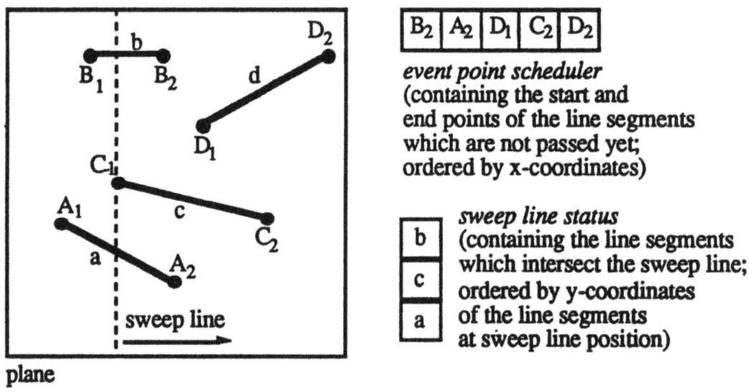


**Fig. 4:** Example of a plane sweep

In order to simplify the following algorithms we assume that different event points have different x-coordinates. This assumption can be cancelled by introducing a more elaborate order criteria (e.g. event points with equal x-coordinates are sorted by their y-coordinates) and by treating vertical edges according to their direction.

### 3.1 Nievergelt and Preparata's algorithm to determine the regions of a self-intersecting polygon

Nievergelt and Preparata's algorithm [NP 82] is a plane-sweep algorithm to determine the regions of a self-intersecting polygon (see figure 5). It is based on an algorithm developed by Bentley and Ottmann [BO 79] which computes the intersecting points of a set of line segments. In Nievergelt and Preparata's algorithm the vertices and computed intersection points are stored in the event point scheduler. The edges intersecting the sweep line are kept in the sweep line status determined by their y-coordinate at the sweep line position. The event points are classified into one of four categories:

1. start point: both edges at the event point are to the right of the point
2. end point: both edges at the event point are to the left of the point
3. bend: one edge is on the right and one is to the left of the event point
4. intersection point: two edges intersect in the event point

For computing the regions between the edges, the sweep line status consists additionally of two pointers to point lists for each edge. The point lists describe the boundaries of the regions. A corresponding list is extended by an event point when the sweep line reaches this point depending on its category. For every region between two edges in the sweep line status there is an identification number (region ID). Figure 5 depicts an example.

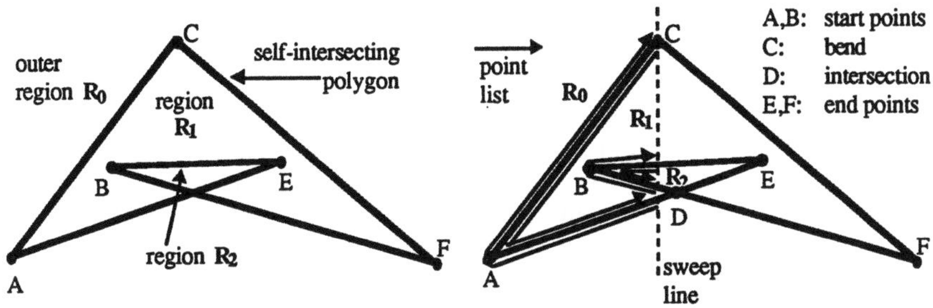


**Fig. 5:** Example of Nievergelt and Preparata's algorithm [NP 82]

## 3.2 Our algorithm for map overlay

In [NP 82] Nievergelt and Preparata mentioned to generalize the classification of the event points for generating an algorithm which intersects arbitrary maps. In the following we present and examine an algorithm for map overlay in detail which uses the algorithm described above as a building block and which realizes the generalization of this classification. The regions evaluated by the algorithm form the areas of the output map. Obviously, not only the determination of regions is needed for an overlay, but the algorithm has to identify the polygons of the input maps covering the new region. These polygons are called *parent polygons*. This step must be done to calculate the value of the thematic attribute of the region by the overlay function.

For handling the point lists correctly, every edge touching an event point must be attached to this point. There are edges starting, ending, or crossing the event point. We consider crossing edges as a combination of starting and ending edges (see figure 6). Therefore, the treatment of event points is processed in two steps:

1. Treat the edges to the left of the event point.
2. Treat the edges to the right of the event point.

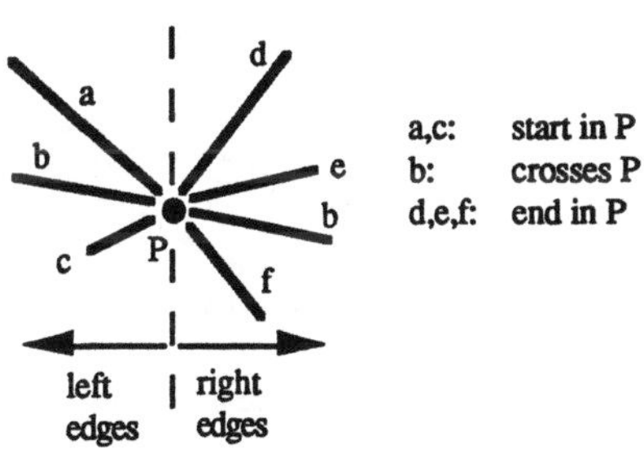


**Fig. 6:** Example of an event point

As long as only regions of one polygon are computed, the classification of the event points is directly possible when the point is inserted into the event point scheduler. If more than one polygon is treated, the vertices or intersection points of different polygons may coincide. Such points must be combined in one event point. The edges attached to one event point are sorted by their gradient.

Two types of point lists describing regions can be distinguished: *Outer point lists* describe for each region the area outside of this region. Outer point lists are directed clockwise. Point lists describing the inner area of a region are called *inner point lists*.

They are directed counterclockwise. The inner point list of a region is the outer one of a neighboring region separated by this list, and vice versa.

### 3.2.1 Treatment of the event points

Let P be the actual event point. Then $s_1$ to $s_n$ ($t_1$ to $t_m$) are the left (right) edges attached to P ordered from top to bottom. For each edge k exist two pointers A(k) and B(k). A(k) points to the tail of the point list above the edge k and B(k) points to the head of the point list below the edge k. The sweep line status is denoted by y.

Initially, we inspect the edges on the left side (see figure 7). According to their number we proceed:

case 1: $n = 0$ A new outer point list is necessary which only consists of the point P. $A(t_1)$ and $B(t_m)$ must point to this list.

case 2: $n = 1$ The two point lists referenced by $A(s_1)$ and $B(s_1)$ are extended by the point P and are attached to the pointers $A(t_1)$ and $B(t_m)$, respectively. $s_1$ is deleted from y.

case 3: $n > 1$ If $m = 0$, the outer point list must be closed. There exist two subcases: If $A(s_1)$ and $B(s_n)$ point to the same list, this list (extended by P) describes a polygonal hole (or an outer region) which is stored in the output map. If $A(s_1)$ and $B(s_n)$ point to different lists, the lists and P are only concatenated.
If $m > 0$, the outer point lists $A(s_1)$ and $B(s_n)$ are extended by P and are attached to $A(t_1)$ and $B(t_m)$ as in case 2.
For all regions between the edges $s_i$ and $s_{i+1}$ exist also two subcases: If $B(s_i)=A(s_{i+1})$, the inner list $B(s_i)$ (extended by P) describes a new region which is stored in the output map. Otherwise $A(s_{i+1})$, P, and $B(s_i)$ are concatenated.
$s_1, \ldots, s_n$ are deleted from y.

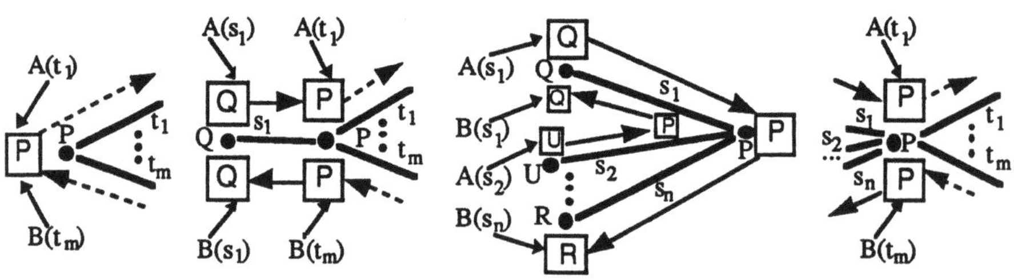


case 1: n=0 case 2: n=1 case 3a: n>1, m=0 case 3b: n>1, m>0

**Fig. 7:** Treatment of left edges

The edges on the right side of the event point are examined in the second step. Additionally, intersection points are determined in this step. A computed intersection point is inserted into the event point scheduler. There are also three cases in the second step:

case 1: $m = 0$ The two regions which are separated by the edges $s_1$ and $s_n$ up to now are identical (see figure 7, case 3a). Therefore the region IDs must be adjusted if they are different. Because $s_1$ to $s_n$ are deleted from the sweep line status y, two edges of y which were not neighbors now are new neighbors and must be tested for intersection.

case 2: $m = 1$ $t_1$ is inserted into y and tested for intersection with its new neighbors.

case 3: $m > 1$ $t_1$ to $t_m$ are inserted into y. Then $t_1$ is tested for intersection with its new neighbor above and $t_m$ is tested with its new neighbor below. For all $i=1,...,m-1$ a new region is starting between the edges $t_i$ and $t_{i+1}$ and receives its own ID. Therefore a new inner point list consisting of P is generated where the two pointers $A(t_i)$ and $B(t_{i+1})$ refer to this list (see figure 8).

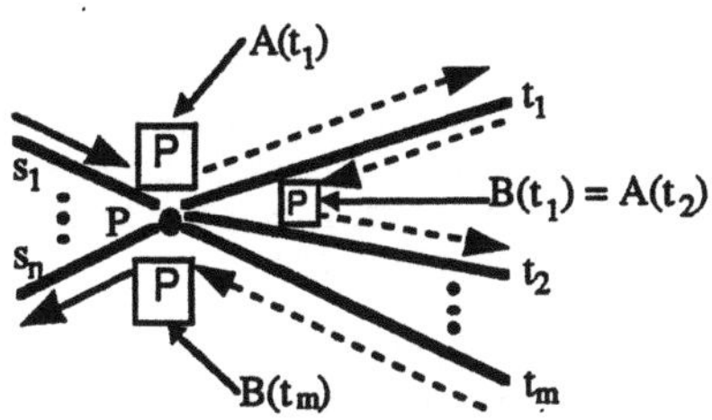


**Fig. 8:** Treatment of right edges

A region which describes a polygon hole can be assigned to the surrounding region by the region IDs. A formal description of the algorithm is presented in the appendix.

### 3.2.2 Identification of the parent polygons

In the overlay operation it is necessary to identify the parent polygons of a region because we need these polygons to determine the thematic attribute of each area in the output map evaluating the overlay function. Other overlay algorithms (e.g. [Fra 83]) are inefficient with respect to determining the parent polygons, e.g. costly point location tests have to be performed. Using the plane-sweep technique our algorithm allows a simple and efficient solution because the neighborhood of the regions intersecting the sweep line is inherently stored by the sweep line status:

In every input map $M_i$, there exists exactly one or no parent polygon for every evaluated region. Therefore, we attach to every edge of an input polygon a lower and an upper mark initially. The mark which is set, indicates the side where the polygon is located and contains a pointer to the polygon. The other mark is empty. While combining event points, coinciding edges are discovered. In this case, also the marks are combined. The set of parent polygons P(R) of a new region R is identified when the region is starting (step 2, case 3). At this point of time, the parent polygons of the region Q lying above R are known and also the marks of the edge e between R and Q:

$$P(R) = ( P(Q) \setminus \text{upper mark}(e) ) \cup \text{lower mark}(e)$$

This approach works also for holes by using exchanged marks.

## 3.3 A suitable coordinate representation for map overlay

The instability of plane-sweep algorithms against numerical errors is an objection being raised. This reproach may be justified if a floating point representation is used to compute the intersection points. However *rational coordinates* are a more suitable representation because they form a vector space. For example, a rational representation of coordinates is used in an implementation of a map overlay in [FW 87].

A more detailed analysis of a representation by rational coordinates leads to the following statements:

1. The coordinates in maps recorded by a GIS can be represented by pairs of integers. This assumption is realistic because both, the described part of the world and the resolution are limited.
2. To compute intersection points, integer coordinates are unsufficient [Fra 84]. But a map overlay with input maps containing integer coordinates only, needs a limited number of digits to represent the intersection points by rational numbers. Let n the number of digits of the integers of the input map then the number of digits of the nominator of the intersection points is smaller than 2*n+4 and the number of digits of the denominator is smaller than 3*n+4 (see [Bri 90]).
3. If the input maps result from an overlay themselves (thus containing rational coordinates), the same number of digits as in statement 2 is sufficient for the representation of the intersection points. This is due to the fact that no line segments connecting intersection points are introduced.

Under realistic assumptions rationals of finite precision are an easy, relative efficient, and numerical exact coordinate representation for the map overlay. The presented algorithm and other plane-sweep algorithms are absolutly robust by this approach. For an efficient use of rational coordinates an adequate support by hardware and system software is desirable.

### 3.4 Further applications of the algorithm

The presented algorithm can be used with little modifications for other problems. The *merge operation* is one of them which is closely related to the map overlay [Fra 87]: Merge neighboring areas in one map representing the same thematic attribute. For example, such maps may result from a classification of the attributes or from an overlay with a non-injective overlay function. The neighboring areas with the same atttribute can be merged by an algorithm similar to the presented overlay algorithm. The merge algorithm does not insert edges which separate areas with the same attribute into the sweep line status. Thus the resulting regions describe the merged areas.

The evaluation of the *intersection, union,* or *difference of polygons (with holes)* is a special case of the overlay problem. It can be done by using the following overlay functions $f_\cap$ , $f_\cup$ and $f_\setminus$. To handle the polygons within the overlay function correctly we assign to each polygon the thematic attribute 1.

$$f_\cap(t_1, t_2) = \begin{cases} 1 & (t_1.A = 1) \text{ and } (t_2.A = 1) \\ r & \text{otherwise} \end{cases}$$

$$f_\cup(t_1, t_2) = 1 \qquad \text{(+ merging of the results)}$$

$$f_\setminus(t_1, t_2) = \begin{cases} 1 & (t_1.A = 1) \text{ and } (t_2.A = r_2.A) \\ r & \text{otherwise} \end{cases}$$

## 4 Analytical and experimental performance analysis

### 4.1 Analytical performance analysis

Let n be the total number of edges of all polygons and k be the total number of intersection points of all edges. In [MS 87] is shown that the problem of computing all intersections between two sets S and T of line segments in the plane, where no two segments in S (similarly, T) intersect, can be executed optimally in O (n*log(n) + k) time. Because overlay algorithms have to compute in any case the intersection points between the input maps this is also a lower bound for map overlay algorithm.

In the following we consider the analytical performance of our overlay algorithm in detail. If no vertices coincide, the maximum number of event points is reached and is equal to n+k. Therefore the operation on the event point scheduler can be performed in O (log(n+k)) = O (log n) (because $k < n^2$). Because n is the total number of edges, the operation on the sweep line status can be executed in O (log(n)) time. The vertices of the polygon can be sorted in O (n*log(n)) time. If we assume that the number of edges attached to one event point is limited by a constant, the overlay can be performed in:

$$t(n,k) = O((n+k) * \log(n)) \text{ time}$$

Thus our overlay algorithm has the same time complexity as Nievergelt and Preparata's algorithm for determining the regions of a self-intersecting polygon [NP 82].

## 4.2 Experimental performance analysis

We implemented our plane-sweep overlay algorithm in Modula-2. To examine the performance experimentally, we ran tests on a SUN workstation 3/60 under UNIX. Because of missing hardware and system software for rational coordinates we used a 8 Byte floating point representation which was supported by the system. The implementation was developed to demonstrate the properties of the algorithm but it was not tuned for speed. Consequently, there is scope to speed up the overlay.

We performed the overlay between two input maps, where n is the total number edges of all polygons of both input maps. We ran four series of tests with maps consisting of (1) a regular net of areas to get a constant proportion p of k / n where k denotes the total number of intersection points, (2) areas covering the map completely which are generated by a tool, (3) tool-generated areas covering only 50 per cent of the map, and (4) real data. Test serie 1 was performed with different proportions p. In test serie 4a two maps of administrative divisions of Italy were overlaid where one was translated by an offset. In test serie 4b the state frontier of Uganda and lakes near by were overlaid. Typical input maps of the series are depicted in figure 9:

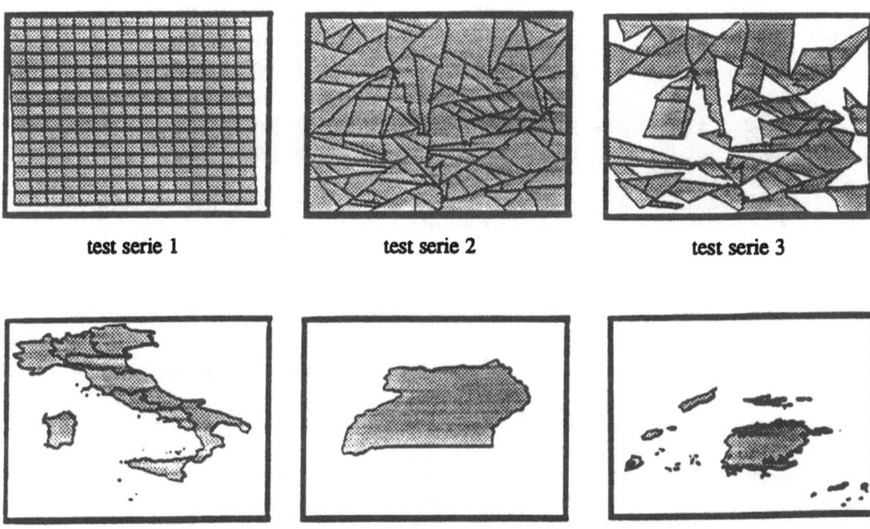

test serie 1 | test serie 2 | test serie 3

test serie 4a (Italy) | test serie 4b (Uganda: state frontier) | test serie 4b (Uganda: lakes)

**Fig. 9:** Input maps of the test series

The results of the test serie 1 are shown in table 1. t is the used CPU time in sec which is needed to perform the overlay. Additionally, we want to determine the constant c of our map overlay algorithm hidden by the O-notation (c = t / (n * ln n)).

**Table 1:** Results of the test serie 1

serie 1a (p = 0.25):

| n | t [sec] | c [msec] |
|---|---|---|
| 2048 | 25 | 1.61 |
| 4608 | 59 | 1.52 |
| 8192 | 112 | 1.51 |
| 10368 | 142 | 1.48 |
| 15488 | 227 | 1.52 |
| 21632 | 313 | 1.45 |
| 25088 | 363 | 1.43 |

serie 1b (p = 0.1):

| n | t [sec] | c [msec] |
|---|---|---|
| 2880 | 29 | 1.26 |
| 5120 | 53 | 1.22 |
| 8000 | 86 | 1.20 |
| 11520 | 130 | 1.21 |
| 15680 | 183 | 1.21 |
| 20480 | 246 | 1.21 |
| 25920 | 307 | 1.17 |

serie 1c (p = 0.033):

| n | t [sec] | c [msec] |
|---|---|---|
| 2160 | 18 | 1.11 |
| 4860 | 44 | 1.06 |
| 8640 | 83 | 1.06 |
| 11760 | 113 | 1.03 |
| 15360 | 147 | 0.99 |
| 19440 | 190 | 0.99 |
| 24000 | 246 | 1.02 |

The test series of table 1 demonstrate how the constant depends on the number of intersection points. An analysis of these tests results in the following function:

$$t(n,k) = c' * (n + 1.75 * k) * \ln(n)$$

The value of c' in the test series 1a to 1c is approximately 1.05 msec. We would like to emphazise that this constant is very small with respect to performance criteria.

In table 2 the test series 2 and 3 are depicted where c' is calculated additionally.

**Table 2:** Results of the test series 2 and 3

serie 2:

| n | p | t [sec] | c [msec] | c'[msec] |
|---|---|---|---|---|
| 13176 | 0.240 | 188 | 1.50 | 1.02 |
| 28837 | 0.154 | 372 | 1.26 | 0.97 |
| 30285 | 0.161 | 413 | 1.32 | 1.01 |

serie 3:

| n | p | t [sec] | c[msec] | c'[msec] |
|---|---|---|---|---|
| 6251 | 0.262 | 101 | 1.85 | 1.21 |
| 14179 | 0.184 | 221 | 1.63 | 1.20 |
| 15260 | 0.188 | 245 | 1.66 | 1.22 |

Series 2 and 3 demonstrate another dependency of the running time: If a large number of edges of the polygons coincide (see figure 9), the running time decreases. The reason is that the algorithm detects such edges and combines them.

Table 3 depicts the results of test serie 4 with files of real data. In serie 4a (Italy) varying administrative divisions and in serie 4b (Uganda) varying resolutions are considered.

**Table 3:** Results of the test serie 4

serie 4a (Italy):

| division | n | p | t [sec] | c [msec] | c'[msec] |
|---|---|---|---|---|---|
| state | 6666 | 0.012 | 67 | 1.14 | 1.12 |
| groups of regions | 9622 | 0.015 | 94 | 1.07 | 1.04 |
| regions | 11542 | 0.014 | 110 | 1.02 | 0.99 |
| provinces | 20378 | 0.023 | 194 | 0.96 | 0.92 |

serie 4b (Uganda, p < 0.004):

| n | t [sec] | c[msec] |
|---|---|---|
| 1852 | 16 | 1.16 |
| 8973 | 86 | 1.05 |
| 17829 | 180 | 1.03 |

The results of test serie 4 demonstrate the validity of the previous results for real data.

# 5 The use of spatial access methods for map overlay

The database system of a GIS must support efficient query processing as well as efficient manipulation and combination of maps. To fulfill these requirements we assume that the database system uses suitable *spatial access methods (SAMs)* for the management of the database. In particular, this allows to extract the relevant parts from the seamless database (maps). In the following we assume that each map is supported by its own SAM because in GISs an efficient access to a map of one topic is desirable, e.g. land utilization or soil pollution.

An often used technique to store areas with SAMs is to approximate them by *minimal bounding boxes (MBBs)*. MBBs preserve the most essential geometric properties of geometric objects, i.e. the location of the object and the extension of the object in each axis. The query processing is carried out in two (or more) steps. MBBs are used as a first filter to reduce the set of candidates. The second step examines those candidate polygons by decomposing them into simple spatial objects such as convex polygons, triangles, or trapezoids [KHHSS 91b]. To test the polygons for intersection with a sweep line or query rectangle, MBBs are a sufficient first filter.

In the following we assume that the SAM organizes the access to the objects of a database using a tree-like directory. Such access methods are adequate in handling non-uniform spatial data [KSSS 89]. The inner nodes are called *directory pages*, the leaves of the tree are *data pages*. The data and directory pages of a SAM define a partition of the data space. In our case a record in a data page consists at least (a) of a MBB, (b) of the value of the thematic attribute, and (c) of a polygon description or of a pointer to such a description depending on the size of the polygon, see [KHHSS 91a].

As mentioned in the introduction, the database and the maps in a GIS may be very large. Therefore it is not useful to keep all overlay maps in main memory, especially not in multi user systems. In systems with a virtual storage manager the efficiency could decline by a large number of page faults.

Instead of processing the maps completely, it is more efficient to partition the maps and to carry out the overlay on these partitions. One approach is to partition the map using a uniform grid like in [Fra 89]. Obviously, this is not the best way because a non-uniform data distribution is not adequately handled by this approach. We will partition the map by using SAMs and the proposed overlay algorithm.

Another important reason to partition the maps is the running time of the plane-sweep overlay which is more than linear. By partitioning we reduce the number of polygons and edges for each overlay. Thus it may be possible to carry out the total overlay faster.

## 5.1 Sweep-line partition

In the presented plane-sweep overlay only those polygons are relevant which intersect the sweep line. Thus we have a criterion for partitioning by the algorithm itself: Only polygons intersecting the sweep line or close to the sweep line, are kept in main memory. In terms of SAMs this means to read data pages from secondary storage as soon as the sweep line intersects them. We call this approach *sweep-line partition*.

### 5.1.1 Sweep-line partition and transformation

The sweep-line partition can be realized by the *transformation technique* [SK 88]. This technique transforms the coordinates of a 2-dimensional MBB into a 4-dimensional point. There are two representations of such points: (1) The *center representation* consists of the center of the rectangle ($c_x$,$c_y$) and the distance of the center to the sides of the rectangle ($e_x$,$e_y$). (2) The *corner representation* stores the lower left ($x_1$,$y_1$) and the upper right corner ($x_2$,$y_2$) of the box. The 4-dimensional points are stored by a suitable multidimensional point access method, e.g. the grid file [NHS 84], PLOP-hashing [KS 88], the BANG file [Fre 87], or the buddy tree [SK 90].

In the following we use the transformation technique with corner representation. For the SAM of each input map an own sweep line is introduced. These sweep lines are oriented at the partition of the SAM. The pages to the left of or intersecting these sweep lines have already been read from secondary storage. Performing the map overlay, we must synchronize the overlay sweep line and sweep lines on SAM level for each input map. When the overlay sweep line overtakes one of the sweep lines of the SAMs, new data pages must be read from secondary storage by the corresponding SAM and the sweep line of the SAM is adjusted accordingly. An example is depicted in figure 10.

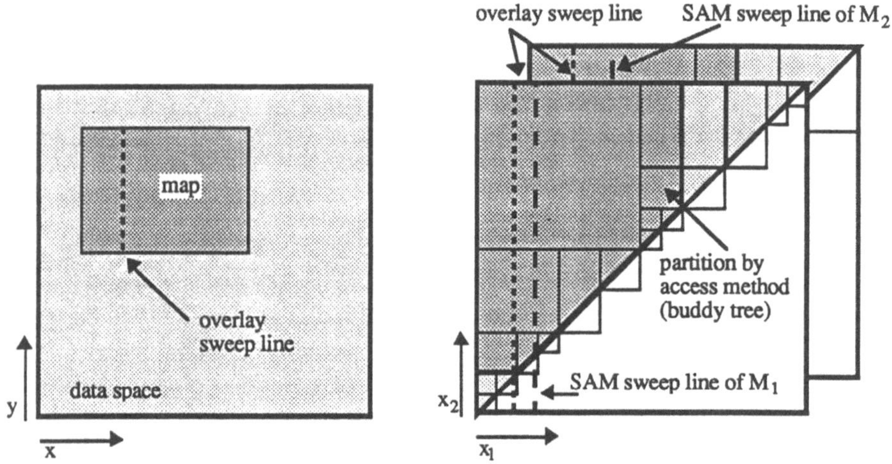


**Fig. 10:** Sweep-line partition and realization by transformation (x-dimensions are shown only)

### 5.1.2 Performance

Using the sweep line partition reduces the number of page faults considerably because only those parts of the maps intersecting the sweep line reside in main memory. Minimizing the number of page faults during the overlay improves the overall performance.

This gain of efficiency is only slightly reduced by the following effect: Without partition every required page is accessed exactly once. The pass through the tree of the SAM according to the sweep line may cause several accesses to the same directory page. However, the number of accessed directory pages is, compared to the total number of read pages, generally very small (considerable less than 1 per cent).

Compared to the complete overlay we observed that the CPU-time used for the partitioned overlay decreases about 10 per cent independent of the number of edges. The reason is the smaller average number of entries in the event point scheduler.

## 5.2 Strip Partition

Contrary to a partition oriented by the sweep line, an orthogonal partition is possible. This idea is used, for example in [Pul 90]. To support the plane sweep, it is sensible to divide the map into strips $S_i$ which extend over the whole map *(strip overlay)*. In the following we assume proceeding from the top strip $S_1$ to the bottom strip $S_M$ (see figure 11) The strip partition shortens the length of the sweep line which decreases running time. The height of the strips may vary to adapt the partitions to non-uniform data distributions.

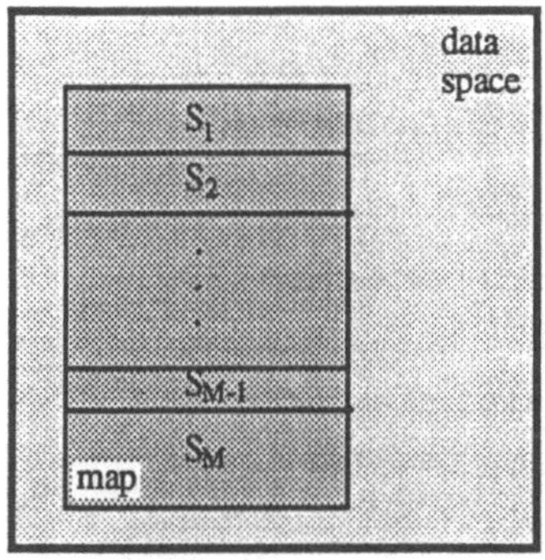


**Fig. 11:** Strip partition

Some areas of a map may intersect more than one strip. To process these areas correctly, computed regions intersecting a strip which is not processed yet must be stored temporarily in a buffer. Such regions are an additional input to the next strip overlay. Thus, every area of an input map can be assigned to exactly one strip $S_i$ and needs to be read from secondary storage only once.

As in section 5.1 we assume that all data pages are completely read. The access to the areas of one strip corresponds to standard region queries which supply all polygons intersecting the strip. There is only one exception: Data pages accessed by previous strips are not read again. We call this kind of query *modified region query*. Such queries are performed very efficiently by the R*-tree [BKSS 90], a variant of the well-known R-tree [Gut 84]. This is caused by the minimization of area, margin, and overlap of directory rectangles in the R*-tree.

### 5.2.1 Generating an optimal strip partition

In the following, we want to describe how an optimal strip partition of the map is generated. An optimal strip partition adapts the strips to the distribution of the areas of the input maps, exploits the size of main memory, and avoids page faults. The strip partition is best supported by using an efficient SAM, such as the R*-tree.

As mentioned, the areas of each input map which are simple polygons with holes, are approximated by minimal bounding boxes, which preserve the location and the extension of the areas. The number of bytes representing an area is assigned to the MBB. This is necessary because we cannot expect in GIS applications that each area is described by the same number of bytes. Each data page represents a set of areas. Thus, for each data page the number of bytes can be calculated which is necessary to store the data of this page in main memory. This information is stored in lowest level of the directory.

In a preprocessing step of the map overlay, we determine the data pages for all input maps which intersect the map. Each data page of a SAM corresponds to a partition of the data space, e.g. the regions of the R*-tree are rectangles. These regions are sorted in descending order according to the highest y-coordinate of the region. Initially, the buffer is empty which stores areas which are not performed completely by a strip overlay. According to the order mentioned above the first k regions are determined where the sum of bytes which are represented by these k regions is smaller than the size of main memory minus the size of the buffer. Thus, the first strip $S_1$ is limited by the highest y-coordinate of the (k+1)st data page. The areas which are not handled completely in the first strip overlay will be stored in the buffer. The above procedure is iterated.

To illustrate this approach we present the following example where the size of main memory is restricted to 8 Megabytes (see figure 12).

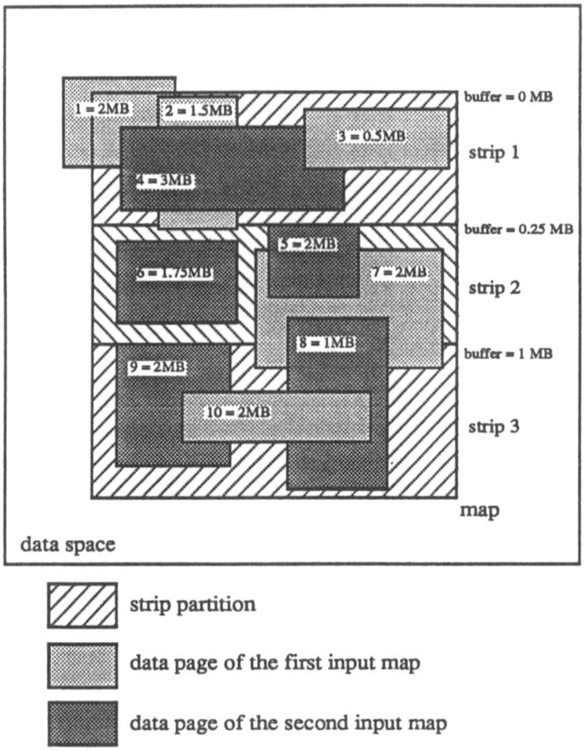


**Fig. 12:** Example of generating an optimal strip partition

The numbers 1 to 10 of the data pages of the input maps indicate the order mentioned above. The size of the data pages 1 to 4 amounts to 7 MB. With the next data page the size of main memory would be exceeded (7MB+2 MB ≥ 8MB) and page faults could occur. Therefore, the first strip ends at the highest y-coordinate of the data page 5. Let us assume that after the first strip overlay 0.25 MB are stored in the buffer. Then the second strip can be extended until 7.75 MB are not exceeded. Thus, the data pages 5 to 8 are associated to the second strip. Finally, the data pages 9 and 10 and the regions stored in the buffer are accomodated in the third strip.

Generating the optimal strip partition is not time intensive because only directory pages are read to get the necessary information, such as the size of the data pages and bytes represented by the data pages. Only if the map overlay is performed data pages with the complete description of the areas are read from secondary storage. The ratio of read directory and data pages is smaller than 1% when performing the map overlay.

# 6 Conclusions

In this paper, we presented an efficient map overlay algorithm based on spatial access methods and computational geometry concepts, in particular on the plane-sweep paradigm. The marriage of these two areas was enabled by the property that the spatial access method supports the plane-sweep paradigm. Since plane-sweep processing generates results in sorted order, the spatial access method must be robust with respect to sorted insertions for storing these results. The operation map overlay was selected because it is the most important analysis operation in a GIS, e.g. Tomlin's map analysis package [Tom 90] is completely based on map overlay. Thus the presented map overlay algorithm provides good analysis performance in GIS. Good analysis and retrieval performance are the most important factors for good user acceptance of a GIS. Thus, in our future work, we will design efficient algorithms based on spatial access methods for all retrieval operations. Performance improvements which exceed those realized in this paper by coupling spatial access methods and computational geometry, are feasible by using processors suitable for rational numbers and by implementing parallel GIS algorithms on parallel multi-processor systems. These issues are important goals in future work.

## Acknowledgement

We thankfully acknowledge receiving real data representing national administrative divisions of the European countries by the Statistical Office of the European Communities. The data of test serie 'Uganda' is taken from the World Data Bank II. Additionally, we would like to thank Holger Horn for making his map generator available to us.

## References

[BKSS 90] Beckmann, N., Kriegel, H.-P., Schneider, R., Seeger, B.: The R*-tree: An Efficient and Robust Access Method for Points and Rectangles. Proc. ACM SIGMOD Int. Conf. on Management of Data, 322-331, 1990

[BO 79] Bentley, J.L., Ottmann, T.A.: Algorithms for Reporting and Counting Geometric Intersections. IEEE Trans. on Computers, Vol. C-28, No. 9, 643-647, 1979

[Bri 90] Brinkhoff, T.: Map Overlay of Thematic Maps Supported by Spatial Access Methods. Master Thesis (in German), University of Bremen, 1990

[Bur 86] Burrough, P.A.: Principles of Geographical Information Systems for Land Resources Assessment. Oxford University Press, 1986

[Fra 83] Franklin, W.R.: A Simplified Map Overlay Algorithm. Proc. Harvard Computer Graphics Conf., 1983

[Fra 84] Franklin, W.R.: Cartographic Errors Symtomatic of Underlying Algebra Problems. Proc. Int. Symp. on Spatial Data Handling, Vol. I, 190-208, 1984

[Fra 87] A.U. Frank: Overlay Processing in Spatial Information Systems. Proc. 8th Int. Symp. on Computer-Assisted Cartography (Auto-Carto 8), 16-31, 1987

[Fra 89] Franklin, W.R. et al.: Uniform Grids: A Technique for Intersection Detection on Serial and Parallel Machines. Proc. 9th Int. Symp. on Computer-Assisted Cartography (Auto-Carto 9), 100-109, 1989

[Fre 87] Freeston, M.: The BANG file: a new kind of grid file. Proc. ACM SIGMOD Int. Conf. on Management of Data, 260-269, 1987

[FW 87] Franklin, W.R., Wu, P.Y.F.: A Polygon Overlay System in Prolog. Proc. 8th Int. Symp. on Computer-Assisted Cartography (Auto-Carto 8), 97-106, 1987

[Gün 89] Günther, O.: The Design of the Cell Tree: An Object-Oriented Index Structure for Geometric Databases. Proc. IEEE 5th Int. Conf. on Data Engineering, 598-605, 1989

[Gut 84] Guttman, A.: R-Trees: A Dynamic Index Structure for Spatial Searching. Proc. ACM SIGMOD Int. Conf. on Management of Data, 47-57, 1984

[KHHSS 91a] Kriegel, H.-P., Heep, P., Heep, S., Schiwietz, M., Schneider, R.: A Flexible and Extensible Index Manager for Spatial Database Systems. Proc. 2nd Int. Conf. on Database and Expert Systems Applications (DEXA '91), 1991

[KHHSS 91b] Kriegel, H.-P., Heep, P., Heep, S., Schiwietz, M., Schneider, R.: An Access Method Based Query Processor for Spatial Database Systems. Proc. Int. Workshop on DBMS's for geographical applications, Capri, May 16-17, 1991

[KS 88] Kriegel, H.-P., Seeger, B.: PLOP-Hashing: A Grid File without Directory. Proc. 4th Int. Conf. on Data Engineering, 369-376, 1988

[KSSS 89] Kriegel, H.P., Schiwietz, M., Schneider, R., Seeger, B.: Performance Comparison of Point and Spatial Access Methods. Proc. 1st Symp. on the Design of Large Spatial Databases, 1989. In: Lecture Notes in Computer Science 409, Springer, 89-114, 1990

[MS 87] Mairson, H.G., Stolfi, J.: Reporting and Counting Intersections Between Two Sets of Line Segments. Proc. NATO Adv. Study Inst. on Theoretical Foundations of Computer Graphics and CAD, 307-325, 1987

[NHS 84] Nievergelt, J., Hinterberger, H., Sevcik, K.C.: The Grid File: An Adaptable, Symmetric Multikey File Structure. ACM Trans. on Database Systems, Vol. 9, No. 1, 38-71, 1984

[NP 82] Nievergelt, J., Preparata, F.P.: Plane-Sweep Algorithms for Intersecting Geometric Figures. Comm. of the ACM, Vol. 25, No. 10, 739-747, 1982

[Oos 90] Oosterom, P.J.M.: Reactive Data Structures for Geographic Information Systems. PhD-thesis, Department of Computer Science at Leiden University, 1990

[PS 88] Preparata, F.P., Shamos, M.I.: Computational Geometry. Springer, 1988

[Pul 90] Pullar, D.: Comparative Study of Algorithms for Reporting Geometrical Intersections. Proc. 4th Int. Symp. on Spatial Data Handling, 1990

[SK 88] Seeger, B., Kriegel, H.-P.: Techniques for Design and Implementation of Efficient Spatial Access Methods. Proc. 14th Int. Conf. on Very Large Data Bases, 360-371, 1988

[SK 90] Seeger, B., Kriegel, H.-P.: The Buddy-Tree: An Efficient and Robust Access Method for Spatial Database Systems. Proc. 16th Int. Conf. on Very Large Data Bases, 590-601, 1990

[Til 80] Tilove, R.B.: Set Membership Classification: A Unified Approach To Geometric Intersection Problems. IEEE Trans. on Computers, Vol. C-29, No. 10, 874-883, 1980

[Tom 90] Tomlin, C.D.: Geographic Information Systems and Cartographic Modeling. Prentice-Hall, 1990

[Whi 78] White, D.: A Design for Polygon Overlay. Harvard Papers on Geographic Information Systems, Vol. 6, 1978


# Appendix: Formal description of the plane-sweep overlay

A(e) (B(e)) are the A(B)-lists of e in the following. * denotes the concatenation of lists.

```
PROCEDURE PlaneSweepOverlay;

  VAR x    : EventPointScheduler (Point,LeftEdges,RightEdges);
      y    : SweepLineStatus (Edge, A-list, B-list, RegionID);
      P    : Actual event point;
      s    : Left edges of P; (* sorted as in section 3.2 *)
      t    : Right edges of P; (* sorted as in section 3.2 *)
      n,m  : Number of entries of s and t respectively;
      h,l  : neighbor edges of s[1] and s[n] in y;

  PROCEDURE TestForIntersection (edge1,edge2);
  BEGIN
     IF (edge1 and edge2 intersect) AND (intersection point right of P) THEN
        Insert intersection point of edge1 and edge2 into x;
     END;
  END TestForIntersection;

  PROCEDURE TestAndReportLists (A-list,B-list,P);
  BEGIN
     IF A-list points to the tail of B-list THEN
        Report polygon B-list * P;
     ELSE
        Concatenate: A-list * P * B-list;
     END;
  END TestAndReportLists;
```

```
BEGIN (* PlaneSweepOverlay *)
    (* Initialization *)
    x := Polygon vertices sorted by x-ordinates with attaching edges of the polygon;
    y := Ø;
    (* Pass *)
    WHILE x ≠ Ø DO
        (P,s,t) := Combined first event point of x;
        Delete (p,s,t) from x;
        n := Number of edges of s;
        m := number of edges of t;
        (* Process left edges *)
        IF n = 0 THEN (* case 1 *)
            A(t[1]) = B(t[m]) := P;
        ELSIF n = 1 THEN (* case 2 *)
            A(t[1]) := A(s[1]) * P;
            B(t[m]) := P * B(s[1]);
        ELSE (* case 3 *)
            IF m = 0 THEN (* case 3a *)
                TestAndReportLists (A(s[1]),B(s[n]),P); (* hole in region *)
            ELSE (* case 3b *)
                A(t[1]) := A(s[1]) * P;
                B(t[m]) := P * B(s[n]);
            END;
            FOR i := 1 TO n-1 DO
                TestAndReportLists (A(s[i+1]),B(s[i]),P); (* region *)
                Delete s[i] from y;
            END;
            Delete s[n] from y;
        END;
        (* Process right edges *)
        IF m = 0 THEN (* case 1 *)
            h := former upper neighbor edge of s[1] from y;
            l := former lower neighbor edge of s[n] from y;
            Adjust IDs of regions between h and l in y;
            TestForIntersection (h,l);
        ELSIF m = 1 THEN (* case 2 *)
            Insert t[1] into y;
            TestForIntersection (t[1], upper neighbor edge of t[1] from y);
            TestForIntersection (t[1], lower neighbor edge of t[1] from y);
        ELSE (* case 3*)
            FOR i := 1 TO m DO
                Insert t[i] into y;
            END;
            FOR i := 1 TO m-1 DO
                Set new region ID between t[i] and t[i+1] in y;
                B(t[i]) = A(t[i+1]) := P;
            END;
            TestFor'ntersection (t[1], upper neighbor edge of t[1] from y);
            TestForIntersection (t[m], lower neighbor edge of t[m] from y);
        END;
    END PlaneSweepOverlay;
```

# How to Split Buckets in Spatial Data Structures*

Andreas Henrich
Hans-Werner Six

We investigate several split strategies for data buckets in spatial data structures. Based on a theoretical analysis and experimental results, we derive their advantages and disadvantages. Furthermore, we illustrate how crucial the choice of the split strategy is for the performance of spatial queries. Most results are independent of the underlying data structure and hence provide deeper insight into parameters determining the performance of spatial data structures.

## 1 Introduction

In non-standard applications such as cartography, CAD and robotics, Database Management Systems have to organize large sets of multidimensional geometric objects on secondary storage such that these objects can quickly be retrieved according to their spatial locations. Typical spatial queries are the retrieval of an object by its coordinates (exact match) and range queries where all objects geometrically intersecting the query region are selected for further processing or presentation on a screen. Since the set of objects varies over time, insertions and deletions have to be performed as well.

In recent years, various efficient data structures for large sets of multidimensional geometric objects have been developed. Most of these structures have been designed for multidimensional points (see e.g. [4, 13–14, 18–20, 22–26, 28]). In typical applications, however, objects are arbitrary geometric, i.e. non-point objects. In many situations, it has been proven to be useful to represent non-point objects by their (minimal) bounding boxes, serving as simple geometric keys. Hence, non-point data structures deal with multidimensional intervals (see e.g. [1, 7, 9, 15, 28–29]). Only the cell tree [6] does not use this approximation.

Data structures which efficiently support spatial access to geometric objects, cluster geometric objects in data buckets according to their spatial locations. With each data bucket a subspace of the data space, the so-called bucket region, is associated containing all objects of the corresponding bucket. Except for the cell tree, bucket regions are multidimensional intervals, i.e. boxes with sides parallel to the axes (and to the bounding boxes of the non-point objects).

Since the performance of exact match and small range queries deteriorates when the overlapping of bucket regions increases, the overlapping of bucket regions should be kept small. Hence, for point objects the bucket regions are pairwise disjoint. In the more complex situation of non-point objects most data structures create overlapping

* This work has also been supported by DFG grant Si 374/1-5.

data buckets in order to ensure that any object is completely contained in one bucket region, such that no object clipping on bucket region borders occurs. However, keeping the overlapping small is important for the performance of these data structures. Again, the cell tree using a partition of the data space is the only exception in this respect.

The second aspect which is crucial for the efficiency of spatial data structures is the utilization of the data buckets. A low bucket utilization causes an unsufficient performance of (at least) larger range queries, where several to many buckets must be accessed.

For a given object distribution both, the overlapping of bucket regions and the bucket utilization, are determined by the split strategy used by the data structure. In other words, the pattern of bucket regions created by the split strategy (and the object distribution) is crucial for the efficiency of spatial queries.

A split strategy has to be applied when a data bucket overflows, i.e. an additional object to be inserted cannot be kept by a bucket. Then all objects of the overflowing bucket including the new object were distributed over two buckets thus creating space for the insertion of further objects.

Main issue of any split strategy is the question of how to distribute the objects stored in one bucket over two buckets, i.e. of how to make two bucket regions out of one.

In this paper we investigate several split strategies analytically and with the help of experimental results and work out their advantages and disadvantages. Most results are independent of the underlying data structure and hence provide deeper insight into parameters determining the performance of spatial data structures.

The paper is organized as follows. Section 2 deals with split strategies for point objects. A theoretical analysis as well as an experimental evaluation based on the LSD tree [9] is provided. In section 3, split strategies for the non-point situation are discussed and experimental results are presented. Section 4 concludes the paper.

## 2 Split strategies for point data structures

Point data structures create bucket regions which are multidimensional intervals and pairwise disjoint. Figure 2.1 shows a possible partition of the data space into nine bucket regions. An efficient spatial data structure establishes a bucket region pattern with high bucket utilization. This is achieved by the split strategy used by the data structure.

The split of an overflowing bucket is performed by introducing a *split line* which divides the corresponding bucket region into two, and distributes the affected objects over two data buckets accordingly. A split line is determined by its *split dimension* and *split position*.

As far as the split dimension is concerned, it has been proven useful to choose the data space dimensions alternately in a cyclic manner. This approach creates bucket regions which tend to be "quadratic" and hence ensure a reasonable (worst case) behaviour for all kinds of range queries.

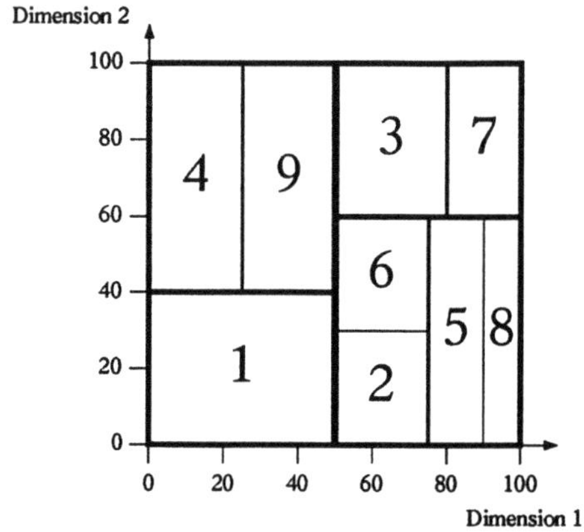


Figure 2.1: Possible partition of the data space into bucket regions

The problems encountered when the split position is to be chosen are more difficult and will now be discussed in more detail. To this end, we start with some definitions being useful in the course of the discussion.

1. For a data bucket $w$ with bucket region $D(w)$, $w_{\lambda_i}$, respectively $w_{v_i}$, denotes the lower, respectively upper, boundary in dimension $i$ ($1 \leq i \leq d$). Hence
$$D(w) = \left(w_{\lambda_1}; w_{v_1}\right] \times \ldots \times \left(w_{\lambda_d}; w_{v_d}\right] \tag{1}$$
2. Let $\left(L_1; U_1\right] \times \ldots \times \left(L_d; U_d\right]$ denote the whole data space. Note that we use half-open intervals so that we need not consider special cases, e.g. when the lower boundary of a bucket region in one dimension coincides with the lower bound of the whole data space.
3. Let $O$ be the set of all $n$ objects to be stored.
4. Let $O_{D(w)}$ be the set of all objects of $O$ contained in $D(w)$
$$O_{D(w)} = \{o \,|\, (o \in O) \wedge (o \in D(w))\} \tag{2}$$

For a perfect split strategy, the split position $s_{Pos}$ must be chosen such that the number of objects in $O$ located in one of the new bucket regions equals the number of objects in $O$ located in the other bucket region. Note that the split position $s_{Pos}$ uniformly divides the objects in $O_{D(w)}$ and not in $D(w)$.

To be more precise, we consider the coordinate value of an arbitrary object $o$ of $O$ in dimension $i$ ($1 \leq i \leq d$) as the realisation of a random variable $X_i$ with probability density function $f_{X_i}$. Let us first assume that the random variables $X_i$ are independent of each other.

For a given split dimension $s_{Dim}$ a split position $s_{Pos}$ should divide the intervall $(w_{\lambda_{s_{Dim}}}; w_{v_{s_{Dim}}}]$ of the bucket region to be split such that the area between $w_{\lambda_{s_{Dim}}}$ and $s_{Pos}$ under the graph of $f_{X_{s_{Dim}}}$ equals the area between $s_{Pos}$ and $w_{v_{s_{Dim}}}$.

This principle is illustrated in figure 2.2.

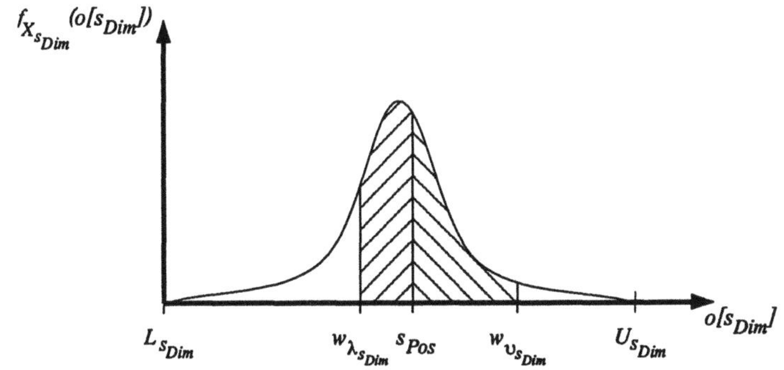


Figure 2.2: Division of a bucket region into two subregions of equal probability

In other words, the probability of an object $o$ of $O_{D(w)}$ to be located to the left of the split line equals the probability to be located to the right.

There are various different split strategies trying to determine a split position fulfilling the condition mentioned above. These split strategies can basically be classified as either *distribution dependent* or *data dependent*.

## 2.1 Distribution dependent split strategies

A distribution dependent split strategy determines the split position independently of the objects that are actually stored in the overflowing bucket. The split position is determined on the basis of an assumption about the distribution of the objects of $O$. More precisely, a certain density function $f_{X_{s_{Dim}}}$ is assumed and $s_{Pos}$ is chosen such that the probabilities of a random object of $O_{D(w)}$ being located on either side of the split line are equal. In figure 2.2, for example, the areas of the two hatched surfaces must be of equal size. To this end, $s_{Pos}$ must be chosen such that the following equation holds:

$$\int_{w_{\lambda_{s_{Dim}}}}^{s_{Pos}} f_{X_{s_{Dim}}}(x)dx = \int_{s_{Pos}}^{w_{\upsilon_{s_{Dim}}}} f_{X_{s_{Dim}}}(x)dx = \frac{1}{2} \int_{w_{\lambda_{s_{Dim}}}}^{w_{\upsilon_{s_{Dim}}}} f_{X_{s_{Dim}}}(x)dx \tag{3}$$

The above equation can be transformed into

$$\int_{-\infty}^{s_{Pos}} f_{X_{s_{Dim}}}(x)dx = \frac{1}{2}\left( \int_{-\infty}^{w_{\lambda_{s_{Dim}}}} f_{X_{s_{Dim}}}(x)dx + \int_{-\infty}^{w_{\upsilon_{s_{Dim}}}} f_{X_{s_{Dim}}}(x)dx \right) \tag{4}$$

For simple density functions, one can derive a formula for computing the split position $s_{Pos}$ directly from equation (4). We will briefly demonstrate this for the case of a uniform distribution. For a uniform distribution of objects in the data space, the density function $f_{X_i}$ for the $i$-th dimension ($1 \le i \le d$) is given by

$$f_{X_i}(x) = \begin{cases} 0 & x \le L_i \\ \frac{1}{U_i - L_i} & L_i < x \le U_i \\ 0 & x > U_i \end{cases} \tag{5}$$

Substituting $f_{X_i}(x)$ in equation (4) we get

$$\int_{L_{s_{Dim}}}^{s_{Pos}} \frac{1}{U_{s_{Dim}}-L_{s_{Dim}}}dx = \frac{1}{2}\left(\int_{L_{s_{Dim}}}^{w_{\lambda_{s_{Dim}}}} \frac{1}{U_{s_{Dim}}-L_{s_{Dim}}}dx + \int_{L_{s_{Dim}}}^{w_{v_{s_{Dim}}}} \frac{1}{U_{s_{Dim}}-L_{s_{Dim}}}dx\right)$$

$$\Leftrightarrow \quad \frac{s_{Pos} - L_{s_{Dim}}}{U_{s_{Dim}} - L_{s_{Dim}}} = \frac{1}{2}\left(\frac{w_{\lambda_{s_{Dim}}} - L_{s_{Dim}}}{U_{s_{Dim}} - L_{s_{Dim}}} + \frac{w_{v_{s_{Dim}}} - L_{s_{Dim}}}{U_{s_{Dim}} - L_{s_{Dim}}}\right)$$

$$\Leftrightarrow \quad s_{Pos} = \frac{w_{\lambda_{s_{Dim}}} + w_{v_{s_{Dim}}}}{2} \tag{6}$$

Figure 2.3 illustrates the effect of such a split strategy.

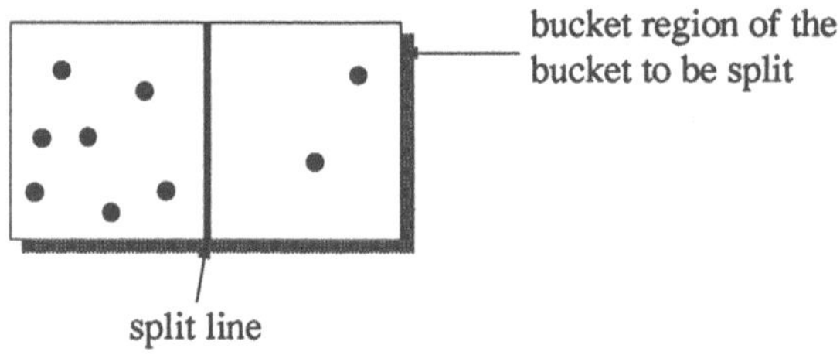


**Figure 2.3: Example of a distribution dependent split strategy under the assumption of a uniform distribution of objects**

Under the assumption of a uniform distribution, finding a distribution dependent split strategy is obviously very easy. However, as soon as this assumption has to be dropped a variety of problems is encountered. On the one hand, there are probability density functions, e.g. the beta distribution (see [16]), that cannot be transformed as demonstrated above. On the other hand, in some cases it may be necessary to work with empirically derived probability densities that are available only in tabular form. We will therefore describe an algorithm for determining $s_{Pos}$ that can be used for any known density function $f_{X_i}$.

For all $i$ ($1 \leq i \leq d$) the quantiles of the density function $f_{X_i}$ are tabulated in a two-dimensional array $V$ with $d$ columns. For each interpolation node $\zeta$ (= row) and each dimension (= column) the value $\int_{-\infty}^{\zeta} f_{X_i}(x)dx$ is stored in $V$. How many interpolation nodes are needed is mainly determined by the shape of the density functions $f_{X_i}$.

In order to determine for a given bucket $w$ the split position $s_{Pos}$ , we start computing the value

$$F = \frac{1}{2}\left(\int_{-\infty}^{w_{\lambda_{s_{Dim}}}} f_{X_{s_{Dim}}}(x)dx + \int_{-\infty}^{w_{v_{s_{Dim}}}} f_{X_{s_{Dim}}}(x)dx\right) \tag{7}$$

with the help of array $V$. If $w_{\lambda_{s_{Dim}}}$ or $w_{v_{s_{Dim}}}$ is not an interpolation node, linear approximation can be used to this end.

Then a binary search is started on $V$ for the interpolation node $\zeta_j$, for which $V[\zeta_j] \leq F < V[\zeta_{j+1}]$ holds, and the split position $s_{Pos}$ is determined by linear approximation as follows:

$$s_{Pos} = \zeta_j + \frac{\zeta_{j+1} - \zeta_j}{V[\zeta_{j+1}] - V[\zeta_j]} \cdot (F - V[\zeta_j]) \tag{8}$$

It is now obvious that computing $s_{Pos}$ can be performed with tolerable effort for arbitrary known density functions. On the other hand, to state a profound hypothesis for the density function may often be impossible or too expensive. Therefore, a uniform distribution will often be assumed, even if one knows, that this assumption is not true. We will show in section 2.3 that this does not cause significant performance problems unless the distribution is extremely uneven.

Up to now, we have discussed the situation where the random variables $X_i$, corresponding to the coordinates of an Object $o$ are independent of each other. Now we will give up this assumption and turn to the case where dependencies between the coordinates come in. Besides various realistic applications the dependency of coordinates occurs in the non-point object situation when the so-called transformation technique is used (see section 3).

In the dependency case the coordinates of an object $o \in O$ can be viewed as realisation of a $d$-dimensional random variable$(X_1, \ldots, X_d)$ with density function $f_{X_1,\ldots,X_d}$. Then for a given bucket $(w_{\lambda_1}; w_{v_1}] \times \ldots \times (w_{\lambda_d}; w_{v_d}]$ and a given split dimension $s_{Dim}$ a split position $s_{Pos}$ has to be chosen such that the following equation holds:

$$\int\limits_{w_{\lambda_1}}^{w_{v_1}} \cdots \int\limits_{w_{\lambda_{s_{Dim}-1}}}^{w_{v_{s_{Dim}-1}}} \int\limits_{-\infty}^{s_{Pos}} \int\limits_{w_{\lambda_{s_{Dim}+1}}}^{w_{v_{s_{Dim}+1}}} \cdots \int\limits_{w_{\lambda_d}}^{w_{v_d}} f_{X_1,\ldots,X_d}(x_1,\ldots,x_d)dx_1\ldots dx_d$$

$$= \frac{1}{2}\int\limits_{w_{\lambda_1}}^{w_{v_1}} \cdots \int\limits_{w_{\lambda_{s_{Dim}-1}}}^{w_{v_{s_{Dim}-1}}} \int\limits_{-\infty}^{w_{\lambda_{s_{Dim}}}} \int\limits_{w_{\lambda_{s_{Dim}+1}}}^{w_{v_{s_{Dim}+1}}} \cdots \int\limits_{w_{\lambda_d}}^{w_{v_d}} f_{X_1,\ldots,X_d}(x_1,\ldots,x_d)dx_1\ldots dx_d + \tag{9}$$

$$\frac{1}{2}\int\limits_{w_{\lambda_1}}^{w_{v_1}} \cdots \int\limits_{w_{\lambda_{s_{Dim}-1}}}^{w_{v_{s_{Dim}-1}}} \int\limits_{-\infty}^{w_{v_{s_{Dim}}}} \int\limits_{w_{\lambda_{s_{Dim}+1}}}^{w_{v_{s_{Dim}+1}}} \cdots \int\limits_{w_{\lambda_d}}^{w_{v_d}} f_{X_1,\ldots,X_d}(x_1,\ldots,x_d)dx_1\ldots dx_d$$

Note that equation 9 corresponds to equation 4. Therefore, one can again use the general method described above to determine split position $s_{Pos}$. However, the array $V$ for the tabulation of the density function increases drastically, because the density function depends no longer only on one variable but on $d$ variables.

To summarize, distribution dependent split strategies do not cause algorithmical problems, even if there are dependencies between the coordinates. Main problem of distribution dependent split strategies is to find a good hypothesis for the density functions. This problem becomes even more complex, if there are dependencies between the coordinates. Hence, in applications where such dependencies occur, a careful examination of these dependencies has to be done in order to find a satisfying distribution dependent split strategy. As mentioned above, a typical example of such an application comes in storing $d$-dimensional intervals as $2d$-dimensional points using the so called transformation technique. This problem will be discussed in section 3.

## 2.2 Data dependent split strategies

As we have seen, for a distribution dependent split strategy the most important problem is to find a profound distribution hypothesis. If nothing is known in advance about the object distribution, it seems natural to determine the split position on the basis of the objects actually stored in the bucket, because these objects can be viewed as a sample of the parent population $O_{D(w)}$. Then an estimator for the split position $s_{Pos}$ should be used, which determines $s_{Pos}$ on the basis of this $b+1$-element sample (here $b$ denotes the *bucket capacity*).

We call split strategies of this type data dependent split strategies. Hence, looking for a good data dependent split strategy means looking for a good estimator for the median of all objects in $O_{D(w)}$.

Naturally, the sample median can be used as an estimator for the median of the parent population $O_{D(w)}$. Figure 2.4 shows the effect of such a split strategy.

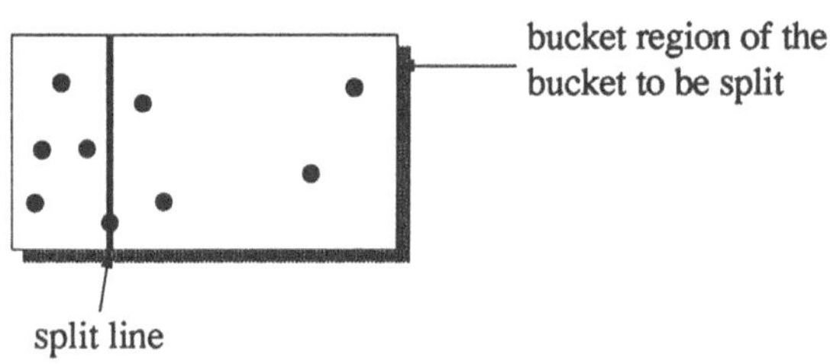


Figure 2.4: Example for the use of the sample median as split position

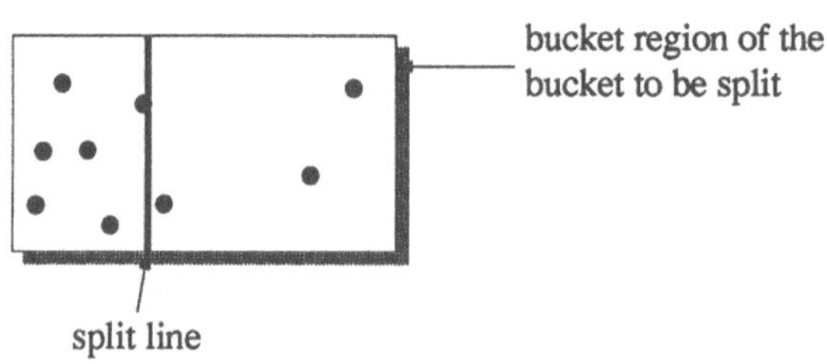


Figure 2.5: Example for the use of the sample mean as split position

Alternatively, one can use the mean over the coordinates in the split dimension of all objects stored in the bucket to be split (see figure 2.5). One advantage of the sample mean over the sample median is, that it can be computed more easily. On the other hand, a disadvantage of the sample mean is that usually an equal distribution of the objects actually stored in the bucket over the two new buckets is not achieved. This becomes obvious from figures 2.4 and 2.5. Whereas the use of the sample median leads to the best possible distribution of 5:4 objects, the sample mean leads to a distribution of 6:3. Hence, at first glance, the sample median seems to be better suited than the sample mean to estimate the median of the parent population. In the following, we will show that this is not necessarily true.

If we want to judge the quality of a data dependent split strategy, we can use the quality measures for estimators known from statistics. According to [30] these are first of all

1. unbiasedness,
2. consistency, and
3. efficency.

An estimator $\hat{\Theta}$ for the value $\Theta$ to be estimated is *unbiased* if the expected value of the estimator equals the value to be estimated, i.e. if $E(\hat{\Theta}) = \Theta$. *Consistency* requires only that $P(|\hat{\Theta}_m - \Theta| > \epsilon) \stackrel{m\to\infty}{\longrightarrow} 0$, were *m* is the sample size. The *efficiency* of an estimator corresponds to its variance. The smaller the variance of the estimator, the more efficent it is.

In the following, we will study the sample mean and the sample median according to their suitability as estimators for the median of the parent population $O_{D(w)}$. Because of space limitations, we refer to literature as far as statistical foundations are concerned. A good overview is, for example, given in [17], where further references can be found.

If we assume that the objects are inserted into the data structure in random order, the objects stored in the bucket to be split are a simple random sample (see [8, p.274]).

Since the mean over the coordinates of all objects in $O_{D(w)}$ with respect to $s_{Dim}$ in general does not coincide with the corresponding median, we directly get, that the sample mean is neither an unbiased nor a consistent estimator for the median of the parent population. Nevertheless, for $s_{Dim}$, the mean over the coordinates of the $b+1$-element sample is an unbiased estimator for the mean over the coordinates of all objects in $O_{D(w)}$.

On the other hand, the sample median is a consistent but not unbiased estimator for the median of the parent population.

We will now turn our attention to efficiency. To this end, we will first determine a confidence interval for the median of the parent population on the basis of the sample median, or to be precise, on the basis of an ordered sample. Thereafter, we will determine a confidence interval for the mean of the parent population on the basis of the sample mean. The smaller the confidence interval, the more efficient the estimator.

If we denote the coordinates of the $b+1$ objects involved in the split in split dimension $s_{Dim}$ by $x_i$ $(1 \le i \le b+1)$, we can sort the objects according to these coordinates:

$$x_{(1)} \leq x_{(2)} \leq \ldots \leq x_{(r)} \leq \ldots \leq x_{(s)} \leq \ldots \leq x_{(b+1)} \tag{10}$$

In the following $u_{1-\frac{\alpha}{2}}$ denotes the $1-\frac{\alpha}{2}$ quantile of the N(0,1)-distribution. If we choose the indices

$$r = \left\lfloor \frac{1}{2}\left(b+1-u_{1-\frac{\alpha}{2}}\sqrt{b+1}\right) \right\rfloor \tag{11}$$

and

$$s = \left\lceil \frac{1}{2}\left(b+1+u_{1-\frac{\alpha}{2}}\sqrt{b+1}\right) \right\rceil, \tag{12}$$

the probability for the median of the parent population to fall out of the confidence interval$[x_{(r)}; x_{(s)}]$ is at most $\alpha$ (see [8, p.236]).

Figure 2.6 illustrates the width of the confidence interval for $\alpha = 0.05$ and multiple values of $b$. The lower bound of the hatched area is given by $r$ and the upper bound by $s$, where the Gaussian brackets are not taken into consideration. Note that the hatched area is determined by the indices $r$ and $s$ instead of the values $x_{(r)}$ and $x_{(s)}$. Hence, the width of the interval $[r; s]$ illustrated in figure 2.6 is smaller than the width of the confidence interval $[x_{(r)}; x_{(s)}]$.

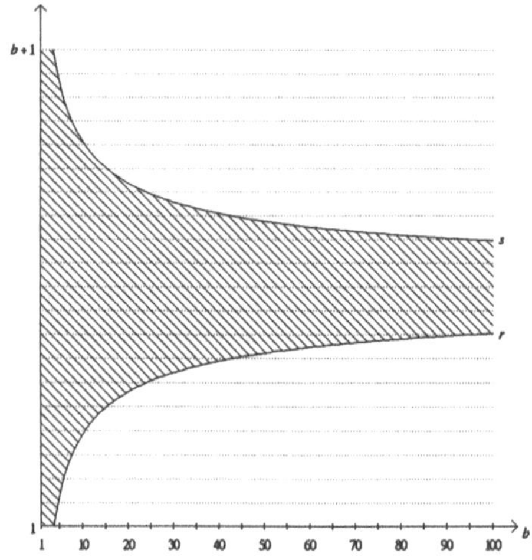


Figure 2.6: Illustration of the confidence interval of the median at a confidence level of 95 %

Let for dimension $s_{Dim}$ $\bar{x}$ denote the mean over the coordinates of all objects in the $b+1$-element sample and $\bar{o}[s_{Dim}]$ the mean over the coordinates of all objects in $O_{D(w)}$. Then the $1-\alpha$ confidence interval for $\bar{o}[s_{Dim}]$ is given by

$$\left[\bar{x} - \frac{t_{b;1-\frac{\alpha}{2}} \cdot \sigma_{o[s_{Dim}]}}{\sqrt{b+1}} \cdot \sqrt{1-\tfrac{b+1}{n}}\ ;\ \bar{x} + \frac{t_{b;1-\frac{\alpha}{2}} \cdot \sigma_{o[s_{Dim}]}}{\sqrt{b+1}} \cdot \sqrt{1-\tfrac{b+1}{n}}\right] \tag{13}$$

where $\sigma_{o[s_{Dim}]}$ denotes the standard deviation of the coordinates of all objects in $O_{D(w)}$ in $s_{Dim}$ and $t_{b;1-\frac{\alpha}{2}}$ denotes the $1-\frac{\alpha}{2}$ quantile of the Student t-distribution with $b$ degrees of freedom (see [8, p.275]).

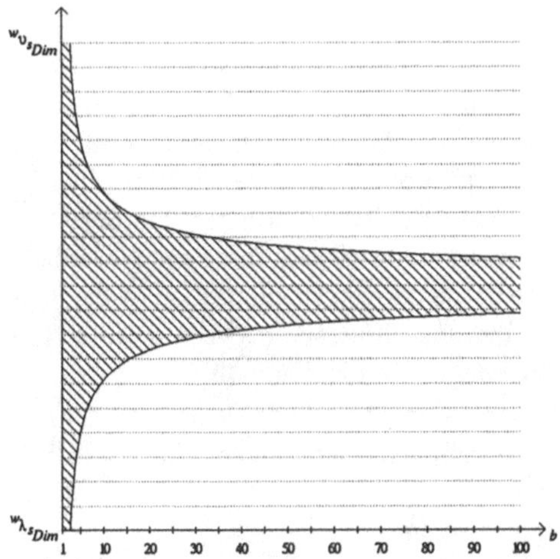


Figure 2.7: 95 % confidence interval of the mean $\bar{o}[s_{Dim}]$ with $\bar{x} = \frac{1}{2}(w_{\lambda_{sDim}} + w_{v_{sDim}})$

The factor $\sqrt{1 - \frac{b+1}{n}}$ in (13) is called finite population correction. An upper bound for this correction is 1. Hence, interval (14) always contains the interval given in (13).

$$\left[\bar{x} - \frac{t_{b;1-\frac{\alpha}{2}} \cdot \sigma_{o[s_{Dim}]}}{\sqrt{b+1}} \; ; \; \bar{x} + \frac{t_{b;1-\frac{\alpha}{2}} \cdot \sigma_{o[s_{Dim}]}}{\sqrt{b+1}}\right] \tag{14}$$

For the sample mean $\bar{x} = \frac{1}{2}(w_{\lambda_{sDim}} + w_{v_{sDim}})$, for example, at a confidence level of 95 % for $\bar{o}[s_{Dim}]$ we get the confidence interval shown in figure 2.7 with respect to the bucket capacity $b$.

Note that in figure 2.7 we use the standard deviation of the uniform distribution for the standard deviation $\sigma_{o[s_{Dim}]}$, because it is an upper bound for the standard deviation of all continuous unimodal distributions on a given interval.

Figure 2.7 can be interpreted as follows: For the sample mean $\bar{x} = \frac{1}{2}(w_{\lambda_{sDim}} + w_{v_{sDim}})$ the probability for $\bar{o}[s_{Dim}]$ to fall outside the hatched area is at most 5 %.

A comparison of figure 2.6 and 2.7 shows, that for the same sample size, the sample mean is a more efficient estimator for the mean of the parent population than the sample median for the median of the parent population. This becomes also obvious from the fact, that the variance of an estimator based on the sample mean is $\frac{\sigma^2}{m}$, whereas the variance of an estimator based on the sample median is $\frac{\pi\sigma^2}{2m}$ ($m$ denotes the sample size). Hence, the variance of the sample median is about 57 % higher than the variance of the sample mean (see [30, p.173]).

To summarize, we have seen several effects:

1. The quality of a data dependent split strategy depends on the bucket capacity $b$.
2. In contrast to the sample mean, the sample median is a consistent estimator for the median of the parent population.
3. The sample mean is a more efficent estimator for the mean of the parent population than the sample median for the median of the parent population.

In order to relate the last two effects, we should be aware of the fact that the mean of the parent population coincides with the median of the parent population for continuous symmetric distribution density functions. However, this does not imply that a split strategy based on the sample mean is unbiased, even if the objects in $O$ follow such a distribution, because the population under concern is not $O$ but the subset $O_{D(w)}$. Only if the objects in $O$ are uniformly distributed, the objects in $O_{D(w)}$ always have a continuous symmetric distribution density function. Nevertheless, the mean over all objects in $O_{D(w)}$ is close to the corresponding median, if the distribution of the objects is not too skew. (Note, that the difference between mean and median is usually used as a measure of the skewness of a distribution.)

The empirical results in section 2.3 will show that even for rather skew distributions, the smaller variance of the sample mean has more influence than its missing consistency. Hence, in most cases the sample mean seems to be better suited to determine a split position than the sample median.

An important point deserves further attention. The considerations about data dependent split strategies are based on the assumption that the objects actually stored in the overflowing bucket are a random sample of all objects in $O_{D(w)}$. In several applications this assumption may be violated, because the objects are somehow sorted. In this case, the use of a data dependent split strategy – based on the sample mean or the sample median – may lead to very bad results. We will discuss this point in more detail in section 2.3.

Hence, using the sample mean as well as the sample median, there are two main parameters crucial for the quality of a data dependent split strategy:

1. *bucket capacity* and
2. *order of insertion*.

At the end of this section, we summarize our considerations about data dependent and distribution dependent split strategies:

1. The greatest advantage of distribution dependent split strategies compared to data dependent split strategies is, that the resulting partition of the data space is independent of the order in which the objects are inserted. Hence, a degeneration caused by an unfavourable order of insertion cannot occur.
2. The drawback of distribution dependent split strategies is, that a hypothesis about the object distribution must be found in advance. Since the determination of the object distribution is unrealistic in many practical applications, a hypothesis which works acceptable for a broad spectrum of object distributions must be found. This drawback is at the same time the greatest advantage of the data dependent split strategies, which work (nearly) independently of the object distribution as far as the objects are inserted in random order.

The effect of these advantages and disadvantages in practical applications will be illustrated by empirical results in the following section.

## 2.3 Experimental results

For the validation of the theoretical analysis in the previous section we have performed a series of experiments. We used the LSD tree [9] as spatial data structure because of its flexibility allowing all kinds of split strategies.

We start with a short description of the LSD tree. For sake of simplicity, we restrict the explanations to the two-dimensional case.

Initially, the whole data space corresponds to one bucket. After a certain number of insertions the initial bucket has been filled, and the attempt to insert an additional object causes the need for a bucket split. To this purpose, a *split line* is determined and the objects on one side of the split line are stored in one bucket, while those on the other side are stored in another bucket. After some further insertions, the capacity of another bucket will be exceeded. In this case, a split line in the corresponding bucket region is determined, thus splitting this region into two subregions. This process is repeated each time the capacity of a bucket is exceeded.

For example, consider the two-dimensional data space in figure 2.1 at the begin of section 2. The first split was made according to coordinate 50 in the first dimension, giving buckets 1 and 2. Then bucket 2 was split according to position 60 of the second dimension, giving buckets 2 and 3, and so on.

The split lines of the LSD tree are maintained in a directory which is a generalized k-d tree [2]. For each split, a new node containing the position and the dimension of the split line is inserted into the directory tree. The leaves of the directory tree reference the buckets in which the actual objects are stored. For example, the partitioning displayed in figure 2.1 results in the directory tree displayed in figure 2.8.

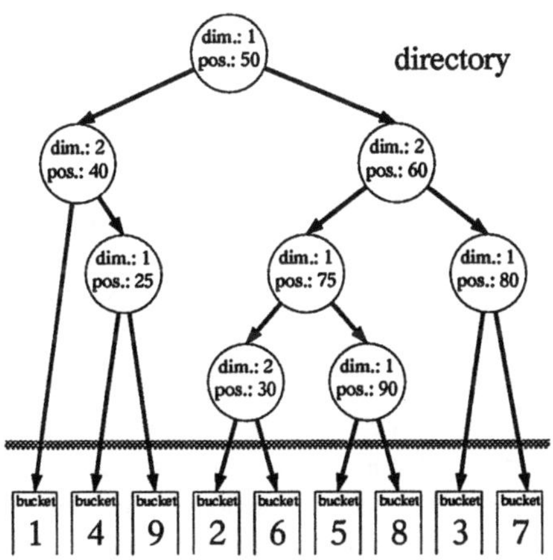


Figure 2.8: The LSD tree corresponding to the partitioning displayed in figure 2.1

Usually, the directory grows up to a point where it cannot be kept in main memory any longer. In this case, subtrees of the directory are stored in external pages on secondary memory, whereas the part of the directory near the root remains in main memory. For the details of the paging algorithm we refer to [10] or [11]. For this paper, it is sufficient to know that an LSD tree uses a partially paged k-d-tree as directory, i.e. the directory consists of an internal k-d tree and external directory pages containing k-d subtrees (see figure 2.9).

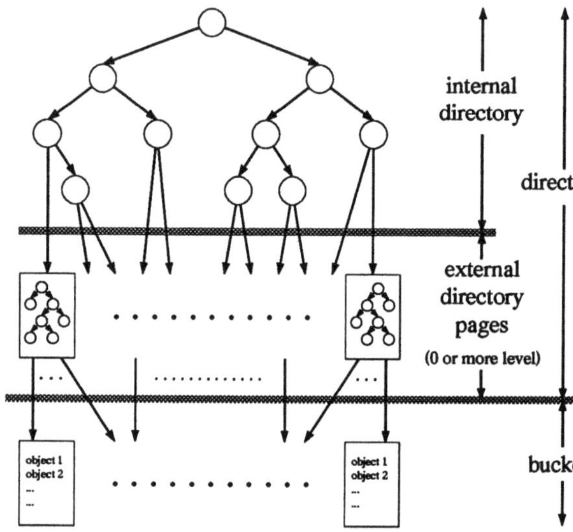


Figure 2.9: Overall structure of the LSD tree

In our experiments we have used five different object distributions:

- ***Uniform distribution:*** The points are uniformly distributed over the data space (see figure 2.10).
- ***Heap distribution:*** The coordinates in both dimensions are beta distributed with parameters $a = 40$ and $b = 100$ (see figure 2.11).
- ***Strip distribution:*** The horizontal coordinates are uniformly distributed, whereas the vertical coordinates are beta distributed with parameters $a = 400$ and $b = 1000$ (see figure 2.12).
- ***Corner distribution:*** The coordinates in both dimensions are beta distributed with parameters $a = 2$ and $b = 0,5$ (see figure 2.13).
- ***Diagonal distribution:*** First we have generated one-dimensional intervals of equal length with uniformly distributed centers. The sum of the length of all intervals equals 0.1 of the length of the whole one-dimensional data space. Then we transformed the intervals into two-dimensional points using the lower and upper bounds of the intervals as dimensions in the image space. This transformation is a well known variant of the transformation technique, which maps $d$-dimensional intervals to $2d$-dimensional points in order to apply point data structures to non-point objects. We will explain the transformation technique in more detail in section 3. Figure 2.14 shows the resulting (image) points.

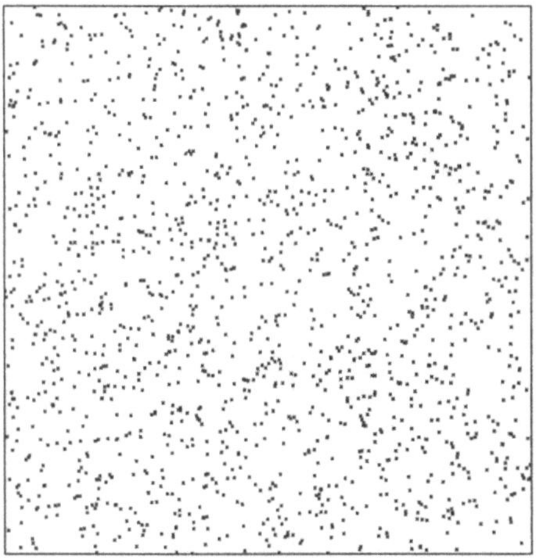

Figure 2.10: 1.600 uniformly distributed points

Figure 2.11: Heap distribution

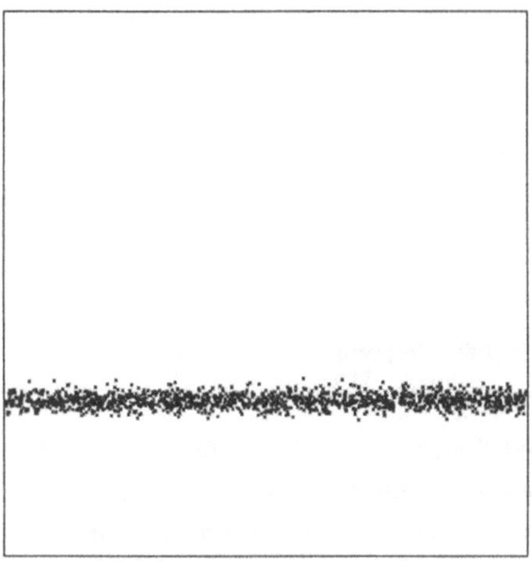

Figure 2.12: Strip distribution

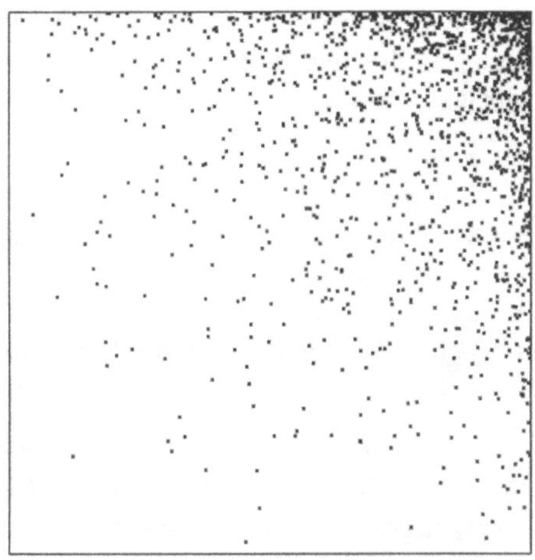

Figure 2.13: Corner distribution

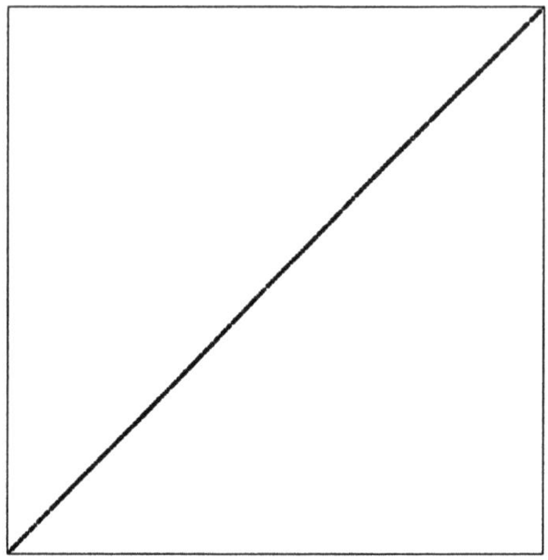

Figure 2.14: Diagonal distribution

First, we have inserted 80000 points distributed according to the five distributions into five LSD trees. For each LSD tree we used a *distribution dependent split strategy* based on the hypothesis of a uniform distribution. The parameters of the experiments are listed below:

- *data structure:* LSD tree
- *maximal number of nodes in the internal directory:* 100
- *directory page capacity:* heap of height 6, i.e. capacity $\leq$ 63 nodes
- *bucket capacity:* 16 objects
- *number of inserted points:* 80000
- *point distribution:* uniform, heap, strip, corner, diagonal
- *split strategy:* distribution dependent, $s_{Pos} = \frac{w_{\lambda_{s_{Dim}}} + w_{v_{s_{Dim}}}}{2}$

Table 2.1 shows the bucket utilization and the characteristics of the directory for the five LSD trees. With respect to the bucket utilization, it has to be noted, that empty pages need not be stored physically, because they can be represented by dummy addresses in the directory.

For all distributions except the diagonal, the LSD tree provides a bucket utilization near the optimal 69 %, although a distribution dependent split strategy had been used, using the hypothesis of a uniform distribution (Note that data structures based on a splitting technique which uniformly distributes the objects of one overflowing bucket over two bucktes yield a bucket utilization of ln 2, i.e. 69 % (see e.g. [3] or [21])). Only for the extreme diagonal distribution the LSD tree tends to degenerate. However, even for this extreme distribution the bucket utilization computed without empty buckets is over 65 %, and hence not bad at all. But the directory size is inacceptably high. Another important point is, that the number of external directory layers does not exceed two for all distributions except for the diagonal distribution, although the maximal number of internal directory nodes is restricted to 100. Even for the extreme diagonal distribution, the external directory height remains three, which is highly acceptable.

| distribution | no. of buckets (incl. empty buckets) | bucket utilization (incl. empty buckets) | no. of empty buckets | bucket utilization (excl. empty buckets) | no. of directory pages | no. of external directory layers |
|---|---|---|---|---|---|---|
| uniform | 7354 | 68.0 % | 1 | 68.0 % | 230 | 2 |
| heap | 7218 | 69.3 % | 10 | 69.4 % | 262 | 2 |
| corner | 7337 | 68.1 % | 1 | 68.1 % | 384 | 2 |
| strip | 7432 | 67.3 % | 81 | 68.0 % | 351 | 2 |
| diagonal | 14615 | 34.2 % | 6975 | 65.4 % | 2848 | 3 |

Table 2.1: Bucket utilization and characteristics of the directory for the LSD trees after 80000 insertions

From table 2.1 it also becomes obvious, that the directory size is more sensitive against skew data distributions than the bucket utilization.

To visualize the behaviour of the LSD tree, we have constructed five LSD trees for five sets with different distributions where each set consists of 1600 points. All parameters, except the number of points, are the same as in the experiments explained above. The following figures show the LSD trees generated and the corresponding partitions of the data space. Paged subtrees are shown by hatched rectangles. The buckets are indicated by small black rectangles at the leaves of the trees.

Figures 2.15 and 2.16 illustrate the situation for uniform distributed points. It comes out clearly, that the LSD tree is well balanced, and that the directory pages are well filled.

Using the heap distribution, broad areas of the data space do not contain any points. In our example, all points are located in the lower left quarter of the data space. If a distribution dependent split strategy based on the assumption of a uniform distribution is used, the first actions of the LSD tree are to encircle the data. Hence, the right son of the root and the right son of the left son of the root refer to empty buckets. After a few splits, however, the LSD tree reaches the populated part of the data space.

For the corner distribution, the maximum of the distribution density occurs in the upper right corner of the data space. Hence, the resulting LSD tree is skew to the right.

For the strip distribution, after a few splits creating empty buckets, the LSD tree reaches the populated part of the data space and shows a rather balanced shape.

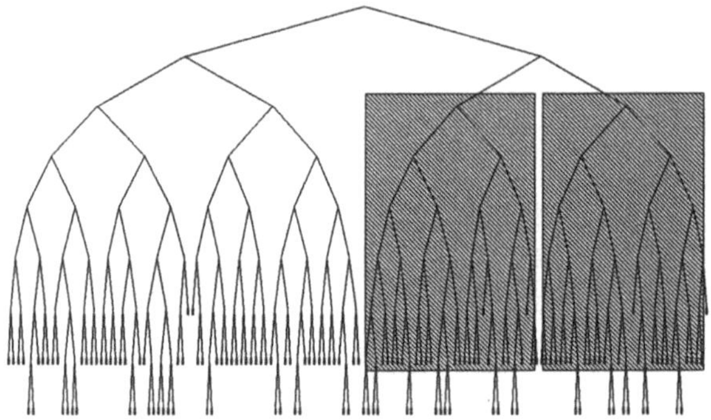

Figure 2.15: Shape of the LSD tree for uniform distributed points using a distribution dependent split strategy based on the hypothesis of a uniform distribution

Figure 2.16: Partition of the data space induced by the LSD tree of figure 2.15

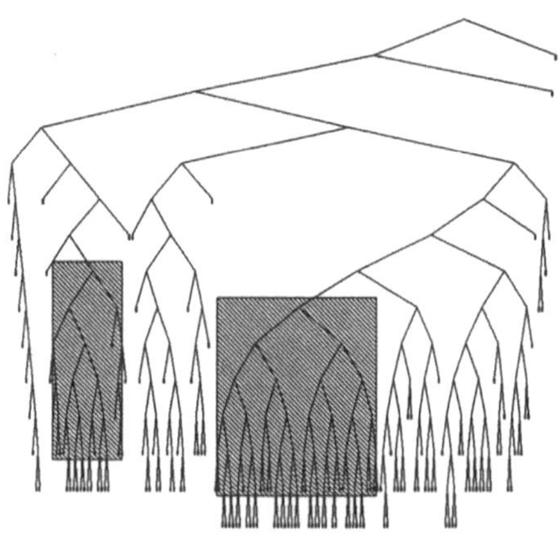

Figure 2.17: Shape of the LSD tree for points distributed according to the heap distribution using a distribution dependent split strategy based on the hypothesis of a uniform distribution

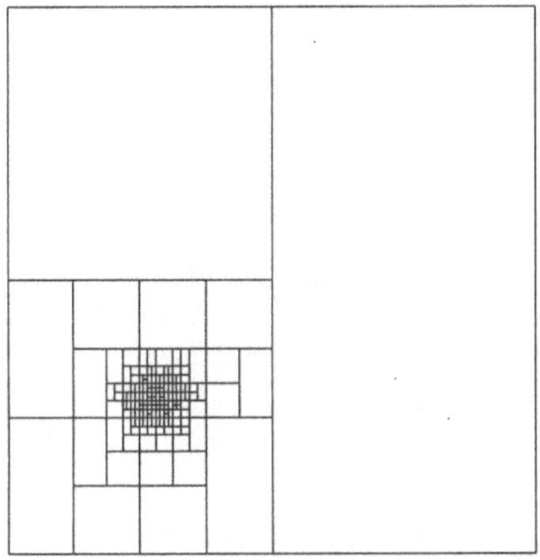

Figure 2.18: Partition of the data space induced by the LSD tree of figure 2.17

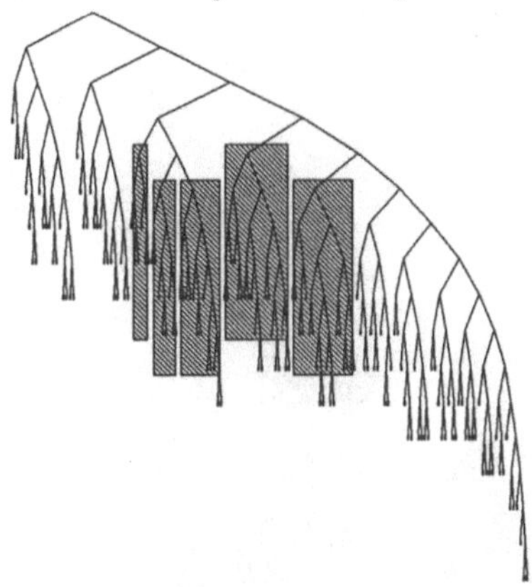

Figure 2.19: Shape of the LSD tree for points distributed according to the corner distribution using a distribution dependent split strategy based on the hypothesis of a uniform distribution

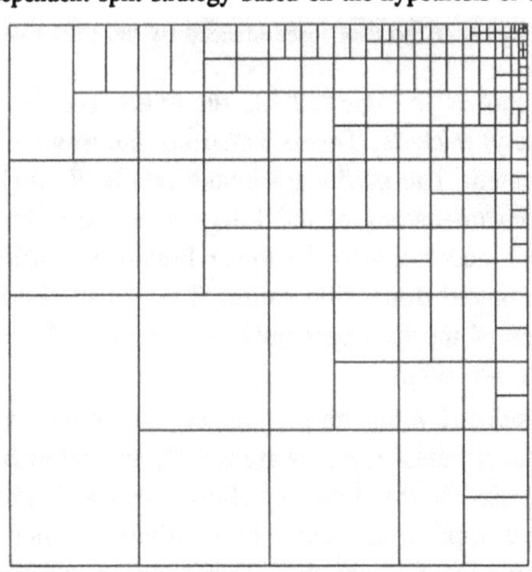

Figure 2.20: Partition of the data space induced by the LSD tree of figure 2.19

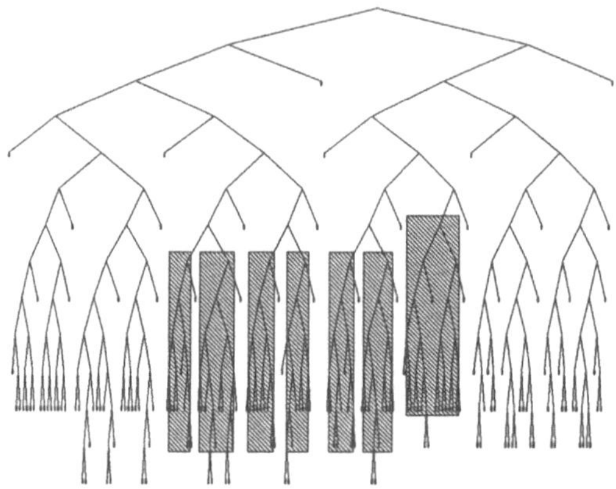

Figure 2.21: Shape of the LSD tree for points distributed according to the strip distribution using a distribution dependent split strategy based on the hypothesis of a uniform distribution

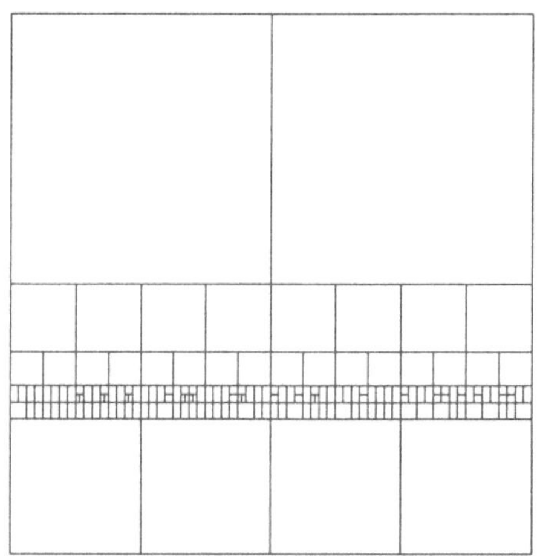

Figure 2.22: Partition of the data space induced by the LSD tree of figure 2.21

In the LSD tree shown in figure 2.23, on every second level nearly half the references point to (empty) buckets. This is a result of the way the diagonally distributed points have been generated. The points are image points of small intervals, and hence the coordinates in both dimensions of the image space are almost equal. Therefore, whenever a split is done according to the lower bound dimension, the following split according to the upper bound dimension causes a very bad distribution of the objects. Here, in most cases one of the two new buckets is empty. This leads to the fact that nearly half the buckets are empty.

If we take into consideration that empty buckets do not have to be stored physically, the bucket utilization nevertheless remains at 65.4 %, an acceptable value. On the other hand, the unacceptable size of the directory shows, that a distribution dependent split strategy based on the assumption of a uniform distribution is not a good choice for this application. Hence, if we want to use the transformation technique to store non-point

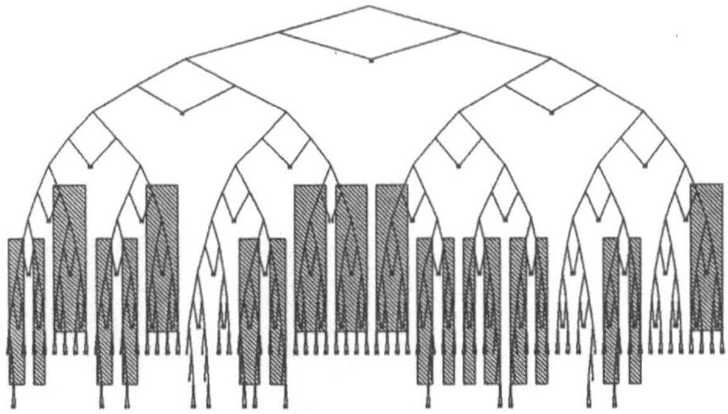

Figure 2.23: Shape of the LSD tree for points distributed according to the diagonal distribution using a distribution dependent split strategy based on the hypothesis of a uniform distribution

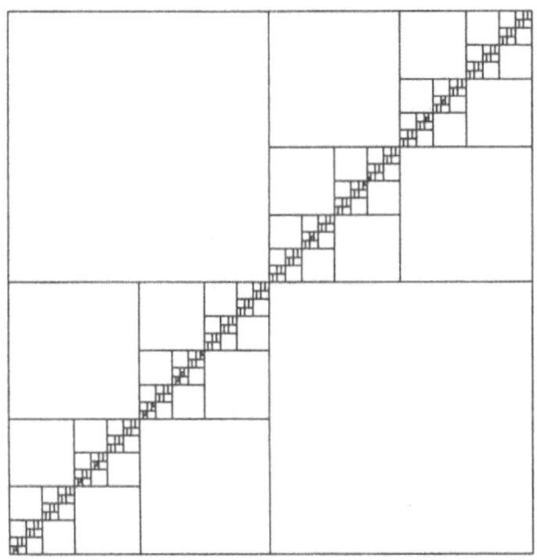

Figure 2.24: Partition of the data space induced by the LSD tree of figure 2.23

objects, we have to develop more sophisticated (distribution dependent) split strategies. We will discuss this in more detail in section 3.

Up to now, we have always used a distribution dependent split strategy. In the following, we will study the behaviour of an LSD tree using a data dependent split strategy. First, we investigate the effects induced by the missing unbiasedness and the high variance. Thereafter, we will examine the impact of the order of insertion.

In section 2.2, we have seen, that a data dependent split strategy can be based either on the sample mean or on the sample median. For the sample mean, the estimator for the median of the parent population is neither unbiased nor consistent for nonuniform distributions, and for the sample median, the variance of the estimator is relatively high.

To examine the consequences of these two aspects, we have inserted 1600 uniformly distributed points into two LSD trees. In the first case, we used a data dependent split strategy based on the sample mean and in the second case, a split strategy based on the sample median. The bucket capacity is again 16. The maximal number of internal nodes is 100 and the external directory pages can store subtrees up to a height of 6.

Since mean and median of the parent population coincide for a uniform distribution, the use of the sample mean should lead to a better balanced LSD tree because of its smaller variance. Figures 2.25 to 2.28 clearly show this effect.

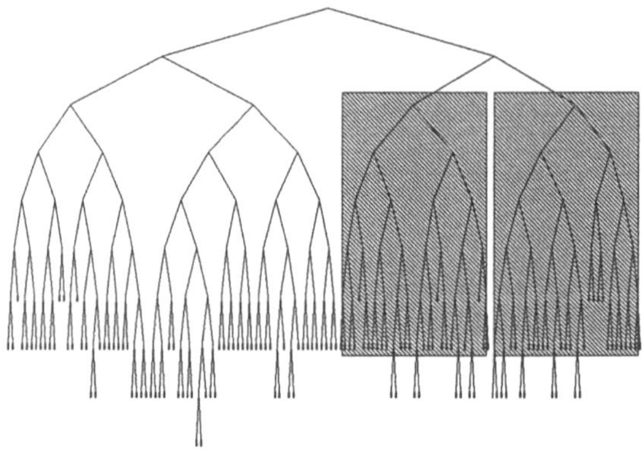

Figure 2.25: Shape of the LSD tree for uniformly distributed points using a data dependent split strategy based on the sample mean

Figure 2.26: Partition of the data space induced by the LSD tree of figure 2.25

To investigate the effect of the missing consistency of the sample mean for nonuniform distributions, we have repeated the same experiment for the heap distribution. Figure 2.29 shows that the split strategy based on the sample mean leads to a "belly" in the shape of the LSD tree. This is a consequence of the difference between the mean and the median of the parent population. However, the sample mean is competitive to the sample median, because the higher variance of the sample median leads to a slightly more unbalanced tree (see figure 2.31).

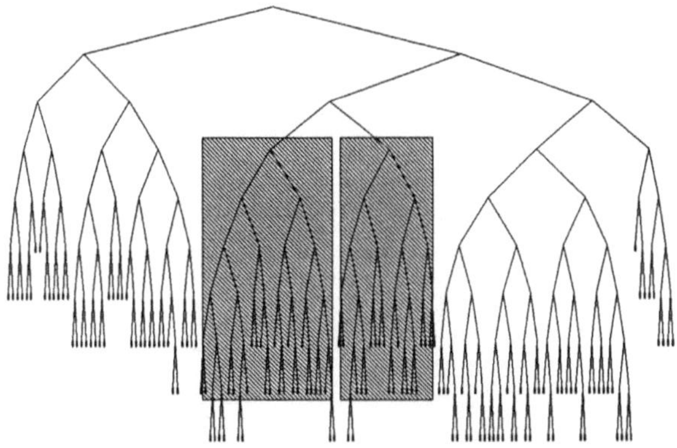

Figure 2.27: Shape of the LSD tree for uniformly distributed points using a data dependent split strategy based on the sample median

Figure 2.28: Partition of the data space induced by the LSD tree of figure 2.27

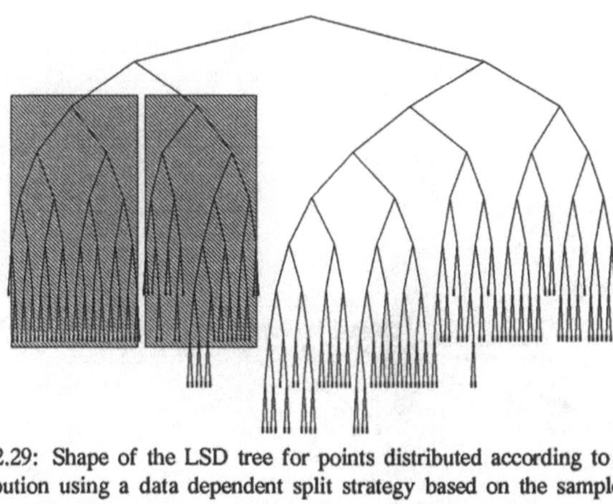

Figure 2.29: Shape of the LSD tree for points distributed according to the heap distribution using a data dependent split strategy based on the sample mean

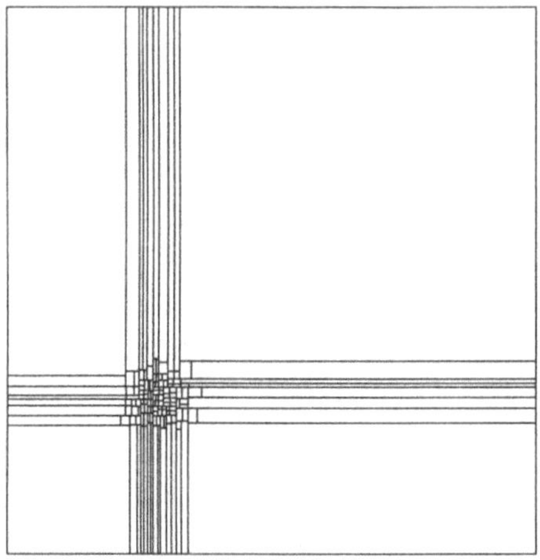

Figure 2.30: Partition of the data space induced by the LSD tree of figure 2.29

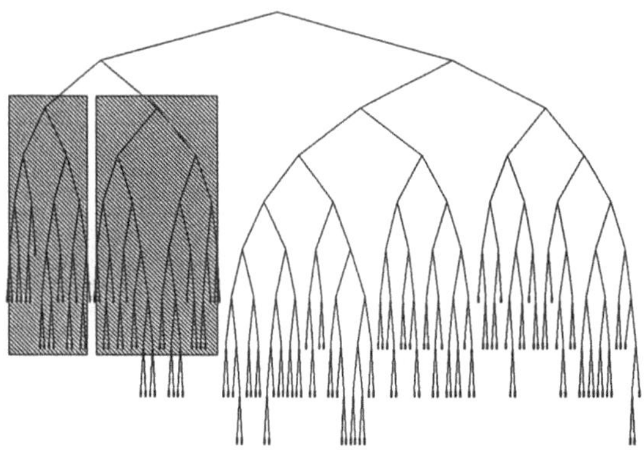

Figure 2.31: Shape of the LSD tree for points distributed according to the heap distribution using a data dependent split strategy based on the sample median

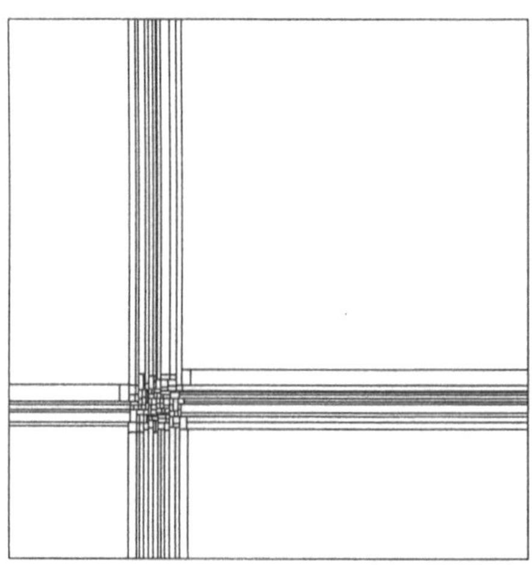

Figure 2.32: Partition of the data space induced by the LSD tree of figure 2.31

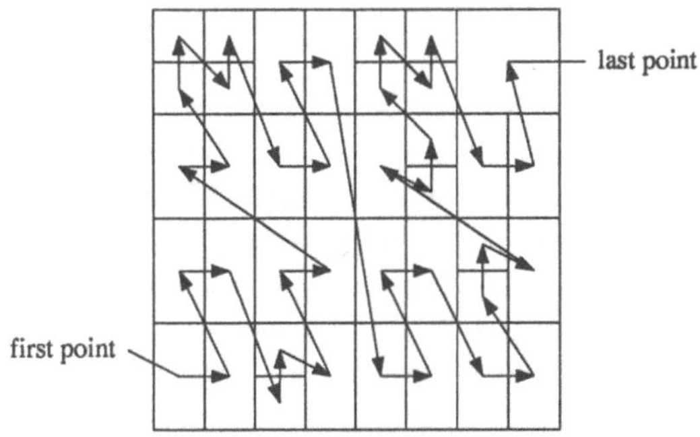


Figure 2.33: Example for the order used for insertions

Summarizing, the sample mean seems to be the better choice for a data dependent split strategy in almost all cases, even for rather skew object distributions.

**The impact of insertions in sorted order**

So far we have seen that the LSD tree is much more robust against skew data distributions when a data dependent split strategy is used instead of a distribution dependent split strategy based on the hypothesis of a uniform distribution (see figures 2.17 and 2.29 for example). But an important precondition for all experiments was that the objects are inserted in random order. Hence, it remains to study the impact of insertions in sorted order. To this end, we have performed experiments where the points have been presorted by the LSD tree itself. The presorting of the input has been done as follows. We inserted 1600 uniformly distributed points into an LSD tree with bucket capacity 1. Then these points were inserted into another LSD tree in the order of a left to right scan through the leaves (= buckets) of the first LSD tree. Figure 2.33 shows an example of the order induced by this sorting technique.

First we inserted the sorted points into an LSD tree using the distribution dependent split strategy. This leads to the LSD tree shown in figure 2.34.

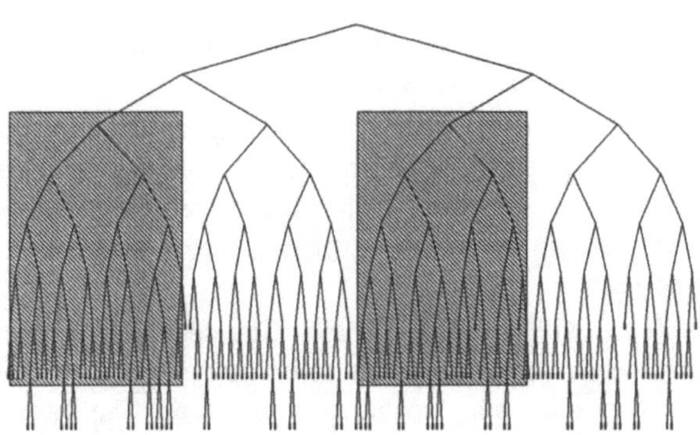

Figure 2.34: Shape of the LSD tree after the insertion of 1600 sorted points in the two-dimensional case using a distribution dependent split strategy

Note, that the underlying binary tree is exactly the same as for the LSD tree of figure 2.15, which was generated by inserting the same input in random order. Only the paged subtrees differ because the nodes in both trees are created in different orders.

Whereas the order of the insertions does not influence the shape of an LSD tree created by a distribution dependent split strategy, the situation differs when a data dependent split strategy is used. We have sorted one-, two- and three-dimensional uniformly distributed points, and inserted them into three LSD trees using a data dependent split strategy based on the sample mean (The sample median does not outperform the sample mean in this situation). The resulting LSD trees are given in figure 2.35.

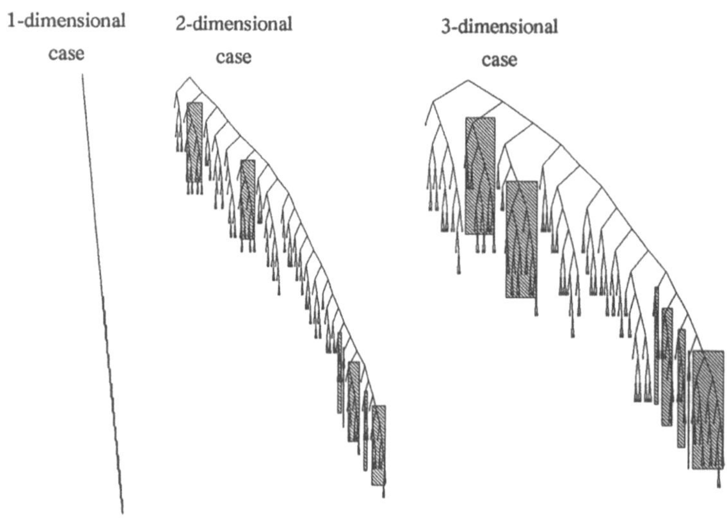


**Figure 2.35: Shape of an LSD tree after the insertion of 1600 sorted points in the one-, two- and three-dimensional case using a data dependent split strategy**

In the one-dimensional presorted case the LSD tree degenerates to a linear list. In the two- and three-dimensional case the situation is somewhat better because the ordering is not unique and hence not that "strong". Nevertheless, here again the LSD tree degenerates in an unacceptable way.

Summarizing our experiences gained from analytical and experimental investigations we suggest a data dependent split strategy based on the sample mean as a good choice in most applications. However, if there are indications of a strong presorting or there is a need for a certain worst case guarantee, then a distribution dependent split strategy should be used.

We close section 2 with a further remark. The degeneration of the LSD tree in certain applications when an inappropriate split strategy has been used is much less prominent compared to some other spatial data structures. The flexibility of the LSD tree allows for each bucket region to establish a new split line independent of earlier split lines. Therefore, bad consequences of deficient split decisions made earlier can be prevented to a certain extent. This does not hold, for example, for the widely used grid file which partitions the data space into grid-oriented bucket regions. Here, a new split position to be created for the split of a certain bucket region must often be chosen such that it coincides with earlier split positions introduced in other bucket regions of the data space.

# 3 Split strategies for non-point data structures

If data structures for points are used to maintain non-point objects, the disjoint partition of the data space leads to an intersection of large objects with several bucket regions. If, as in the so-called clipping approach, an object is stored in any data bucket whose corresponding bucket region it intersects, bucket utilization is low and results in an inefficient range query performance.

Two reasonable approaches which seem to be completely different at first glance try to overcome the disadvantages of the clipping technique. Both provide for bucket regions in a way that every object is completely contained in a bucket region.

The first approach uses the so-called transformation technique [12, 27] were $d$-dimensional intervals are transformed into $2d$-dimensional points which are then maintained by a point data structure. Prominent representative of this kind is the LSD tree for non-point objects. The second approach uses overlapping bucket regions in the original data space such that no object intersects a bucket region boundary. A prominent representative of this kind is the R-tree [7].

We first turn our attention to the transformation technique. As usually, we assume arbitrary $d$-dimensional objects to be approximated by their minimal bounding boxes, i.e. $d$-dimensional intervals. The transformation technique maps $d$-dimensional intervals into $2d$-dimensional points. Two different types of transformations can be distinguished. In the first transformation, the so-called *corner representation*, the lower bound of a rectangle in dimension $i$ $(1 \le i \le d)$ is used as coordinate in dimension $2i$-1 in the image space, and the upper bound in dimension $i$ as coordinate in dimension $2i$. Formally, this transformation $\psi$ can be described as follows:

$$\psi([l_1; u_1], [l_2; u_2], \dots, [l_d; u_d]) \stackrel{\text{def}}{=} (l_1, u_1, l_2, u_2, \dots, l_d, u_d) \tag{15}$$

Figure 3.1 shows a one-dimensional example for the corner representation.

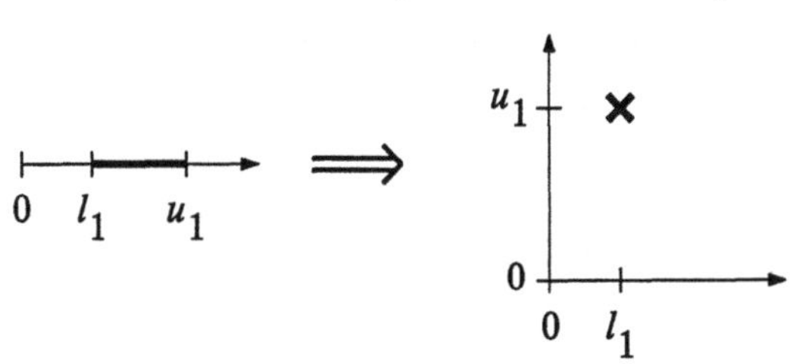


Figure 3.1: One-dimensional example for the corner representation

The second transformation, the so-called *center representation*, uses instead of the lower and upper bounds the center and the halved extension of the interval. More formally

$$\tau([l_1; u_1], \dots, [l_d; u_d]) \stackrel{\text{def}}{=} \left(\frac{l_1 + u_1}{2}, \frac{u_1 - l_1}{2}, \dots, \frac{l_d + u_d}{2}, \frac{u_d - l_d}{2}\right) \tag{16}$$

Figure 3.2 shows a one-dimensional example for the center representation.

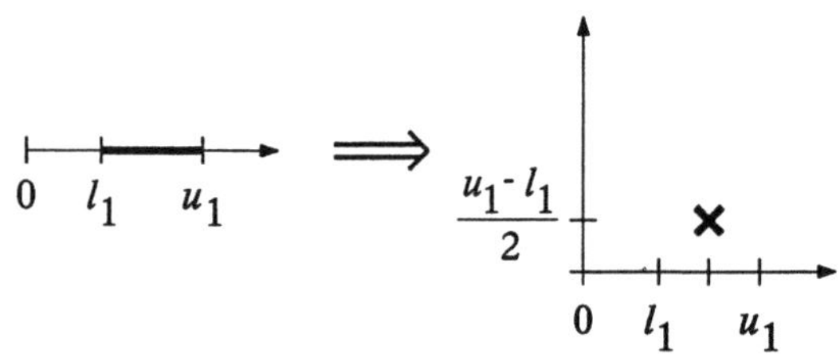


Figure 3.2: One-dimensional example for the center representation

The image points of the transformation technique can then be stored in a normal point data structure. But two problems arise using this approach.

1. Since the transformation technique causes a skew distribution of the image points, we have to design a sophisticated split strategy in order to avoid a performance degeneration. (Note that for any dimension the upper bounds of $d$-dimensional objects always exceed the lower bounds and hence the corresponding $2d$-dimensional image points are located in the upper left triangular subspace of the image space. Furthermore, since in most applications the objects are small compared to the data space, the image points are located in a small strip above the diagonal.) Remember, for example, the diagonal distribution in section 2, which leads to a performance degeneration.
2. In a range query, we have to search with a $d$-dimensional query region of arbitrary shape in a $2d$-dimensional data structure. It is not clear, how to handle this situation.

In the following, we will briefly describe a solution for the second problem using the LSD tree as an example. A detailed description can be found in [11]. First we need the definition of a *data region*.

- The data region of a bucket is its bucket region.
- The data region of a directory node is the union of the data regions of its sons.

The $2d$-dimensional data regions of the data structure are transformed back into $d$-dimensional data regions, which contain all rectangles, whose image points are located in the $2d$-dimensional data region. The query region is then compared with these $d$-dimensional data regions in the original data space. In addition, for each bucket the minimal box containing all objects of the bucket is stored together with the bucket reference in the LSD tree directory (in a compressed form) to minimize the number of bucket accesses needed.

We now return to the first problem, i.e. the design of a sophisticated split strategy. An important difference between corner representation and center representation is, that the latter is better suited for the design of a sophisticated split strategy, because here we can clearly distinguish between splits according to the location of the objects and splits according to the size of the objects. If the split dimension is odd, the split concerns the centers of the objects, i.e. their location. If the split dimension is even, the split concerns the halved extensions of the objects, i.e. their size.

Now we are in a position to describe a sophisticated split strategy for the center representation. In many experiments, this split strategy has been proven most efficient compared to various other split strategies (see [11]).

Splits concerning the location of objects are performed distribution dependent assuming a uniform distribution for the centers of the objects. This works well on the average and prevents from a possible degeneration in case of sorted inputs. Splits concerning the size of objects are performed data dependent, because a uniform distribution hypothesis is too unrealistic in this case. A possible, nevertheless unlikely input where the objects are sorted according to their size does not cause a degeneration of the data structure because splits concerning the object size are rare compared to splits concerning the location. This stems from the strategy of choosing the split dimension which will be explained now.

Obviously, it is not a good choice to use the dimensions of the $2d$-dimensional image space in a cyclic manner. We propose, to use the dimensions of the original $d$-dimensional space in a cyclic manner. Then for each dimension, we have to choose between a split according to the location of the objects (dimension $2i$-1) and a split according to the size of the objects (dimension $2i$). The solution is to choose the dimension for which the size of the two minimal rectangles containing all objects in the two new buckets is minimal. Here it comes out clearly, how the split strategy tries to control the overlapping of the two bucket regions.

If the LSD tree has been created using this split strategy it turns out that the performance of range queries based on the technique described is approximately the same as for range queries in the point situation. The following experiment validates this statement.

We inserted 100000 objects into LSD trees with bucket capacity 10. We used one-, two and three-dimensional objects, and points as well as intervals. For the intervals, the sum of the areas of all intervals was 2.5 times the area of the whole data space. Then we performed range queries of different sizes, and measured the hit ratio defined as follows:

$$\text{hit ratio} \stackrel{\text{def}}{=} \frac{\text{no. of found objects}}{\text{no. of bucket accesses} \times \text{bucket capacity}} \tag{17}$$

Figure 3.3 shows the hit ratio for points, and figure 3.4 shows the hit ratio for the rectangles.

An interesting aspect of figures 3.3 and 3.4 is, that there is almost no difference in performance between points and intervals, whereas there are great differences between one-, two- and three-dimensional objects. The reason for this phenomenon is illustrated

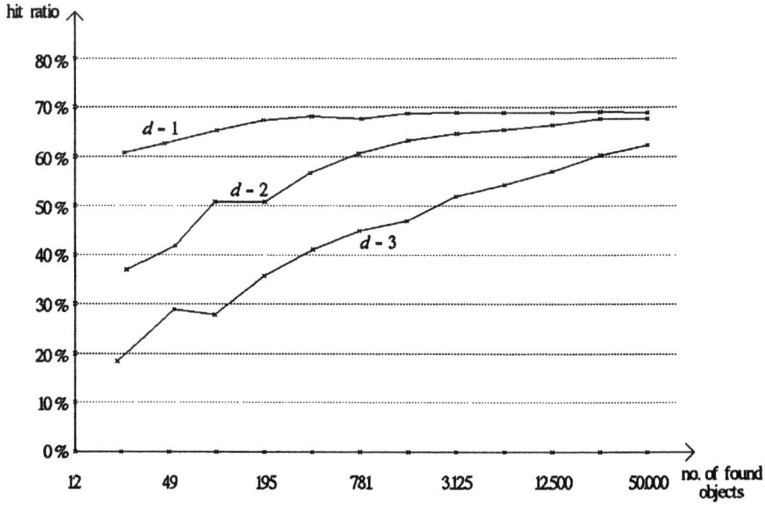


Figure 3.3: Hit ratio for point objects

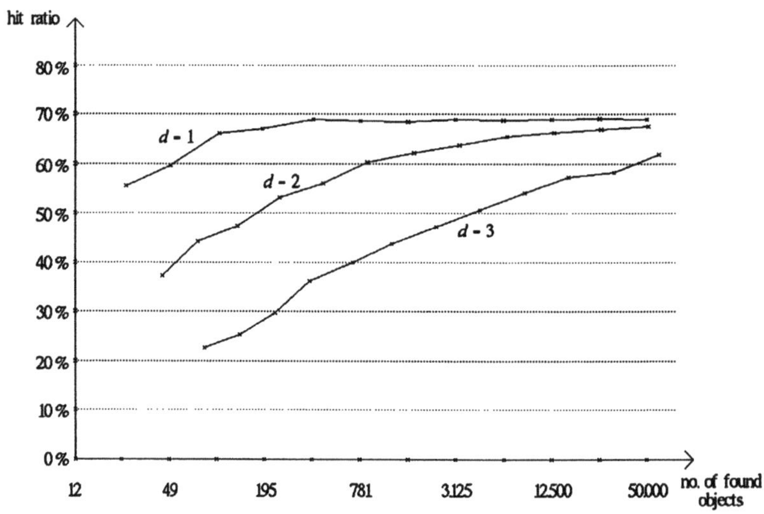


Figure 3.4: Hit ratio for intervals

in figure 3.5. We have partitioned three data spaces into eight bucket regions. If we look at the range query example, whose query region has the shape and the size of a bucket region, in the one-dimensional case, the query region intersects two bucket regions, in the two-dimensional case, the query region intersects four bucket regions, and in the three-dimensional case, the query region intersects eight bucket regions.

To validate the general applicability of the split strategy described for the LSD tree using the center transformation, we implemented this split strategy in an R-tree [7] and compared the performance with the quadratic cost split algorithm proposed by Guttman. The quadratic cost split algorithm has two phases. First, it searches for

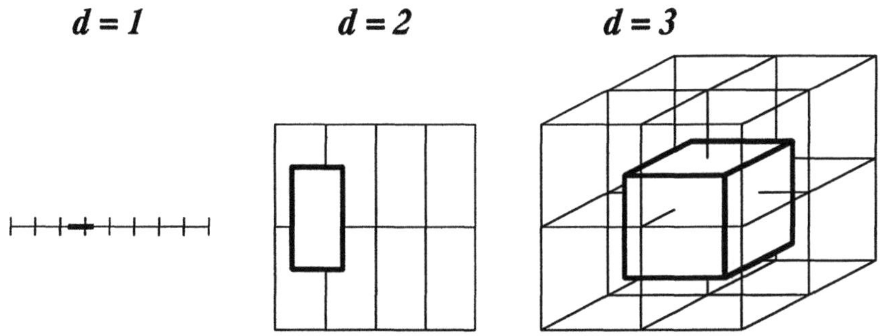


Figure 3.5: A range query with the shape of a bucket region in the one-, two- and three-dimensional case

the pair of objects, for which the area of a rectangle covering both objects, minus the areas of the objects themselves, becomes maximal. These objects are stored in different buckets. The remaining objects are then assigned to the buckets, one at a time. At each step the area expansion required to add each remaining object to each bucket is calculated, and the object assigned is the one showing the greatest difference.

To evaluate both split strategies, we inserted 100000 uniformly distributed rectangles into both R-tree variants. The sum of the areas of all intervals was again 2.5 times the area of the whole data space. The bucket capacity was 10. Table 3.1 shows the bucket utilization, the directory utilization and the average overlapping of the bucket regions for both data structures.

| split strategy | Guttman's quadratic split | our split |
|---|---|---|
| no. of buckets | 14590 | 14312 |
| bucket utilization | 68,54 % | 69,87 % |
| no. of directory pages | 995 | 878 |
| directory page utilization | 62,65 % | 69,20 % |
| average overlap of the bucket regions | 8,70 | 2,44 |

Table 3.1: Some measures for two R-tree variants

Main difference between Guttman's split strategy and our split strategy is the average overlapping of the bucket regions. The larger overlapping of Guttman's split strategy causes performance problems especially for small range queries. To demonstrate this, we have performed range queries of four different sizes. For each size, we have executed at least ten queries at randomly generated locations of the data space. The performance results are shown in table 3.2.

In comes out clearly, that our split strategy outperforms Guttman's split strategy. This stems from the fact that Guttman does not take into consideration that the geometric key of the set of objects stored in a bucket is a bounding box, i.e. a rectangle with sides parallel to the axes. The following consideration explains the situation:

| | query region size | no. of found objects | Guttman's quadratic split | our split |
|---|---|---|---|---|
| *bucket accesses* | | | | |
| | 0 % | 2.7 | 8.7 | 2.4 |
| | 0.1 % | 109.6 | 41.3 | 24.2 |
| | 1 % | 1063.7 | 221.9 | 175.7 |
| | 10 % | 10207.9 | 1690.3 | 1536.5 |
| *hit ratio* | | | | |
| | 0 % | 2.7 | 3.11 % | 11.43 % |
| | 0.1 % | 109.6 | 26.53 % | 45.29 % |
| | 1 % | 1063.7 | 47.94 % | 60.54 % |
| | 10 % | 10207.9 | 60.39 % | 66.47 % |
| *directory page accesses* | | | | |
| | 0 % | 2.7 | 13.7 | 5.5 |
| | 0.1 % | 109.6 | 18.5 | 8.5 |
| | 1 % | 1063.7 | 42.0 | 24.1 |
| | 10 % | 10207.9 | 168.7 | 124.3 |
| *geometrical tests between two rectangles* | | | | |
| | 0 % | 2.7 | 274.0 | 107.4 |
| | 0.1 % | 109.6 | 572.7 | 291.3 |
| | 1 % | 1063.7 | 1680.1 | 915.2 |
| | 10 % | 10207.9 | 5343.9 | 2811.6 |

Table 3.2: Range query performance of the R-tree

The first pair of objects determined by the quadratic cost split algorithm most likely consists of two objects located in opposite corners of the bucket region. If this object pair is used as starting point for the second phase, this often leads to a distribution, where the objects in one bucket are located above the diagonal and the objects in the other bucket below the diagonal. This in turn causes bucket regions with high overlapping. Figure 3.6 shows this situation.

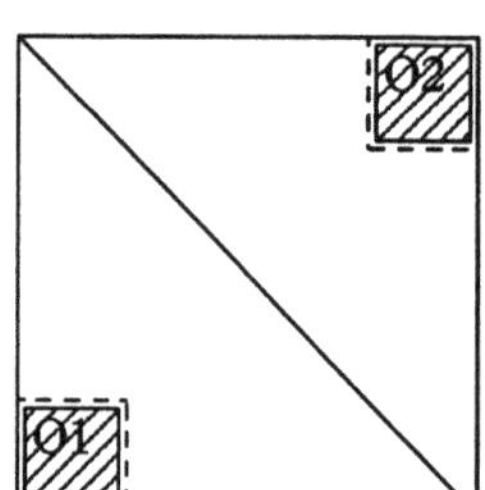


Figure 3.6: The object distribution induced by Guttman's quadratic cost split algorithm

In this section we have outlined that an efficient split strategy for non-point data structures has to take into account the bucket utilization as well as the overlapping of bucket regions. A first validation of this claim has been given by the split strategy presented which can be used with advantage in two prominent representatives of non-point data structures.

## 4 Conclusion

We have – analytically and experimentally – investigated several split strategies for spatial data structures and have worked out their advantages and disadvantages. As far as the point situation is concerned, we have been able to state clear recommendations. However, an important precondition for an efficient performance is the use of a flexible data structure in the sense that new split lines can be placed independent of the placement of earlier split lines.

As far as non-point data structures are concerned, the situation is much more complicated because two parameters, namely bucket utilization and overlapping of bucket regions, must be optimized at the same time. Here we provide first insights, but further investigations are necessary.

## References


1. Beckmann, N., Kriegel, H.-P., Schneider, R., Seeger, B.: 'The R*-tree: an efficient and robust access method for points and rectangles', Proceedings of the ACM SIGMOD Int. Conf. on Management of Data, 1990
2. Bentley, J.L.: 'Multidimensional Binary Search Trees Used for Associative Searching', Communications of the ACM, Vol. 18, No. 9, 509-517, 1975
3. Fagin, R., Nievergelt, J., Pippenger, N., Strong, H.R.: 'Extendible Hashing — A Fast Access Method for Dynamic Files', ACM Transactions on Database Systems 4, 315–344, 1979
4. Freeston, M.: 'The BANG file: a new kind of grid file', Proc. ACM SIGMOD Int. Conf. on Management of Data, 260-269, 1987
5. Greene, D.: 'An Implementation and Performance Analysis of Spatial Data Access Methods', Proc. IEEE 5th Int. Conf. on Data Engineering, 606-615, 1989
6. Günther, O., Bilmes, J.: 'The Implementation of the Cell Tree: Design Alternatives and Performance Evaluation', University of California, Santa Barbara, Department of Computer Science, TRCS88-23, October 1988
7. Guttman, A.: 'R-Trees: A Dynamic Index Structure for Spatial Searching', Proc. ACM SIGMOD Int. Conf. on Management of Data, 47-57, 1984
8. Hartung, J.: 'Statistik: Lehr- und Handbuch der angewandten Statistik', 3. Aufl., R. Oldenbourg Verlag, München, 1985
9. Henrich, A., Six, H.-W., Widmayer, P.: 'The LSD tree: spatial access to multidimensional point and non-point objects', Proc. 15th Int. Conf. on VLDB, 45-53, 1989
10. Henrich, A., Six, H.-W., Widmayer, P.: 'Paging binary trees with external balancing'. In: Nagl (Ed.), Proc. 15th Int. Conf. on Graph-Theoretic Concepts in Computer Science (WG '89), Lecture Notes in Computer Science, vol. 411, 260-276, Springer, Berlin, 1989
11. Henrich, A.: 'Der LSD-Baum: eine mehrdimensionale Zugriffsstruktur und ihre Einsatzmöglichkeiten in Datenbanksystemen', Dissertation, FernUniversität Hagen, Fachbereich Mathematik und Informatik, 1990
12. Hinrichs, K.: 'The grid file system: implementation and case studies of applications', Dissertation Nr. 7734, ETH Zürich, 1985

13. Hutflesz, A., Six, H.-W., Widmayer, P.: 'Globally Order Preserving Multidimensional Linear Hashing', Proc. IEEE 4th Int. Conf. on Data Engineering, 572-579, 1988
14. Hutflesz, A., Six, H.-W., Widmayer, P.: 'Twin Grid Files: Space Optimizing Access Schemes', Proc. ACM SIGMOD Int. Conf. on Management of Data, 183-190, 1988
15. Hutflesz, A., Six, H.-W., Widmayer, P.: 'The R-File: An Efficient Access Structure for Proximity Queries', Proc. IEEE 6th Int. Conf. on Data Engineering, 372-379, 1990
16. Kleiter, G.D.: 'Bayes-Statistik: Grundlagen und Anwendungen', Walter de Gruyter, Berlin New York, 1980
17. Kotz, S., Johnson, N.L. (Hrsg.): 'Encyclopedia of Statistical Sciences', Vol. 5: 'Lindeberg Condition to Multitrait-Multimethod Matrices', John Wiley & Sons, New York, 1985
18. Kriegel, H.-P., Seeger, B.: 'Multidimensional Order Preserving Linear Hashing with Partial Expansions'. In: Ausiello/Atzeni (Eds.), Proc. Int. Conf. on Database Theory, Lecture Notes in Computer Science, vol. 243, 203-220, Springer, 1986
19. Kriegel, H.-P., Seeger, B.: 'PLOP-Hashing: A Grid File without Directory', Proc. IEEE 4th Int. Conf. on Data Engineering, 369-376, 1988
20. Krishnamurthy, R., Whang, K.-Y.: 'Multilevel Grid Files', IBM Research Report, Yorktown Heights, New York, 1985
21. Larson, P.: 'Dynamic Hashing', BIT 18, 184–201, 1978
22. Lomet, D.B., Salzberg, B.: 'A Robust Multi-Attribute Search Structure', Proc. IEEE 5th Int. Conf. on Data Engineering, 296-304, 1989
23. Nievergelt, J., Hinterberger, H., Sevcik, K.C.: 'The Grid File: An Adaptable, Symmetric Multikey File Structure', ACM Transactions on Database Systems, Vol. 9, No. 1, 38-71, 1984
24. Ooi, B.C., McDonell, K.J., Sacks-Davis, R.: 'Spatial kd-Tree: An Indexing Mechanism for Spatial Databases', IEEE COMPSAC, 433-438, 1987
25. Otoo, E.J.: 'Balanced Multidimensional Extendible Hash Tree', Proc. 5th ACM SIGACT / SIGMOD Symposium on Principles of Database Systems, 100-113, 1986
26. Robinson, J.T.: 'The K-D-B-Tree: A Search Structure for Large Multidimensional Dynamic Indexes', Proc. ACM SIGMOD Int. Conf. on Management of Data, 10-18, 1981
27. Seeger, B., Kriegel, H.-P.: 'Techniques for Design and Implementation of Efficient Spatial Access Methods', Proc. 14th Int. Conf. on VLDB, 360-371, 1988
28. Seeger, B.: 'Entwurf und Implementierung mehrdimensionaler Zugriffsstrukturen', Dissertation, Universität Bremen, Fachbereich Mathematik und Informatik, 1989
29. Six, H.-W., Widmayer, P.: 'Spatial Searching in Geometric Databases', Proc. IEEE 4th Int. Conf. on Data Engineering, 496-503, 1988
30. Yamane, T.: 'Statistik: Ein einführendes Lehrbuch', Band 1, Fischer Taschenbuch Verlag, Frankfurt am Main, 1976

# A Guaranteed Approximation Algorithm for On-line Computing Quadtree Border(*)

Paolo Giulio Franciosa
Enrico Nardelli

**Abstract**

In this paper we face the problem of *on-line* computing the border of a bidimensional object represented by a quadtree. The quadtree is supposed to be given one level at a time and the border is found by computing a series of guaranteed approximations, each one obtained from the previous one.

Computing the border of an object is the first step toward obtaining a vectorial representation of the object.

The whole series of approximations converges to the real border of the object and is computed within the same time bounds of known approaches for computing the boundary of a quadtree, but our algorithm allows to:

- obtain approximated solutions as soon as the necessary informations are available;
- evaluate the meaningfulness of approximated solution;
- stop the computation as soon as a solution with sufficient precision is obtained;
- reducing space resource required for computing the new approximation.

Our algorithm is shown to be competitive with respect to algorithms that off-line compute the exact boundary.

The proposed algorithm works building and maintaining some links between the approximated border nodes and their neighbours.

First experimental results show that the characteristics of this algorithm are particularly useful in order to solve proximity queries on shapes typical of geographical applications.

## 1. Introduction

In the last decades several data structures based on a hierarchical decomposition of the geometric space have been proposed in order to represent raster spatial objects (a very extensive list of references can be found in [11]). One of the most popular is the *region quadtree* [9, 11], firstly introduced by Klinger and Dyer [5], that consists of a quaternary tree in which each node is associated to a square-shaped region and each node whose region partially intersects the object has four sons, each associated to a square-shaped subregion contained in its father's one (see fig. 1). Nodes associated to regions contained in (disjoint from) the object are usually referred to as *black* (*white*) nodes; other nodes are referred to as *grey* nodes.

Quadtrees are well known to provide a compact representation for objects with "regular" shapes, but in several cases we are interested in computing only the border of an object [3, 4, 12, 1].

The problem of computing the boundary of a quadtree has been extensively studied in [3, 7, 4, 12, 11]. All approaches are based on a depth first traversal of the quadtree,

(*) Work partially supported by: ESPRIT II Basic Research Action Programme of the EEC, projects “ALCOM” and “Basic GOODS” and Italian CNR Research Programme “Sistemi Informatici e Calcolo Parallelo”, project “Multidata”.

and the main operation is the identification of the neighbour nodes of the node currently examined; this is generally done by climbing the tree up to the nearest common ancestor of current node and its neighbour.

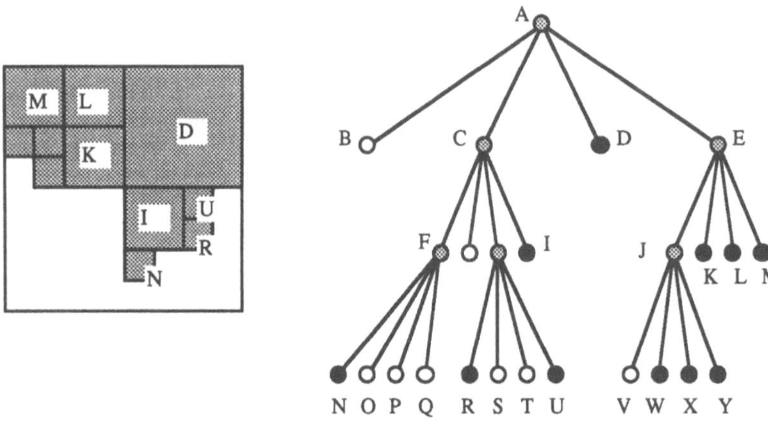


**Fig 1:** An object and its quadtree

Quadtrees are well suited for representing successive approximations of spatial objects, just considering the first $k$ levels of the representation and "resolving" possible grey nodes into black or white nodes.

On this approximated representation it is possible to compute an approximated border, where by *border* of a quadtree we mean the set of its border nodes, i.e. nodes intersecting the boundary line (see fig. 1, where all nodes are border nodes).

A different approach is possible, by introducing the concept of *unsolved* border at level $k$, that is the set of grey nodes and black nodes in levels not greater than $k$ for which it is not possible, considering only the first $k$ levels, to decide whether they belong to the border or not.

We define the $k$-approximated border as the union of the unsolved border at level $k$ and the set of black nodes in levels not greater than $k$ which are known to belong to the border.

In this paper we focus on the problem of computing a series of guaranteed approximations of the border of a given quadtree, which converges to the real border. The quadtree can be either completely known at the beginning or fed into the algorithm one level at a time.

Computing the approximated border of a spatial object can be of great interest in several contexts. For example in several motion planning problems, or while solving proximity queries on spatial data bases, it is very useful to compute only the roughest approximation necessary to answer the query. A similar argument applies in real time image processing, when the object shape is not exactly known but for example the observer is approaching the object, hence better approximations of the border are built as soon as a more precise representation of the object is available.

Moreover computing the approximated border is a first step toward obtaining a vectorial representation of the object.

More generally we propose an algorithm that can be helpful in various practical situations, such as:

- when the object is not a priori known but it is given as a series of increasingly closer approximations of the exact object. In this case it can be useful to start working as soon as a partial knowledge of the object is available;
- when a good approximated solution must be produced within a not a priori fixed amount of time: this can be the case of real time applications, where it is required to produce the best possible approximated solution before an unpredictable event takes place;
- when the approximation required is not predictable, in the sense that after producing a certain approximated solution a better one could be required.

In order to achieve the exposed characteristics it is not helpful, even when it can be done, to naively convert known boundary detection algorithms, typically based on a depth first traversal of the quadtree or on following the contour of the object, into an algorithm that visits the quadtree one level at a time, because this would heavily affect its time performances.

We are interested in designing on-line *competitive* algorithms, in the sense defined in [2, 6], i.e. algorithms that given a sequence of queries $q_1, q_2, \ldots, q_k$, compute the whole sequence of query solutions $S_1, S_2, \ldots, S_k$, (where $S_i$ is computed knowing only the initial part $q_1, q_2, \ldots, q_i$, of the sequence of queries), within the same time bound of the best off-line algorithm, which computes the same solutions knowing a priori the whole sequence $q_1, q_2, \ldots, q_k$.

We propose an algorithm that, given a quadtree representing a spatial object, computes a series of guaranteed approximations of its border. The $k$-th approximation is derived from the $(k-1)$-th one in worst case time $O(g(k))$, where $g(k)$ is no greater than the number of grey nodes on level $k$. The series of approximations converges to the real border in $T$ steps, where $T$ denotes the depth of the quadtree, and is computed in worst case time $O(G)$, where $G$ is the total number of grey nodes.

The total space required is bounded by $\max_k(N(k))$, where $N(k)$ is the size of the approximated border at level $k$. Note that this is optimal, since this is the size of the output for some $k$.

Our algorithm is shown to be competitive with respect to algorithms that off-line compute the exact boundary. Therefore it can be advantageously used whenever an on-line approach is preferable.

Some experiments are being performed, and the first results confirm that while solving proximity queries on geographical objects, or while checking geometrical relations between "regular" shaped objects is often sufficient to stop the approximation process early: thus our approach seems to be encouraging.

In the following section the main ideas are sketched, and in section 3 a detailed description of the algorithm is given. Its correctness is proved in section 4, while in sections 5 and 6 we show the algorithm performances in terms of approximation characteristics and asymptotic time and space bounds. In section 7 some considerations on the first experimental results are shown.

## 2. Main ideas

The border of an object is well defined only if we first define an adjacency relation between the regions associated to the nodes of the quadtree representing the object. In the following, when it does not lead to ambiguity, we will identify a node with the associated region.

There are two natural ways for defining an adjacency relation: two regions (nodes) are *adjacent*

- if and only if they touch on a side (4-adjacency);

or

- if and only if they touch on a corner (8-adjacency).

For example region *I* in fig 1 is 8-adjacent to *N*, *R*, *U*, *D* and *K* but is not 4-adjacent to *K*.

It is simple to see that if we want the border of a simply connected object to be connected according to the 4-adjacency (8-adjacency) relation, we must define the border of an object as the set of black nodes 8-adjacent (4-adjacent) to some white node. In fig. 2 the middle (right) picture shows the border according to the 4-adjacency (8-adjacency) between black and white nodes.

In the sequel we will consider the 4-adjacency on the object, hence the border of the shape on the left in fig. 2 is the rightmost one.

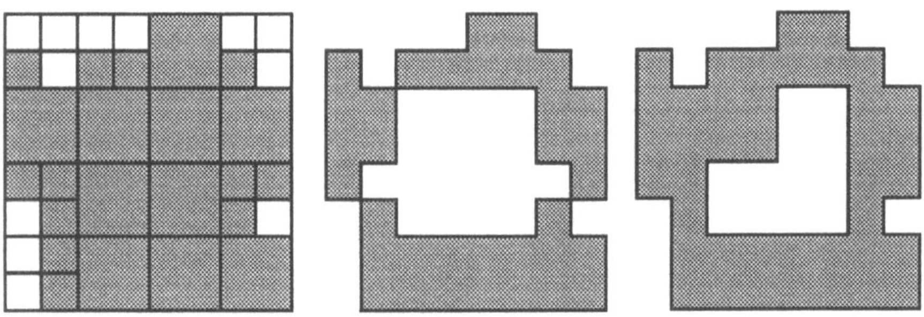

**Fig. 2:** border obtained using 4-adjacency and 8-adjacency between black and white nodes.

We can now formally define the *k*-approximated border.

Let us define the *strong border*(*k*) and the *weak border*(*k*) as the following sets:

- *strong border*(*k*) is the set of black nodes on level not greater than $k$ which are 8-adjacent to a white node on level *i*, with *i* not greater than *k*;
- *weak border*(*k*) is the set of grey nodes on level *k*, plus the black nodes on level not greater than $k$ which are both 8-adjacent to a grey node in level *k* and not 8-adjacent to any white node in level *i*, with *i* not greater than *k*.

An example of the above defined sets is shown in fig. 3.

The *k-approximated border* is defined as the union of strong border(*k*) and weak border(*k*). An example of the series of approximations is shown in fig. 4.

In the sequel we will always consider objects embedded in the interval $[0,1]\times[0,1]$, hence the root of the quadtree always represents the unit square.

An interesting feature of the *k*-th approximated border of an object is that it consists of all the nodes whose $L^\infty$ distance from the boundary line(*) is less than $2^{-k}$. Hence the

(*) Recall that the $L^\infty$ metric defines the distance between points $(x_1, y_1)$ and $(x_2, y_2)$ as $\max(|x_1 - x_2|, |y_1 - y_2|)$. Moreover the distance between two objects *a* and *b* is defined as the minimum distance between any couple of points respectively belonging to *a* and *b*.

$k$-approximated border is well suited to represent an approximation of the contour of the object in terms of a subset of its quadtree nodes. Note that, since we consider all adjacent nodes (both corner adjacent and side adjacent ones) to be logically at the same distance, it is reasonable to adopt the $L^{\infty}$ metric to measure distance between nodes.

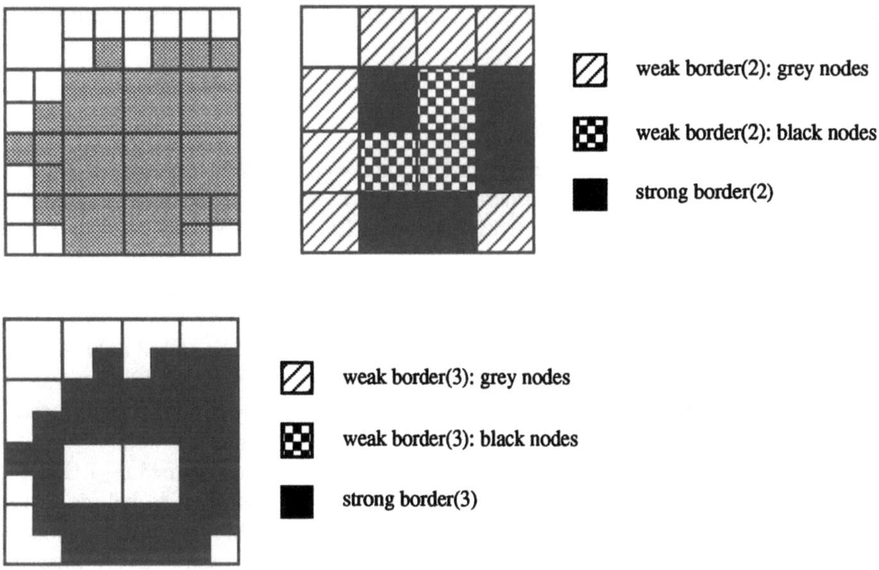


**Fig. 3:** Approximated border on level 2 and 3

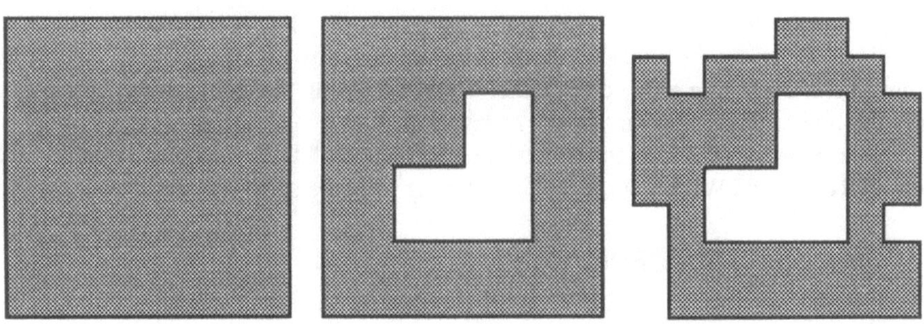

**Fig. 4:** $k$ approximated borders of the object in fig. 2 for $k = 1, 2, 3$.

Moreover, a guaranteed measure of the meaningfulness of the $k$-approximated solution is available as the ratio of the areas of strong border($k$) and weak border($k$). In fact it is simple to see that, while increasing $k$, the area of strong border($k$) can not decrease and the area of weak border($k$) can not increase.

The basic operation for border computation is the access to the 8-adjacent nodes of a given node. Note that, in general, a node can be 8-adjacent to more than 8 nodes, but we will see in the next section that a carefully designed access strategy can limit to a constant the number of accessed nodes.

Our approach is based on a breath first traversal of the quadtree, so that level $k$ is completely examined before starting to work on level $k$+1.

The main idea is to derive the $k$+1-approximated border from the $k$-approximated border by exploding grey nodes on level $k$ and then refining weak border($k$) and strong border($k$).

## 3. Algorithm description

We do not need to represent the strong border($k$) and weak border($k$) sets for each $k$. It is sufficient to maintain the strong border($k$) and weak border($k$) for the current value of $k$ by means of two variables, respectively named STRONG and WEAK, and to update them in moving from level $k$ to level $k+1$.

Each node is represented by means of a record structure, with a temporary link to its father and some links towards its adjacent nodes.

An adjacency link from a node $x$ to an adjacent node $y$ will be stored only if $y$ is a black node larger than $x$ or $y$ is a grey node on the same level of $x$; and a link toward a white node will be represented by a nil pointer. This guarantees that we only need a constant number of adjacency links (at most eight) from each node.

The record used for representing each node contains two flags, named IS_BORDER and IS_INNER. When IS_BORDER is true the node is confirmed to belong to the current strong border set, when IS_BORDER is false and IS_INNER is true the node is confirmed to not belong to the current approximated border, when both are false the node belongs to the current weak border. Obviously these flags are only meaningful for black nodes, since grey nodes always belong to the weak borders.

The basic operation is the access to adjacent nodes: knowing which nodes are adjacent to a node $q$ in level $k$ allows us to know in constant time which is the smallest node (in levels not greater than $k+1$) adjacent to any son of $q$ in any direction. It is therefore possible to build in constant time each adjacency link at level $k+1$.

Since we know the exact border if there are not grey nodes, the basic step of the algorithm is exploding grey nodes in their sons and setting the nature, when possible, of the sons and of their adjacent "weak" nodes.

Here below a sketch of the procedure REFINE_BORDER is shown, which builds sets STRONG and WEAK for level $k+1$, starting from the same sets for level $k$.

```
procedure REFINE_BORDER(WEAK, STRONG);
begin
    OLD_WEAK := WEAK;
    WEAK := { };
    for each grey node v in OLD_WEAK do
        OLD_WEAK := OLD_WEAK - {v};
        for each son w of v do
            PROCESS_NODE(w, OLD_WEAK, WEAK, STRONG);
```

```
        for each node p in OLD_WEAK do
              SELECT(p, WEAK, STRONG);
    end;
```

Procedure REFINE_BORDER is called by the main program APPR_BORDER, which also sets the initial values for STRONG and WEAK, until the real border is obtained.

```
    program APPR_BORDER(root);
    begin
        CREATE(root);
        case root.color of
        grey:       WEAK := {root};
                    STRONG := { };
        black:      WEAK := { };
                    STRONG := {root};
        endcase
        while WEAK ≠ { } do
              REFINE_BORDER(WEAK, STRONG);
    end.
```

Procedure PROCESS_NODE is called by REFINE_BORDER in order to process the sons of grey nodes in OLD_WEAK. When we discover a white son, say $x$, we can definitively confirm that the nodes in weak border($k$) adjacent to $x$ are in strong border($k+1$).

Discovering a grey son, say $y$, we can temporarily exclude that the nodes in weak border($k$) adjacent to $y$ are inner nodes. Moreover node $y$ becomes a node of weak border($k+1$).

```
    procedure PROCESS_NODE(w, OLD_WEAK, WEAK, STRONG);
    begin
        case w.color of
        white:      for each black node p adjacent to w in OLD_WEAK do
                          p.IS_BORDER := true;
        grey:       CREATE(w);
                    WEAK := WEAK + {w};
                    for each black node p adjacent to w in OLD_WEAK do
                          p.IS_INNER := false;
```

```
        black:      CREATE(w);
                    for each node p adjacent to w in level not greater than k+1 do
                         case p.color of
                         white:      w.IS_BORDER := true;
                                     w.IS_INNER := false;
                         grey:       w.IS_INNER := false;
                         endcase;
                    SELECT(w, WEAK, STRONG);
        endcase
    end;
```

Procedure CREATE creates a record for representing a new node and sets the initial values for flags IS_BORDER and IS_INNER (recall that these flags are meaningful for black nodes only).

```
    procedure CREATE(w);
    begin
        create a node representing w;
        w.IS_BORDER := false;
        w.IS_INNER := true;
    end;
```

Procedure SELECT puts the current node either in strong border($k+1$) or in weak border($k+1$) on the basis of the value of the flags set by procedure PROCESS_NODE. Moreover it reinitializes the flags for nodes put in WEAK, i.e. nodes remaining in the weak border($k+1$).

```
    procedure SELECT(x, WEAK, STRONG);
    begin
        if x.IS_BORDER
            then STRONG := STRONG + {x}
            else if not x.IS_INNER
                then WEAK := WEAK + {x};
                     w.IS_BORDER := false;
                     w.IS_INNER := true;
                else dispose(x);
    end;
```

## 4. Algorithm correctness

We will prove the algorithm correctness by first showing in theorem 4.1 that each grey or black node is examined by the algorithm, a black node is discarded only if it is not adjacent to any grey or white node, and a black node adjacent to a white node becomes part of the strong border set.

The following theorems hold:

**Theorem 4.1**: the $k$-th iteration of procedure REFINE_BORDER correctly computes the $k$-approximated border, starting from the $(k-1)$-approximated border.

**Proof**: the $k$-approximated border is the union of weak border($k$) and strong border($k$), which correspond respectively to set WEAK and set STRONG at the end of the $k$-th iteration.

It can be seen that each node in weak border($k$) or in (strong border($k$) - strong border($k-1$)) is either a black node in weak border($k-1$) or a son of a grey node in weak border($k-1$). Hence "new nodes" are generated only from nodes in OLD_WEAK. Let us examine what happens to grey nodes in OLD_WEAK: each of them generates four sons: the grey ones obviously belong to weak border($k$), and therefore are correctly put in WEAK; black ones belong to strong border($k$) if and only if they are adjacent to a white node, while they belong to weak border($k$) if and only if they have at least one grey adjacent and do not have any white adjacent: it is simple to see that both cases are correctly treated in procedures PROCESS_NODE and SELECT.

As far as it regards the black nodes in OLD_WEAK, the inductive hypothesis assures that each of them has at least one grey adjacent and does not have any white adjacent. Hence a black node $p$ in OLD_WEAK only can change its status thanks to the explosion of one of the grey nodes adjacent to it. Three cases are possible:

- if the explosion of adjacent grey nodes creates at least a white node adjacent to $p$ then $p$ belongs to strong border($k$): this is correctly implemented in code for processing white nodes in procedure PROCESS_NODE;
- if the explosion of adjacent grey nodes creates only black nodes adjacent to $p$, then $p$ does not belong either to strong border($k$) of weak border($k$): in fact the IS_INNER and IS_BORDER flags, which are reinitialized at the beginning of the $k$-th iteration, are never modified; hence SELECT correctly processes node $p$;
- otherwise, since no white nodes adjacent to $p$ and at least one grey node adjacent to $p$ are created, $p$ belongs to weak border($k$); hence IS_BORDER is left to *false* and IS_INNER is set to *false*: therefore SELECT correctly processes $p$.

Hence the $k$-approximated border is correctly computed from the $(k-1)$-approximated border.

Q.E.D.

**Theorem 4.2**: program APPR_BORDER correctly computes the series of $k$-approximated borders.

**Proof**: trivial from theorem 4.1, and from the fact that the 0-approximated border is obviously correctly computed by the assignments in program APPR_BORDER.

Q.E.D.

Let us denote the region associated to a set of nodes $S$ as $R(S)$.

**Theorem 4.3**: the following relation holds for each $k$:

$$R(\text{strong border}(k)) \subseteq R(\text{real border}) \subseteq R(k\text{-approximated border}).$$

**Proof**: the left inequality derives from the fact that each node $p$ in strong border($k$) is, by definition, adjacent to a white node, hence $p$ is contained in the real border. The right inequality is proved by observing that for each $k$, each node in the real border either belongs to $k$-approximated border or derives from the explosion of a grey node in weak border($k$).

Q.E.D.

**Theorem 4.4**: program APPR_BORDER correctly computes in STRONG the real border.

**Proof**: when program terminates WEAK is empty and $k$-approximated border coincides with strong border($k$). Since STRONG represents strong border($k$) and given theorem 4.3, the thesis follows.

Q.E.D.

## 5. Approximation characteristics

In the following we characterize the quality of the approximation in terms of guaranteed distance of the approximated border nodes from the boundary line of the object and in terms of the ratio between the approximated border area and the real border area. This result is stated by the following theorems:

**Theorem 5.1**: each node in the $k$-approximated border has an $L^\infty$ distance less than $2^{-k}$ from the boundary line.

**Proof**: by definition we know that a node $p$ belonging to the $k$-approximated border is adjacent to either a grey or a white node. If $p$ is adjacent to a white node $q$ then there is a portion of the boundary line between $p$ and $q$ and the thesis is proved. If $p$ is adjacent to a grey node $r$ then certainly $r$ contains a portion of the boundary line; but, since we are computing the $k$-approximated border, $r$ has edges of size $2^{-k}$. Hence, due to the $L^\infty$ metric, the thesis is proved.

Q.E.D.

The above theorem gives us an upper bound for the distance from the real boundary line of the approximated border obtained if we stop the computation process after $t$ iteration steps. This can be used as a quality measure in order to decide whether to stop the approximation refinement process or not.

Let us denote the area of a set of nodes $S$ as $A(S)$.

**Corollary 5.1**: the following relation holds for each $k$:

$$A(\text{strong border}(k)) \leq A(\text{real border}) \leq A(k\text{-approximated border}).$$

**Proof**: directly derives from theorem 4.3.

Q.E.D.

**Corollary 5.2**: the following relations hold for each $k$:

$$A(\text{strong border}(k+1)) \geq A(\text{strong border}(k))$$
$$A((k+1)\text{-approximated border}) \leq A(k\text{-approximated border})$$

**Proof**: trivial.

Q.E.D.

Theorem 4.3, theorem 5.1 and corollary 5.1 assure that our definition of approximated border, based on weak border($k$) and strong border($k$), provides upper and lower bounds for the real border which become tighter while increasing $k$, and therefore is a good approximation technique. Moreover corollary 5.2 shows an additional quality measure that can be used when deciding whether to stop the approximation refinement process or not.

## 6. Algorithm complexity

In the following time and space bound for the proposed algorithm are given.

**Theorem 6.1**: the $k$-th iteration of REFINE_BORDER can be performed in $O(g(k))$ worst case time, where $g(k)$ is the number of grey nodes in weak border($k$).

**Proof**: the first external cycle in procedure REFINE_BORDER examines each grey node exactly once, and it is trivial to check that both the inner cycle and all cycles in PROCESS_NODE are performed in constant time. As far as it regards the second external cycle in REFINE_BORDER, it only examines black nodes in OLD_WEAK, but since they have to be adjacent to grey nodes of not smaller size, their number can not be greater than $8 \cdot g(k)$.

Q.E.D.

**Theorem 6.2**: procedure REFINE_BORDER is executed exactly $T$ times, where $T$ is the depth of the quadtree.

**Proof**: the $k$-th iteration of procedure REFINE_BORDER takes away from OLD_WEAK all grey nodes in level $k-1$, and does not put in WEAK any grey node in level less than $k$. Hence after the $T$-th step WEAK will contain no grey nodes and therefore, by definition of weak border($k$), no nodes at all.

Q.E.D.

**Theorem 6.3**: the first $L$ $k$-approximated borders are computed in worst case time $O(G(L))$, where $G(L)$ is the total number of grey nodes on levels not greater than $L$.

**Proof**: directly derives from theorems 6.1 and 6.2.

Q.E.D.

**Theorem 6.4**: program APPR_BORDER uses $O$(size($L$-approximated border)) total worst case space to compute the first $L$ $k$-approximated borders.

**Proof**: in program APPR_BORDER space is required to store sets WEAK and STRONG (we can neglect the space required to store OLD_WEAK, which corresponds to set WEAK in the preceding iteration of REFINE_BORDER).

At step $k$, the size of STRONG is not greater than the size of set STRONG at step $L$.

Consider now the set WEAK: it consists of grey nodes and black nodes. Each grey node in WEAK at step $k$ will generate at least one node in the $L$-approximated border, and the number of black nodes in WEAK is not greater than $8\,g(k)$ (see proof of theorem 6.1).

Hence at step $k$ the size of $k$-approximated border is $O$(size($L$-approximated border)).

Q.E.D.

The above results show that our algorithm is competitive, in the sense defined in [2, 6] with respect to the algorithms proposed in [7, 4, 10]. In fact all these algorithms takes $O(G)$ to compute the exact border.

As far as it regards the space complexity of our algorithm, it is interesting to note that theorem 6.4 assures that our algorithm is optimal, since the space used asymptotically matches the output size.

## 7. Experimental results

Some experiments are being performed, both on real shapes and on randomly generated ones, where for real shapes we mean simply connected shapes or shapes typical of geographical applications.

Shapes used in the experiments have been generated as the union of polygons extracted from a fixed set of basic shapes, varying their area and position.

Two kinds of operations have been performed: proximity queries and shape recognition. The first results confirm that while solving proximity queries on geographical objects, or while checking geometrical relations between the above described "regular" shaped objects it is often sufficient to stop the approximation process early: thus our approach is very promising.

A limit in our experimental approach is that results are influenced by the choice of the set of basic shapes. A possible improvement is considering simple randomly generated basic shapes.

Quantitative results will be given in the full paper.

## 8. Conclusions

We proposed an algorithm that on-line computes the border of a bidimensional object represented by a quadtree; the border is defined as the set of black nodes adjacent to at least one white node.

The border is found by computing a series of guaranteed approximations, each one obtained from the previous one.

The series of approximations converges to the real border of the object and is computed within the same time bounds of known approaches for computing the boundary of a quadtree, but our algorithm allows to:

- obtain approximated solutions as soon as the necessary informations are available, by visiting one level of the quadtree at a time;
- evaluate the meaningfulness of each approximated solution;
- stop the computation as soon as a solution with sufficient precision is obtained;
- reducing space resource required for computing the new approximation.

Our algorithm is shown to be competitive with respect to algorithms that off-line compute the exact boundary both in space and time complexity.

## Acknowledgements

We wish to thank Maurizio Talamo for the nice discussions and the very useful hints.

## References

[1] H. Atkinson, I. Gargantini, T. R. S. Walsh, *"Counting Regions, Holes, and Their Nesting Level in Time Proportional to the Border"*, Comp. Graph. and Image Proc., 29, 2 (Feb. 1985), pp. 196-215.

[2] A. Borodin, N. Linial, M. Saks, *"An Optimal On-line Algorithm for Metrical Task Systems"*, Proc. of the $19^{th}$ Ann. ACM Symp. on Theory of Comp., 1987, pp.373-382.

[3] R. Dyer, A. Rosenfeld, H. Samet, *"Region Representation: Boundary Codes from Quadtrees"*, Comm. ACM, 23, 3 (Mar. 1980), pp.171-179.

[4] L. Jackins, S. L. Tanimoto, *"Quad-Trees, Oct-Trees, and K-Trees: A Generalized Approach to Recursive Decomposition of Euclidean Space"*, IEEE Trans. on Patt. An. and Mach. Intel., 5, 5 (Sept. 1983), pp. 533-539.

[5] Klinger, C. R. Dyer, *"Experimentts in Picture Representation Using Regular Decomposition"*, Comp. Graph. and Image Proc., 5, 1 (Mar. 1976), pp. 68-105.

[6] S. Manasse, L. A. McGeoch, D. D. Sleator, *"Competitive Algorithms for On-line Problems"*, Proc. of the 20$^{th}$ Ann. ACM Symp. on Theory of Comp., 1988, pp.322-333.

[7] H. Samet, *"Computing Perimeters of Images Represented by Quadtrees"*, IEEE Trans. on Patt. An. and Mach. Intel., 3, 6 (Nov. 1981), pp. 683-687.

[8] H. Samet, *"Neighbor Finding Techniques for Image Represented by Quadtrees"*, Comp. Graph. and Image Proc., 18, 1 (Jan. 1982), pp. 37-57.

[9] H. Samet, *"The Quadtree and Related Hierarchical Data Structure"*, Comp. Survey, 16, 2 (Jun. 1984), pp. 187-264.

[10] H. Samet, *"A Top-Down Quadtree Traversal Algorithm"*, IEEE Trans. on Patt. An. and Mach. Intel., 7, 1 (Jan. 1985), pp. 94-98.

[11] H. Samet, The Design and Analysis of Spatial Data Structures, Addison-Wesley, Reading, MA, 1990.

[12] H. Samet, R. E. Webber, *"On Encoding Boundaries with Quadtrees"*, IEEE Trans. on Patt. An. and Mach. Intel., 6, 3 (May 1984), pp. 365-369.

# Parallel Processing of a Class of Geographical Queries

Martin L. Kersten

Carel A. van den Berg

Effective exploitation of parallel database management systems for geographical databases is strongly influenced by the inherent parallelism in the queries being posed. The processing skew encountered in solving subproblems and the sheer size of the databases require novel query processing techniques. The dynamic query optimization and processing scheme described here indicates a new and challenging research area for advanced architectures for such next-generation database management systems.

## 1 Introduction

Exploitation of distributed processing capacity provided by large multi-processor systems remains one of the cornerstones to improve the performance of database management systems. To illustrate, several large research groups have produced research prototypes that attack the problem from different angles, such as Bubba [Bea90], Gamma [DGS+90], and PRISMA [KAH+88]. The key to achieve this goal lies in the integration between system architecture and the query optimizer. Although each prototype has demonstrated performance gains over centralized query processing, they are mostly based on the same assumption. Namely, that the runtime query processing cost can accurately be determined at query compile time.

An alternative approach, researched in this paper, is to exploit domain specific optimization rules and to determine for each query at run-time whether a parallel execution pays off, i.e. outweights the overhead incurred by managing fragmented and replicated database structures.

Consider, for example, a large geographical database, such as a database with detailed maps of Europe, and additional descriptive information, such as travel guide information. Our hypothesis is that effective handling of queries against such databases involves two steps. First, given a geographical query derive a program that solves it under the assumption that the database fits in main memory. Second, derive a query specific scheduler to process the database portions at minimal cost.

Our approach is based on the observation that with fast processors and cheap large main memories running queries against databases of 10-100 Megabyte do not pose severe problems to a DBMS designer. One merely loads the database structures

The work reported in this paper is partly supported by the Stichting Informatica Onderzoek, Nederland (SION)

into main memory for the duration of the session and optimizes the queries by massaging the query plan to reduce CPU resource consumption. Techniques can be borrowed from symbolic program optimization, such as data flow analysis, as well as traditional query optimization techniques, such as algebraic transformations, to improve response time and storage for intermediate results.

Remains the problem to handle large geographical databases. They can be seen as a collection of small databases. Thus, we should derive a database processing schedule for each query to minimize IO cost. Moreover, such a query specific scheduler can replicate the query program over a processor pool, it can control the distributed loading of database portions, and it can integrate the partial results produced at each processing node.

This approach can be classified as a dynamic query processing scheme, because the query specific scheduler adapts to the system load distribution, the processing skew of the query subtasks, and the (initial) data distribution. Once a processor in the pool has solved a partial problem it can be assigned a new task. Furthermore, the scheduler can direct subtasks to the data location or, reversely, moves data around to minimize the total IO cost.

In the remainder of this paper we will show that this approach naturally fits the handling of a class of geographical queries, coined *shortest path queries with boundary conditions*. The benefits from the dynamic query processing scheme stem from two aspects. First, the query task scheduler uses a hypothetical answer to guide the loading of database portions and, second, the response time is improved by handling query batches with automatic (replicated) caching of the relevant portions. The net result is that query processing adapts to the inherent parallelism. Simple subqueries lead to non-parallel executions, while for complex subqueries the response time is minimized by adding resources until the overhead incurred in their synchronization outweights their benefits.

Furthermore, we describe the Goblin system architecture built around this query processing scheme. We indicate the results obtained from a queuing model simulation and an actual implementation of the query processing framework on a 100-node multiprocessor [KAH+88].

In Section 2, we introduce a real-world geographical query and analyse the options to parallelize execution. Section 3 describes the components of our dynamic query processing scheme. Section 4 gives an overview of the Goblin object-oriented database system and the results obtained for the query processing scheme. Section 5 describes the current state of the Goblin project and indicates directions for future research.

# 2 The Tourist Problem

In this section we introduce a class of geographical queries through a real-world example. It is used to indicate options and limitations to parallize geographical queries.

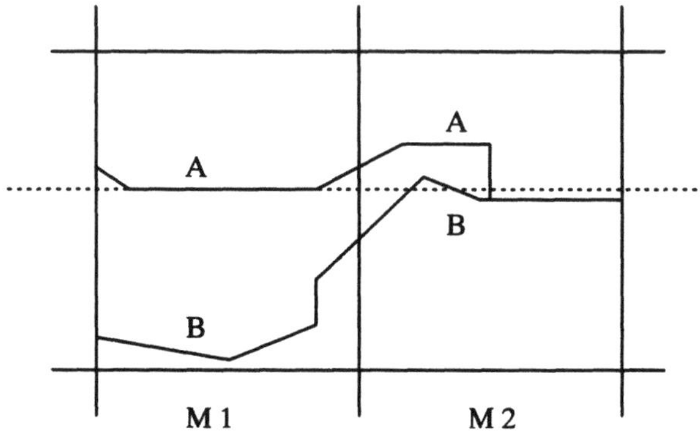


Figure 1: A hypothetical map

## 2.1 Narrative description

Consider the planning of a summer camping trip in Europe, let's say we are interested in the shortest trip from Amsterdam to Vienna. During this trip we need camping accommodation with hot showers. It is unlikely that the camp sites lay on the shortest path, which means that we should allow for a detour each day of, say, 25 km. Furthermore, we are fanatics of Baroque churches, for which we would like a list of candidates to visit. Again, we allow for a detour of 25 km from the shortest path.

This hypothetical holiday trip does not pose severe processing problems in real-life. A workable algorithm runs as follows. Take the box with maps of Europe (with possibly different resolution and topological information), the camping guides, and the tourist guides. Then draw a straight line from Amsterdam to Vienna, the ideal route. Then, retrieve all European maps crossed by the straight line using the map boundary information. Since each guide is associated with a geographic area, we can temporarily drop those that do not intersect with the straight line.

Once these ingredients have been collected we can solve the tourist problem by repeatedly solving a related query for each map involved. Namely, for each map we know the ideal route to our destination. We enter at one edge and leave at another edge (except for the first and last map). So we can locate the closest approximation of the partial route. Thereafter, we can associate the descriptive information using, probably, a single camping and tourist guide per map.
Under the assumption of an idealized collection of maps, i.e. they do not overlap and they are of the same resolution, the outcome of the query is obtained by glueing the partial results.

The latter requires some care, because local optimal solutions may not coincide with the global solution. For example, consider the two 'maps' in Figure 1. The dotted line indicates the ideal route, while the other lines depict local approximations. The optimal local approximations are M1.A and M2.B. Clearly, the combination does not form a proper route. Rather, the optimal is the path M1.A+M2.A. Thus, the algorithm should be refined to obtain a candidate list of routes for each map and join these lists in the final stage, such that the combined path is again minimal. This global join can be based on tuples that represent each map and path identity, start and end point, and the length of the path only. That is, a kind of semi join algorithm suffices. Alternatively, when the amount of partial information gathered fits in main memory, we can derived a 'hypothetical' map and solve the original query against this new one directly. That is, calculate the shortest path through the database of partial solutions.

## 2.2 Gaining speed by parallel processing

This query can be seen as a specific case of the class of shortest path queries with boundary conditions. Although this geographical query is quite simple, it illustrates some phenomena relevant for exploitation of parallelism in a database setting.

The planning of this trip can be speed up by getting all family members involved. Therefore, we merely distribute the maps and guides, such that each person is assigned the task to perform the partial planning. We might even replicate the guides or tear them apart to avoid concurrency problems among the members. The integration of partial results again should be controlled by a single person, who might use an incremental algorithm to work in parallel with the partial planning.

Furthermore, a short trip involving a single map should not be attacked with a parallel evaluation scheme unless the query itself can be broken into independent subqueries. For example, one can recognize that this query consists of three related queries; the shortest path selection, the camp site selection, and the Baroque church selection. Thus, one might start with the shortest path selection first and use the result to drive the two remaining queries in parallel.

In either case, the solution is found by generation of a hypothesis and processing a derived query over a small portion of the database. The hypothesis is that the shortest route to a destination is also the closest approximation to the straight line. The maps being crossed are taken as input to the query scheduler.

Furthermore, the algorithm does not prescribe an evaluation order. One may start planning with any map, provided it is being crossed by the ideal route. In computer terms, one can start processing with whatever map happens to be already loaded in the DBMS buffers. Once a route cannot be found between two maps, we also know which map to look at such that the chances are high to find it.

Breaking the query into related tasks also simplifies managing their differences, such as their resolution, topological information, and data representation. Each kind of map can be handled with the specific kind of query program. Integration of results is under control of the query scheduler, which converts the cross-map data to a common resolution and representation. Never do we have to re-scale complete maps.

Finally, the processing time involved with each map differs greatly. That is, the processing speed of the family members is highly skewed (touristic interesting areas require more time than others), and it is not always cheap to replicate a piece of information quickly (such as copying the travel guides).

To summarize, a query processing scheme for geographical database systems should:

- break a query into independent queries to avoid re-scaling and data conversion,
- adapt itself to the skew in processing time to handle subqueries,

- keep related information together so as to avoid copying,
- and control the amount of replication in the distributed file cache buffers.

## 2.3 Current parallel DBMSs'

As mentioned in the introduction, several parallel database systems have been designed over the last decade. They exploit parallelization by declustering data over the processor pool and they mostly apply a dataflow driven query evaluation scheme. A common mechanism is to decluster the data randomly. This improves query batches with a statistical homogeneous access pattern. It won't work beyond the level of individual maps, because then we are forced to continuously rebuild maps at some site. Instead, a spatial distribution mechanism is preferred.

Query processing in current database machines is generally based on the combination of two ideas: operation pipelining and static scheduling. Operation pipelining is the process of mapping the operations of the operator tree for a query onto virtual processors. In this model the data flows from the leaf processors to the processor associated with the top of the operator tree. Before query execution starts, the virtual processors are allocated to physical processors using a (heuristic) scheduling algorithm. This algorithm uses assumptions based on the data distribution, communication delays, and the selectivity of the operations to come up with an optimal allocation, where optimal means with a minimal response time. This approach is widely used in database machine architectures like Bubba [Bea90], Gamma [DGS+90], and PRISMA [KAH+88].

Although operation pipelined query processing results in a natural and elegant processing model, it has also some disadvantages. An immediate consequence of pipelined execution is that a communication delay is introduced between each operation in the tree. Furthermore, for each operation of pipelined execution it is difficult to obtain a fair load distribution among the processors involved. The prime reason being the data reduction taking place within the tree. The processors at the top of the tree will therefore remain underutilized for most of the time.
Moreover, the assumptions about the data distribution for which an optimal allocation is determined may not hold at the query execution time. The order in which the operations are performed may thus be badly chosen, and lead to inefficient use of the available processors.

The shortest path required in the tourist problem also poses severe problems for the current query optimizers. Either, one implements this algorithm at the level of the spatial data structure, or one uses a transitive closure operator provided by the query language. The disadvantage of the first approach is that this code interferes with the caching policy of the query processor. Moreover, optimization of the algorithm becomes hard wired code, which complicates their integration with a rule driven query optimizer. The second approach cannot exploit the hypothesis of the ideal trip, unless the spatial relationships can be expressed.

These disadvantages related to static scheduling have already been identified in the literature [Ape83, GW89, Mur89]. Apers [Ape83] compared the influence of a static scheduling and dynamic scheduling method on a distributed database system. Graefe [GW89] focussed at improving the performance on a uniprocessor by choosing the access method or join algorithm at run-time. For this purpose the query optimizer prepares a query evaluation plan, which contains enough information to make the choice dynamically. Murphy [Mur89] also concentrates on parallel query execution, where the join processing is decomposed into a join operation on pages (segments), and an optimal schedule for the page join operations is searched. The main difference with our approach is that we determine the scheduling at run-time, taking the actual execution times for the subtasks into account.

# 3 A Dynamic Query Processing Scheme

In this section we have a closer look at the compilations required to arrive at a dynamic query processing scheme. The presentation illustrates work in progress on the design of rule driven query processing.

## 3.1 The Query Language

Manipulation of a geographical database requires a powerful query language including recursion. Several such languages have been proposed [Güt89, Her88, BB81]. The query language developed for Goblin [MK90] is an SQL-like language for object-oriented databases that supports application-domain specific types. The corresponding data structures and operators are either defined in terms of the core database programming language or they are directly implemented in the host language (i.e. C). We assume familiarity of SQL-like query languages and merely focus on the query processing aspects for solving the tourist problem.

The database is organized around three object classes, i.e. the camping guide, the tourist guide, and the map. A partial description of the schema is shown in Figure 2. We assume that the tourist guide only contains monuments. Furthermore, the database is considered perfect, i.e. all locations mentioned in the guides correspond with points of the map. The map is organized as a large graph, centered around node information. A node optionally bears a symbolic name. For simplicity, we assume that each city is represented by a single point, the location of the city center.

Formulation of the query over the database requires some care. One might be tempted to construct a single query to specify the required information as concisely as possible. This would aid the query optimization performed by a system, such as the detection of common subexpression. Although such a track is often possible, it complicates both their formulation and the interpretation of the result. In particular, the result consists of a collection of heterogeneous objects or one ends-up with a universal relation with many null values.

Instead, the tourist problem should be formulated as a batch of related queries. Then the components of the batch are easier to grasp and the results are not cluttered with null values. The first query retrieves the shortest path between Amsterdam and Vienna. The result is converted into a new polyline using an aggregate operator (like avg() in SQL), which takes a list of edges and builds a polyline. The remaining queries collect the boundary information. The details on the detours are ignored.

```
select polyLine(Z.location)
from X in map, Y in map, Z in  map
where distancePoly(Z.location, shortestPoly(X.location,Y.location))=0

select C.name,C.location
from C in campguide, X in map, Y in map
where X.name='Amsterdam' and Y.name='Vienna' and C.shower='hot'
and distancePoly(C.location, shortestPoly(X.location,Y.location))<25

select T.name,T.location
from T in touristguide, X in map, Y in map
where T.kind='Baroque church'
and distancePoly(T.location, shortestPoly(X.location,Y.location))<25
```

```
create class campguide( name string, shower string, location point, ... ) END CLASS
create class touristguide( name string, kind string, location point, ...) END CLASS
create class map(name string, node point,
            neighbours set of  (node point, distance integer), ....) END CLASS
```

Figure 2: The tourist database schema

## 3.2 The Query Program

The first step in solving the tourist problem is to construct a query program that solves it under the assumption that all relevant information can be kept in main memory. That is, we retrieve the polylines and descriptive information for each map separately.

An optimization decision to be taken here is whether we should construct a single query program that handles all three query blocks (sequentially/ parallel light-weight processes), or that we construct three independent query programs. The latter case requires at least three processors to work in parallel, while the cost involved in loading the map data on all processors should not outweight the gain in response time.

For each query program containing multiple query blocks, one can further optimize their processing using optimizing techniques for centralized databases. Since the database fits in main memory, intermediate results can be represented in a compact manner. For, one can build *markings*, i.e. sets of oids, rather than construct the intermediate classes. Furthermore, one can keep these intermediates around for subsequent query blocks. In the tourist problem, the intermediate result that represents the shortest path between Amsterdam and Vienna can be kept available for re-use in the two remaining blocks.

Therefore, we included a rule driven query optimizer in Goblin along the lines described in [BG89, Güt89]. Several aspects of this compilation scheme needed adjustment to work in our scheme.

The executional algebra should be tuned towards efficient main memory processing. In particular, the intermediate results should be kept around to speed-up subsequent steps in the query process and the cost model should be refined to deal with cpu processing cost. Another issue is the role of the projection operator in a main-memory environment. Projection need not be pushed down the query tree unless an aggregate operator is being applied. In most cases, one can postpone projection until tuples are being formed for transport out of the query process.

Finally, the rule set should be extended with semantic query optimization rules. Such rules should model the approach described above to use a hypothesis to drive the query evaluation scheme. To illustrate, the first subquery should be reformulated into a parameterized query to select all polyLines from one map border to another.

```
function partial(P,Q point)
{       return select shortestPoly(Z1.location,Z2.location))
        from Z1 in map, Z2 in map
        where shortestPoly(Z1.location,Z2.location)) is not null
                and (Z1.location.X=P.X or Z1.location.X=P.Y)
                and (Z2.location.X=Q.X or Z2.location.X=Q.Y) }
```

In this query, the ideal route is given as runtime parameters. Furthermore, the routine shortestPoly returns **null** if no such path exists. The body of the qualification tests the shortest path for crossing the map boundaries.

From this query we can derive the subquery

```
select polyHead(P), polyTail(P), polyLength(P)
from P in partial(center("Amsterdam"),center("Vienna"))
```

to collect the information needed for integration of the partial results.

## 3.3 The Query Scheduler

In general, the main memory resources will be insufficient to hold the necessary data. Thus, a strategy to load portions of the database at minimal cost is needed.

This problem can be reformulated as re-using the index structures maintained in most spatial database structures. To illustrate, if the spatial data structure is a grid, such as in our real-world example, we can easily determine the grid-points being crossed by the ideal route. This leads to the candidate grid elements for processing. Furthermore, once we cannot directly find a shortest path, we know in which direction to search for a solution. That is, we can assign a loading priority to each grid element based on the assumption that the path should be close to the first guess. This is illustrated at the left in Figure 3.

This scheduling algorithm can also be applied to common spatial data structures such as quadtrees, grid-files,BSP trees (See [Nie89]). One should merely re-interpret the directory information being maintained, as illustrated in Figure 3.b..

One of the major research challenges is to find a rule driven query processing scheme, which enables the automatic construction of such schedules. The techniques applied to centralized query optimization require major enhancements, because such schedules involve symbolic reasoning over the query and the directory information. This may require the extension of the [BG89] optimizer with a theorem proving capability.

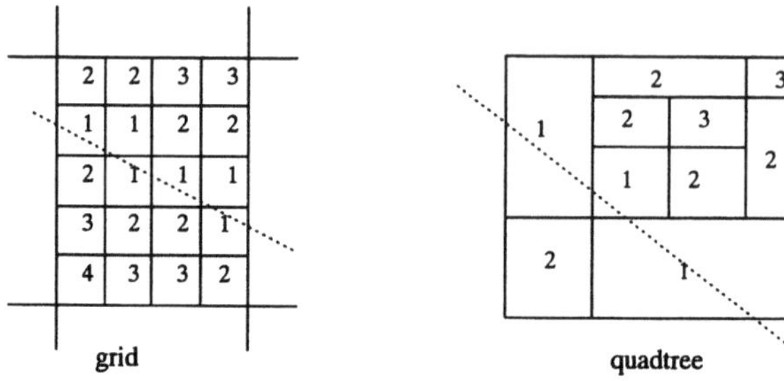


Figure 3: The schedule priority map

# 4 System Architecture

The architecture of the Goblin object-oriented database system is shown in Figure 4. It is based on two different kinds of processes: query dependent and schema dependent processes. The query dependent processes are called the Query Scheduler (QS) and the Query Program (QP). The schema dependent processes are the Class Managers (CM). The query dependent processes are generated for a specific query. The schema dependent processes are generated for a specific class or view. In the following we will briefly discuss their functionality as it pertains to our query processing scheme.

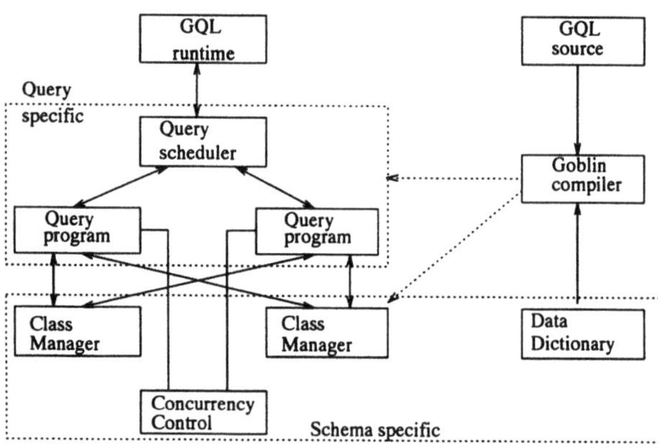


Figure 4: The Goblin architecture

## 4.1 The Class Manager

The Class Managers provide a persistent storage and retrieval facility for a single class. The class storage is composed of a number of segments. Upon receiving a *select* request from the Query Scheduler, the CM retrieves the segments from disk, stores it in its cache and acknowledges the QS with a *cached* message that it has successfully read a segment. The CM cannot only retrieve the raw segments from disk, but it can also apply selection, projection and partitioning operations on the data. When the segment data has been cached, a Query Program can subsequently retrieve this data by sending the *get* message to the CM. The CM then transports the data over the network to the QP. In the initial prototypes, each class manager implements a relational table, c.f. a One-Fragment Manager of PRISMA [KAH+88].

## 4.2 The Query Program

The Query Programs form the engine of the query evaluation process. The Query Scheduler translates the query over the classes into a batch of queries over their segments. Then each Query Program solves a query for a given combination of segments. The final query result is obtained by taking the union of the partial results produced at the QS.

In contrast with a pipelined execution model the execution order of the individual operations in the query is not fixed at compile time. It depends on the availability of the segments. For instance, if the QP is requested by the Query Scheduler to calculate $R_1 \bowtie S_2 \bowtie T_1$, and segments $S_2$ and $T_1$ arrive first, it will first calculate $S_2 \bowtie T_1$, and store the intermediate result in its cache for further use. When segment $R_1$ arrives, it completes the join operation, and informs the QS that it has reduced the $(R_1, S_2, T_1)$ segment combination.

## 4.3 The Query Scheduler

The Query Scheduler, finally, coordinates the distribution of work over the QPs'. It consists of three major components: a task filter, a task allocator, and the task integrator. The QS maintains a task table containing all the segment combinations to be reduced by the QPs'. Figure 5 illustrates the communication pattern between the components.

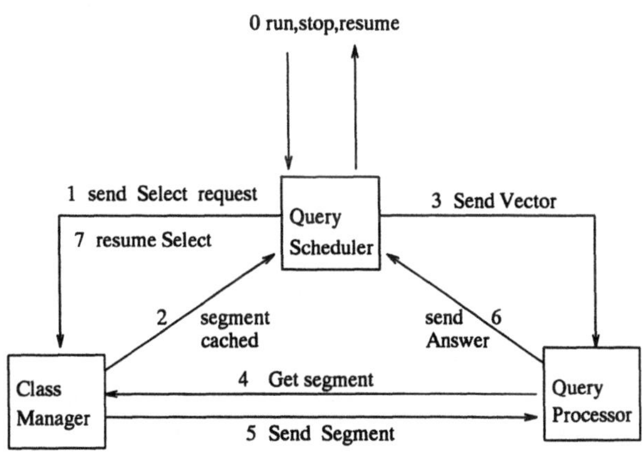


Figure 5: The Goblin communication pattern

The task filter removes those tasks from the task table that do not contribute to the final query result. For instance, this will occur when the relations $R$ and $S$ are hash partitioned on their join attributes. Another possibility is to exploit feedback information available from the QPs. For example the QP can detect which segment combinations produce empty partial join results, say $S_3 \bowtie T_4$. The task filter can then remove the remaining $(R_i, S_3, T_4)$ tasks from its task table.

The task allocator coordinates the execution of the query by selecting an idle QP for each task and removing it from the task table. Currently we consider three different schedule algorithms: *random* scheduling, *cache* scheduling, and *pipeline* scheduling. Clearly, the algorithm to select a task has a great influence on the efficiency. The *random* scheduling method selects randomly a new task from the task table. The consequence is that for practically each task, the QP has to obtain the required segments from the CMs. For the *cache* scheduling method, the QS keeps track of the segments being cached by the QPs. The QS minimizes the IO by selecting a task for which (almost) all the segments are already available. In the *pipelined* scheduling method, the QS also keeps track of the intermediate results available in the QP. It will send those tasks to the QP, which require the same intermediate results, thereby reducing both communication and processing time for a task.

The task integrator collects the partial results and produces the aggregated information, such as the global shortest path for our tourist problem. Such a task integrator uses a query specific data representation and algorithm, which is derived from the original query using domain specific optimization rules.

In our performance studies, we have focussed on the least complex scheduling method, namely the random scheduling method. If this technique is already profitable, the other approaches can only further improve the performance.

We expect from this architecture that the performance of the system does not degrade as the number of QP is increased. We rather expect that the throughput converges towards a maximum value determined by the cost for moving segments around, but not because of scheduling overhead. Our experiments described below asses whether these assumptions holds for our architecture.

## 4.4 Performance simulation

Since the construction of a full-fledged extensible database system is a large enterprise, we have take an evolutionary approach. First, we developed a queuing model of the proposed architecture taking a three-way relational join as a reference point. We choose this operation because joins are commonly used to related tables (the three object classes) and it poses severe problems in communication overhead. It is also easily parallelized. Furthermore, it captures the three-way complex join over the relations in our tourist problem.

The simulation was focussed on acquiring insight in the system utilization and the speed up factor as a function of number of processors, i.e. QPs'. We also expected that the query scheduler might become a bottleneck in the system, which would render the whole dynamic query processing scheme useless.

Two different system architectures have been investigated. The basic architecture directly models the system shown in Figure 5 and an extended model where the QPs' act as alternative segment caches. That is, the memory in the system components together form a large distributed segment cache. A segment can then be retrieved either from a Class Manager or from another Query Program where it might have been dropped for producing a partial result. We expected from this extension that the load is better distributed among the Class Managers and Query Programs.

The simulations showed that for a large processor pool, such as the PRISMA system [KAH+88], the utilization of the processor pool is still rather low, due to the network communication cost. Furthermore, a significant speed up was obtained by the adaptive replication of database fragments. In both cases the central scheduler did not become a bottleneck. The linear speed-up curve flattens before the scheduler becomes overloaded. That is, the speed up from parallelism becomes neglectable before the QS becomes saturated. A detailed account of these experiments are presented in [MKvdB90a, MKvdB90b].

## 4.5 The Validation Study

For a simulation experiment there is a tradeoff between the level of detail modelled and the implementation effort. To asses whether the architecture has been modelled in enough detail, we have also performed a validation study for the three-way join operation.

We have made a prototype implementation of the architecture on the POOMA machine in POOL-X. The POOMA architecture and operating system were specially designed to support the language POOL-X. The machine consists of 100 MC68020 processors interconnected through a dedicated packet switching network.

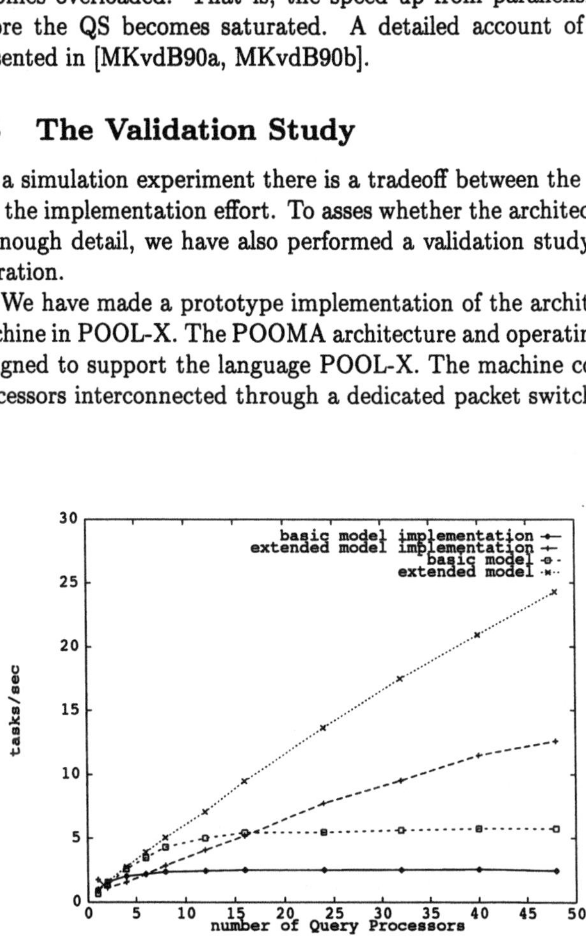


Figure 6: Task throughput for the simulation and the prototype

In the validation experiment we ran the versions with and without segment exchange, changing the number of Query Programs from one to fifty. The processes were all allocated on different physical nodes. We used the hardware timers provided by the processing nodes, to measure the total time spent in the query evaluation.

The validation experiment only measures the total time for a query. Figure 6 shows the measured task throughput for the runs, and the task throughput obtained through simulation. The model and the prototype display a similar behavior. In the range of 1 to 50 Query Programs the extended model and its prototype implementation show a linear speed up upto 25 QPs. The basic model and its implementation reaches its maximum throughput at 15 Query Programs.

On the whole, the performance for the simulated model is better than for the implementation. For the extended model this effect is caused by the simplifying assumption that the cache of the QP is unbounded. In the actual implementation cache misses do occur, which result in additional segment requests for the CM.

A close examination of the implementation showed an unexpected relation between processing speed and communication activity. This effect can explain the difference between the cost measured for reading a segment from disk in isolation and in the actual running system. We think that the constant discrepancy between the basic model and its prototype is caused by this effect. A more detailed account of this implementation and the results obtained can be found in [vdBKSA91].

# 5 Summary

Exploitation of parallel systems poses severe problems to the designers of database systems, because the attainable speed up is strongly related to the application domain and the query complexity. In this paper we have analysed a large class of geographical queries, called *shortest path queries with boundary conditions*, for options to parallelize their execution to improve response time.

We have described an alternative approach to query optimization and query processing which adapts to the potential parallelism in the query and the hardware resources available. The technique is based on splitting the query into a memory resident database problem and a query scheduling problem. Since the scheduler is query specific, it can better control the amount of replication of database fragments and it can better control the processing order using the spatial index structures.

This dynamic query processing scheme is part of the Goblin object-oriented database system currently under development at our institute. Preliminary performance studies have shown that the scheduler itself does not become an immediate bottleneck for highly parallel queries. Furthermore, linear speed up has been attained for parallel join queries where the data itself was not a priory declustered. The compilation techniques for deriving efficient query programs and schedulers is currently being addressed. For both, we extend the rule-driven optimizer described in [BG89]. The final system will run under the Amoeba distributed system [MvRT+90].

The design of query compilers is an open ended track. Besides optimization of the query programs on memory resident databases, much work has to be done on parameterized query schedulers. In particular, feed-back and feed-forward information from the query processors and class managers, respectively, can lead to substantial reductions in the number of query subtasks to process.

# References

[Ape83] P.M.G. Apers. *Query Processing and Data Allocation in Distributed Database Systems.* PhD thesis, Free University Amsterdam, 1983.

[BB81] R. Barrera and A. Buchmann. Schema definition and query language for a geographical database system. *IEEE Trans. on Computer Architecture*, Nov. 1981.

[Bea90] H. Boral and et al. Prototyping bubba, a highly parallel database system. *IEEE Transactions On Knowledge and Data Engineering*, 2(1), March 1990.

[BG89] L. Becker and R.H. Güting. Rule-based optimization and query processing in an extensible geometric database system. Technical report, Fachbereich Informatik Universität Dortmund, 1989.

[DGS+90] D. J. DeWitt, S. Ghadeharizadeh, D.A. Schneider, A. Bricker, H. Hsiao, and R. Rasmussen. The gamma database machine project. *IEEE Transactions On Knowledge and Data Engineering*, 2(1), March 1990.

[Güt89] R.H. Güting. Gral: an extensible relational database system for geometric applications. In *Proc. 15-th Int. Conf. on Very Large Databases*, pages 33–44, 1989.

[GW89] Goetz Graefe and Karen Ward. Dynamic query evaluation plans. In *Proc. SIGMOD*, 1989.

[Her88] J. Herring. Extensions to the sql language to support spatial analysis in a topological data base. In *GIS/LIS'88*, Nov 1988.

[KAH+88] M.L. Kersten, P.M.G. Apers, M.A.W. Houtsma, E.J.A. van Kuyk, and R.L.W. van de Weg. A distributed, main-memory database machine; research issues and a preliminary architecture. In M. Kitsuregawa, editor, *Database Machines and knowledge Base Machines*, pages 353–369, 1988.

[MK90] C.A. vd Berg M.L. Kersten, M.H. vd Voort. The goblin database programming language. In preparation, 1990.

[MKvdB90a] S. Shair-Ali M.L. Kersten and C.A. van den Berg. Performance analysis of a dynamic query processing scheme. In *Proc. of PRISMA Workshop*, pages 239–254, September 1990.

[MKvdB90b] S. Shair-Ali M.L. Kersten and C.A. van den Berg. Performance analysis of a dynamic query processing scheme. Report cs-r9050, CWI Amsterdam, The Netherlands, 1990.

[Mur89] M.C. Murphy. Effective resource utilization for multiprocessor join execution. In *Proc. of the fifteenth VLDB Conference*, pages 67–75, August 1989.

[MvRT+90] S.J. Mullender, G. van Rossum, A.S. Tanenbaum, R. Renesse, and H. van Staveren. Amoeba - a distributed operating system for the 1990s. *IEEE Computer Magazine*, May 1990.

[Nie89] J. Nievergelt. 7 +- 2 criteria for assessing and comparing spatial data structures. In *Design and Implementation of Large Spatial Databases*, LNCS 409, pages 3–28. Springer-Verlag, nov 1989.

[vdBKSA91] C.A. van den Berg, M.L. Kersten, and S. Shair-Ali. Dynamic parallel query processing. Report cs-9112, CWI, Amsterdam, 1991.

# An Access Method Based Query Processor for Spatial Database Systems*

Hans-Peter Kriegel, Peter Heep, Stephan Heep, Michael Schiwietz, Ralf Schneider

## Abstract:

The management of spatial data in applications such as graphics and image processing, geography as well as computer aided design (CAD) imposes stringent new requirements on socalled spatial database systems, in particular on efficient query processing of complex spatial objects. In this paper, we propose a multi-level, multi-representation query processor where the various levels correspond to various representations with different accuracies of approximating complex spatial objects and the answers are, comparable to using increasingly finer filters, more and more reduced to the final set of answers. The efficiency of spatial query processing is improved by an order of magnitude using the following two design paradigms: first, divide and conquer, i.e. decomposition of complex spatial objects into simple spatial objects (components) such as convex polygons, triangles or trapezoids, and second, application of efficient and robust spatial access methods for simple spatial objects. The most powerful ingredient in our approach is the decomposition. Applied to the last and lowest level of the query processor, it removes complex computational geometry algorithms from query processing, replacing it by simple and fast computational geometry algorithms for simple spatial objects. Following a classification and a rigorous study of alternative filter techniques we present one concrete and already implemented occurrence of our multi-level query processor. First experiences with our implementation indicate that the performance of our approach is clearly superior to traditional query processing strategies for complex spatial objects.

**Keywords:** Access Methods, Query Processing, Spatial Database Systems

## 1 Introduction

The demand for using database systems in application areas such as graphics and image processing, computer aided design (CAD) as well as geography and cartography is considerably increasing. The important characteristic of these applications is the occurrence of spatial objects. The management of spatial objects imposes stringent new requirements on socalled spatial database systems. One of the most challenging requirements is efficient query processing of complex spatial objects.

The problem of designing efficient query processing facilities is not yet solved by designing efficient spatial access methods for complex spatial objects, say arbitrary polygons. Assume a spatial application is restricted to the class of arbitrary polygons. Let us further assume that these polygons are approximated by their minimum bounding boxes and that an efficient spatial access method is available for minimum bounding boxes. Then, using the spatial access method supplies a set of candidate polygons on

---

* This work was supported by grant no. Kr 670/4-3 from the Deutsche Forschungsgemeinschaft (German Research Society)

which complex and time-intensive computational geometry algorithms have to be performed yielding overall bad query performance. This example already implies two important design goals for efficient query processing of complex spatial objects:

- A filter mechanism should be used to reduce the set of candidates on which computational geometry algorithms have to be performed.
- No complex computational geometry algorithms should be used in query processing, i.e. complex spatial objects should be decomposed into simple components for which simple and fast computational geometry algorithms are used.

Thus our approach of query processing complex spatial objects is based on a multi-level query processor where the different levels correspond to different accuracies of approximating complex spatial objects and the answers are, comparable to using increasingly finer filters, more and more reduced to the final set of answers. In order to fulfill design goal no.2, we apply the algorithmic paradigm 'divide and conquer' and use on the lowest level of the query processor a structural decomposition of complex spatial objects into simple components such as convex polygons, triangles, or trapezoids. In other words, by decomposition in a single preprocessing step at object insertion time, we reduce the query processing time for large amounts of queries, as we will see, by an order of magnitude.

In the next chapter we analyse spatial objects, spatial queries as well as their support by spatial access methods in spatial database systems. Chapter 3 presents a rigorous study and classification of spatial object representations coupling spatial access methods and decomposition techniques. In chapter 4 we propose a multi-level query processor using these representations and present a first performance comparison of this query processor. The paper concludes with a summary that points out the main contributions of our proposal and gives an outlook to future activities.

# 2 Manipulation of Spatial Objects and Queries Using Spatial Access Methods

## 2.1 What Spatial Objects to Support ?

In applications such as CAD/CAM, GIS, environmental or geographic database systems, two- or three-dimensional spatial objects are an essential object type. Objects such as points, lines, or rectangles are known as simple spatial objects. Semantically complex objects where the complexity is specific to the application, such as contour lines, limits of lots, and contours of CAD objects are shaped in the form of simple polygons. Complexity properties of such polygonal objects, as e.g. the shape, the number of vertices, or the smoothness of the contour are difficult to predict. Additionally holes as a general property of polygons have to be taken into account for a general handling of objects occurring in geographic information systems, e.g. lake areas surrounded by an area of land. For supporting the above type of applications, a special facility managing simple polygons is fundamental to a spatial database system. The types of spatial objects we consider is the class of simple polygons with holes (SPH, see figure 1). A polygon is simple if there is no pair of nonconsecutive edges sharing a point. A SPH is a simple polygon where simple polygonal holes may be cut out. From our experience, the class of SPH is adequate for GIS applications and most 2D CAD/CAM applications.

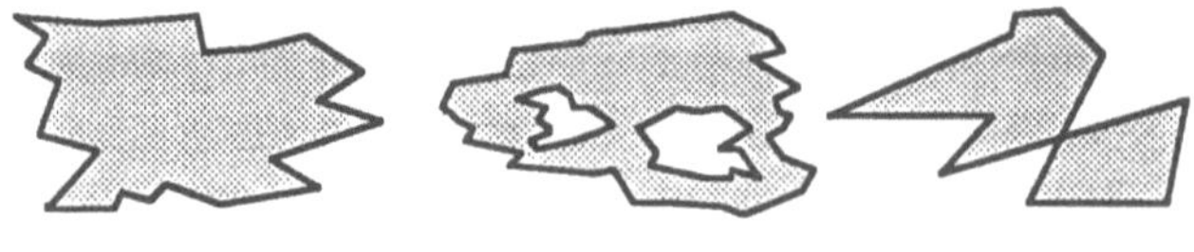

**Fig. 1 :** simple polygon simple polygon with holes non-simple polygon

## 2.2 What Spatial Queries to Support ?

A query in the applications mentioned above generally refers to spatial and nonspatial data. Spatial data can be classified into geometric and topological data whereas nonspatial data is alphanumerical data related to spatial entities. Additionally, spatial queries not only retrieve data, but also construct new objects. Usually these objects are displayed and not necessarily stored in the database.

Geometric data describes properties such as spatial location, size and shape of spatial objects. In a spatial query the spatial location is supported by, and therefore used in combination with a spatial access method. Topological data describes properties such as connectivity, adjacency and inclusion modelling the relationship between geometric data. Examples are: 'which lakes are in a given country' (inclusion), 'which areas are neighbors of each other' (adjacency), 'which polylines are connected and form a network of roads' (connectivity). By the way it is not absolutely necessary to store topological data explicitly, because basically it can be derived from geometric data.

In a spatial database system the following not complete list of spatial queries should be supported. Most application specific queries, as for example described in [Oos 90], [SV 89], can be decomposed into a sequence of such basic spatial queries.

- *PointQuery*:
  Given a point $p \in E2$, find all SPHs in the database where $p \in SPH$.
- *RegionQuery*:
  Given a $SPH^* \subseteq E2$, find all SPHs in the database where $SPH^* \cap SPH \neq \emptyset$.
- *EnclosureQuery*:
  Given a $SPH^* \subseteq E2$, find all SPHs in the database where $SPH^* \supseteq SPH$.
- *ContainmentQuery*:
  Given a $SPH^* \subseteq E2$, find all SPHs in the database where $SPH^* \subseteq SPH$.
- *IntersectionQuery*:
  Given a $SPH^* \subseteq E2$, find the intersection of SPH* with all SPHs in the database.

## 2.3 Spatial Clustering by Spatial Access Methods

In database systems the notion *clustering* is used when objects - which are often required jointly by queries - are stored physically close together. Analogously we use the notion *spatial clustering* if the spatial objects - which are often required jointly by spatial queries - are stored physically close together. Because spatial queries refer to the location of spatial objects in data space, the vicinity of spatial objects has to be mapped on secondary storage in such a way that efficient processing of the spatial queries is supported.

The basic building block for efficient query processing are access methods. In the literature efficient point access methods (PAMs) have been proposed. A survey can be found in [SK 88]. As also discussed in [SK 88], there are three basic techniques to modify these point access methods to support spatial clustering of spatial objects: clipping, where objects are divided along the partitioning lines of the underlying point access method; overlapping regions, where each partition of the access method may contain any object which it overlaps; and transformation, where simple spatial objects are mapped into higher-dimensional points. Figure 2 presents some of the well known spatial access methods (SAMs) and their affiliation to the techniques depicted. The R-tree [Gut 84], for example, uses the technique of overlapping regions, the grid file [NHS 84] transformation and the Cell-tree [Gue 89] uses clipping. The variants of the Buddy-tree [SK 90] (Buddy-tree, Buddy/over, and Buddy/clip) shows the possibility of applying all three techniques to one multidimensional point access method. Performance comparisons for some of these methods are outlined in [KSSS 89], [Ooi 88], [Fre 89], [GB 90], and [SG 90] .

One important property of SAMs based on clipping should be mentioned here: in contrast to the other two techniques clipping may induce an unpredictable amount of redundancy, depending on the size and shape of the stored spatial objects. This fragmentation problem of clipping methods is also emphasized in [SW 88] and [GB 90]. By an increasing number of overlapping spatial objects this fragmentation problem worsen in such a way that clipping has to use overflow mechanisms, as described in [Gue89]. In such a case clipping is to our opinion not an adequate technique.

Another important characteristic of SAMs is the type of spatial objects they are able to handle directly. All spatial access methods proposed up to now, are restricted to the storage of simply shaped object types such as cells of a fixed grid (grid cells for short) [OM 86], rectilinear rectangles (with their sides parallel to the axis, rectangles for short) [Gut 84], [NHS 84], [SK 90], [BKSS 90], [KSSS 90], spheres [Oos 90] or convex cells (convex polygons) [GB 90], [Cla 89].

Figure 2 shows a classification of proposed SAMs according to the two characteristics:

- which 'simple spatial object' type is supported by the SAM
- which technique is applied to handle these types

There are two combinations of object type and technique which are precluded. These methods, belonging to the shaded elements of the matrix, seem to be no competitive alternatives:

- The combination of spheres with a clipping technique leads to arbitrary complex regions induced by the intersection of spheres, which are hard to manage.
- The problem of finding a transformation, given an arbitrary number of parameters (for the convex polygons), cannot be adequately solved.

All other combinations lead to possibly competitive SAMs. Applying the ZB+-tree to grid cells exploits the linear ordering given by a transformation of extended two-dimensional objects (grid cells) into one-dimensional bitstrings. However, a SAM directly storing the two-dimensional grid cells can be developed, using one of the techniques, overlapping regions or clipping. It is obvious that SAMs which support rectangles have been of main interest. In the case of the TR-method (Transformation and Rectangle) due to space limitations only a few well known representatives are shown in figure 2. More complete studies of PAMs respectively SAMs can be found in [See 90], [KSSS 90] and [Oos 90].

| technique | grid cell 'G' | rectangle 'R' | sphere 'S' | convex 'C' |
|---|---|---|---|---|
| clipping 'C' | CG-method: | CR-method:<br>EXCELL [Tam 82]<br>R+-tree [SRF 87]<br>Buddy / clip [SK 90] | | CC-method:<br>Cell-tree [Gue 89] |
| overlapping region 'O' | OG-method: | OR-method:<br>R-tree [Gut 84]<br>R*-tree [BKSS 90]<br>R-file [HSW 90]<br>Buddy / over [SK 90] | OS-method:<br>Sphere-tree [Oos 90] | OC-method:<br>KD2B-tree [Cla 89]<br>P-tree [KS 90] |
| trans-formation 'T' | TG-method:<br>ZB+-tree [OM 86] | TR-method:<br>Buddy-tree [SK 90]<br>Grid-file [NHS 84]<br>etc. | TS-method: | |

'simple spatial object' type

Fig. 2 : Classification of spatial access methods

## 2.4 Handling Complex Spatial Objects by Container Representations

In order to adapt efficient access methods to complex spatial objects, a brute force approach was applied up to now. In this approach, any spatial object is placed in a rectilinear rectangle or convex polygon of minimum shape forming a 'container' for that object, yielding a 'conservative' approximation. A simple spatial object is called a container iff any point inside the contour of the complex spatial object is contained in the simple spatial object. Containers are suitable to be handled by well known spatial access methods.

As simple containers provide only conservative approximations, query processing strategies based on spatial access methods have to procede in a two-step manner. The first step, the socalled *filter step*, reduces the entire set of objects to a subset of candidates. The filter step is based on spatial access methods using the following property: if the container does not fulfill a query, so the object does not fulfill the query[1]. The candidates selected by the filter step have to be examined in the second step which is called *refinement step*. This step is based on complex algorithms known from the field of 'computational geometry' and detects all false hits from the filter step. By the way, if no spatial clustering is supported, i.e. if no filter step based on a spatial clustering of the objects is applied, every single object has to be examined with such complex computational geometry algorithms yielding high processing cost. To make spatial query processing more efficient there are two main goals: first, the accuracy of the filter step should be improved and second, complex computational geometry algorithms should not occur in query processing. The first goal may be achieved using multiple container representations to improve the approximation of each object, i.e. dead space should be minimized. According to the algorithmic paradigm 'divide and conquer' the second goal may be achieved by using complex computational geometry algorithms in a preprocessing step, i.e. at object insertion time the complex spatial object is decomposed into a set of simple spatial components. Query processing is completely performed on those simple components. Chapter 3 describes in detail efficient solutions to these two goals.

# 3 Representation of Spatial Objects Coupling SAMs and Decomposition Techniques

As outlined before the exact evaluation of a spatial query, e.g. 'PointQuery' or 'IntersectionQuery', using a filter step based on container representation requires specific algorithms testing the exact spatial object in a refinement step. These algorithms are based on techniques known from the field of computational geometry. The complexity of computational geometry algorithms is commonly superlinear in object complexity, i.e. in the number of vertices[2], and therefore quite costly, in particular for complex spatial objects. For that reason improving the filter step is important to further reduce the set of candidates to be refined. An optimal filter would exactly retrieve those objects, fulfilling the query. However, for the sake of simplicity, the representation of a container is bound to a low number of parameters, e.g. spatial location and extension. Thus it generally cannot provide an exact spatial object representation and therefore no exact query evaluation. An essential goal for the filter step is to minimize the dead space between the spatial object and the container to minimize the number of objects passing the filter step.

In this section we focus on different techniques used for a container representation. In particular we examine alternative representations gaining in accuracy of the approximation and the quality of the refinement step by introducing redundancy. Finally we generalize the approach of container representation and redundancy with respect to access methods.

[1] Let us notice, that also a 'kernel'-representation is available: if the kernel of an object fulfills a spatial query the same does the object itself. Container and kernel representation may be applied in parallel treating different kinds of queries. Any query may be supported by the appropriate form of representation used for filtering.

[2] point in simple polygon: O(n).
intersection of two polygons: O( (n+k) log n), where k is the number of intersection points.

### 3.1 Single Container Approximation

Known approaches representing complex spatial objects using spatial access methods are based on the concept of object approximations by single containers. Two well known techniques typically used for container representation are minimum bounding boxes (MBB) or bounding convex polygons (figure 3). Another less known container technique is the concept of minimum bounding spheres [Oos 90]. These approximations are based only on the object structure neglecting its spatial location, i.e. a spatial translation of the object would preserve the - as well translated - container. Such type of approximation we call *structural approximation*. The quality of a structural approximation in terms of the additionally covered area in comparison to the area of the spatial object considerably depends on the shape of this object.

Another type of object approximation is based on partitioning spatial objects using a predefined grid upon the object space. Therefore, this type of approximation is called 'grid approximation'. Any spatial object using a grid approximation is represented by an unique grid cell, namely the minimally shaped cell completely encluding the object. The main advantage of this approximation is the simplicity and efficiency of storing the cells in the form of bitstrings representing the hierarchical grid partitioning. Spatial queries such as a 'spatial join', see [OM 86], can be efficiently performed on this representation. A main disadvantage of a predefined grid resolution is the inadequate approximation of the object.

**Fig. 3** : grid approximation, minimum bounding box, convex hull, and bounding sphere

### 3.2 Multiple Container Approximations

The approximations described up to here use one single container and therefore provide a non-redundant approximation of spatial objects. However, there is no possibility for approximating to a given accuracy, because the minimum amount of covered container area is strictly bounded for structural as well as for grid approximations. In order to decrease the container area, redundancy can be introduced. Redundancy is introduced by approximating a spatial object not by a single container, but by a set of containers. Different types of redundancy can be applied for grid as for structural approximation.

#### 3.2.1 Decomposition Techniques Improving the Accuracy of the Filter Step

#### a) SizeBound and ErrorBound

One approach inducing redundancy based on grid approximation is proposed by Orenstein [Ore 89]. A spatial object is represented by a set of disjoint grid cells, socalled elements, completely covering its shape. That approximation provides a leeway of the number of elements controlling the amount of redundancy. In the socalled 'SizeBound(c)' approximation the number of elements is restricted to c. The amount of area covered by the set of elements depends on the shape and location of the object approximated. An 'ErrorBound(c)' approximation generates a set of elements approximating a spatial object according to the fixed grid resolution $2^c$. The number of elements and therefore the amount of redundancy is not restrictable in advance and may seriously increase.

In comparison to single object approximations of spatial objects an approximation based on sets of grid cells usually results in a smaller object covering. This reduces the number of candidates to be refined, as a smaller set of objects fulfills the filter step. However, increasing redundancy affects the performance of the filter step which has to manage a higher number of containers. Additionally, we have to avoid redundant refinement operations on the same objects. Thus the amount of redundancy induced has to be handled with care. This type of enhancement, aiming at a reduction of the objects to be refined, we call accuracy improvement of the filter step. In grid approximations providing accuracy improvement, however, the structural complexity of the candidates resulting from the filter step may be the same as or even higher than that of the original spatial objects. Therefore the refinement step will not profit from a qualitative improvement, because there is no reduction in object complexity, which would effect in more efficient, respectively simple, computational geometry algorithms.

**b) Rectilinear SizeBound**

Analogously to the SizeBound decomposition it seems to be a promising approach to decompose a spatial object into rectilinear rectangles, because the research in access methods has concentrated on efficient spatial access methods for such simple spatial objects. The goals in the development of such a decomposition technique, called 'rectilinear size bound' decomposition (RSB), are the number of generated components should not exceed a predefined number of components (RSB(c)) and the accuracy should be best possible. Because finding a covering of a SPH with a constant number c of rectangles and maximum accuracy is NP-hard, a heuristic algorithm was chosen, which is based on the improvement of the MBB approximation by cutting off rectilinear rectangles. The resulting rectilinear polygon is decomposed by an algorithm yielding an optimum number of components. A reduction in object complexity cannot be achieved by this approach.

### 3.2.2 Decomposition Techniques Improving the Quality of the Filter Step

**a) Edge Quadtree**

The 'edge quadtree' [Sam 90] is based on a grid approximation where the complexity of the object components represented by a grid cell is fixed. As shown in figure 4 any element is forced either to be fully placed within the object, or to contain a part of exactly one edge of the object, or to contain one vertex and parts of the two starting edges. In contrast to the other grid approaches the accuracy of the approximation hardly depends on the shape of the object. Therefore no forecast can be made concerning the number of resulting elements.

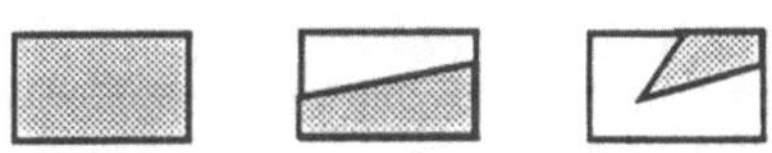

**Fig. 4** : elements of the edge quad tree

**b) Structural Decomposition Techniques**

We suggest a new and more flexible approach that aims at accuracy and qualitative improvement of the filter step based on the concept of a structural object decomposition. Due to the qualitative aspect the refinement step will profit from simpler object types. Complex spatial objects may be decomposed into simple components such as convex polygons, triangles, trapezoids, or rectangles. A structural decomposition of objects consists of a single preprocessing step at object insertion time. This preprocessing simplifies query processing in the refinement step because simple computational geometry algorithms can be applied to simple components. This approach transfers the complex

computational geometry algorithms from query processing to object representation. Structural decomposition techniques replace querying a single complex spatial object by querying a set of simple objects.
The type of object components is defined by the decomposition method applied. Depending on the decomposition algorithm homogenous or heterogenous sets of components are supplied. As in previous approaches these components are aimed to be handled by a spatial access method. If the spatial access method cannot handle the components directly, these components have to be approximated by a suitable type of container. Adequate containers are minimum bounding boxes, convex polygons, or even grid cells as outlined before. Thus the process of object decomposition may result in a set of containers covering the components and therefore the objects theirselves. The main advantage of a structural decomposition is that refinement will restrict to simple components and thus will profit from more efficient algorithms. If the components due to their simplicity are suitable to be managed by an access method without container approximation, already the filter step evaluates a spatial query exactly and no refinement step is necessary.

The quality of a structural decomposition of a spatial object mainly depends on the following properties: the number and the complexity of the components as well as the quality of container approximation. We have examined various types of structural decompositions differing in the generated component types. In the following we will give a brief overview of decomposition methods (figure 5).

*Convex decomposition*: As geometric algorithms for convex polygons generally are more efficient than those for arbitrary SPHs, we consider as a first approach the decomposition of an SPH into convex polygons. A property of the convex decomposition is, that the number of points for a component is not fixed and therefore is not the component complexity. For gaining in accuracy when using a representation by an rectilinear bounded container, convex decomposition algorithms should force the components consisting of horizontal respectively vertical edges as far as possible.

*Triangle decomposition*: One approach of decomposing arbitrary SPHs into triangles ('triangulation') is given by an algorithm based on the Delaunay triangulation [KHS 90], where a minimum number of triangles is generated. Triangles are simple spatial objects with a fixed length description. In contrast to other triangulation algorithms this algorithm restricts to the vertices of the SPH, i.e. no additional vertices, called 'Steinerpoints', are used for the components. Generally, there is no guarantee for triangles to contain horizontal or vertical edges, which may reflect in the quality of a container approximation.

*Trapezoid decomposition*: An algorithm decomposing SPHs into trapezoids is introduced by [AA83]. The number of components generated by that algorithm is less than or equal to the number of vertices of the original object. The components are trapezoids where two edges are parallel to one axis and therefore are well suited for approximation by rectilinear containers. Such as triangles, trapezoids have a fixed length description.

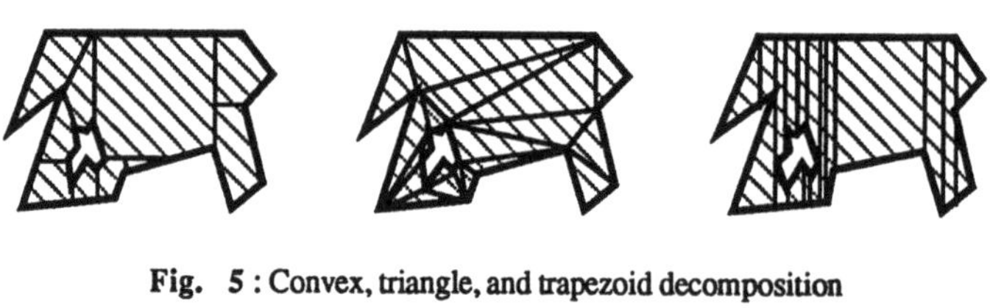

**Fig. 5** : Convex, triangle, and trapezoid decomposition

Decomposing a spatial object into a set of components induces some type of redundancy by definition. The amount of redundancy is one major criterion defining the suitability of a given decomposition method, which holds for structural as well as for grid approximations. Therefore some research on the influence of redundancy to query processing is fundamental. As there has been no systematic study concerning that problem, we will focus on comparing object representations by the amount of redundancy in the next section.

### 3.4 Characterization and Visualization of Container Approximations

As described in the last section query processing can be improved by introducing redundancy into the object representation used by the filter step. Redundancy is dependent to the technique used. As a result of the used technique the accuracy of the filter step as well as the quality of the refinement step can be improved. As a general approach to the role of redundancy we present a matrix (figure 6) which associates different approximation techniques to different redundancy classes. The approximation techniques are classified into grid-oriented and structure-oriented techniques. For each technique we present the resulting object class with respect to the degree of redundancy. The redundancy classes are defined as follows:

- *1-redundancy*: A spatial object is not decomposed at all but it is completely enclosed by a single container.
- *c-redundancy*: A spatial object is decomposed into a constant (predefinable) number of components, like SizeBound techniques do.
- *n-redundancy*: A decomposition technique is affiliated to n-redundancy iff a given spatial object with n vertices is decomposed in O(n) components. A triangulation, for example, decomposes such a spatial object in (n-2) components. In general most decompositions, which are based only on the structural (and topological) characteristics of spatial objects, reside in this class.
- *d-redundancy*: A decomposition technique is affiliated to d-redundancy iff O(n) is not an upper bound for the number of components generated by the decomposition. This is the case if the redundancy is induced by geometric and metric characteristics of the spatial object, e.g. a grid partitioning with a maximum resolution (ErrorBound technique) or a structural decomposition taking additional metric aspects into account.

| Redundancy | Grid-Oriented | Structure-Oriented |
|---|---|---|
| d | Edge quadtree ---> k-gon, k=3..7<br>ErrorBound ---> Set of SPH | Finite element method ---> Triangles<br>Triangulation with Steinerpoints ---> Triangles<br>Structural decomposition ---> Triangles + Rectangles |
| n | SizeBound(n) ---> Set of SPH | Structural decomposition ---> Triangles<br>---> Trapezoids<br>---> Convex Polygons |
| c | SizeBound(c) ---> Set of SPH | RectilinearSizeBound ---> Set of SPH |
| 1 | SizeBound(1) ---> SPH | MBB, MBS, ConvexHull ---> SPH |

Approximation

**Legend:**
- SPH = Simple Polygons with Holes (explained in chapter 2)
- MBB = Minimum Bounding Box
- MBS = Minimum Bounding Sphere

**Fig. 6** : Approximation techniques and redundancy

To give an indication of the approximation techniques we depict some visualizations in figure 7 which are taken from implementations performed by our research group.

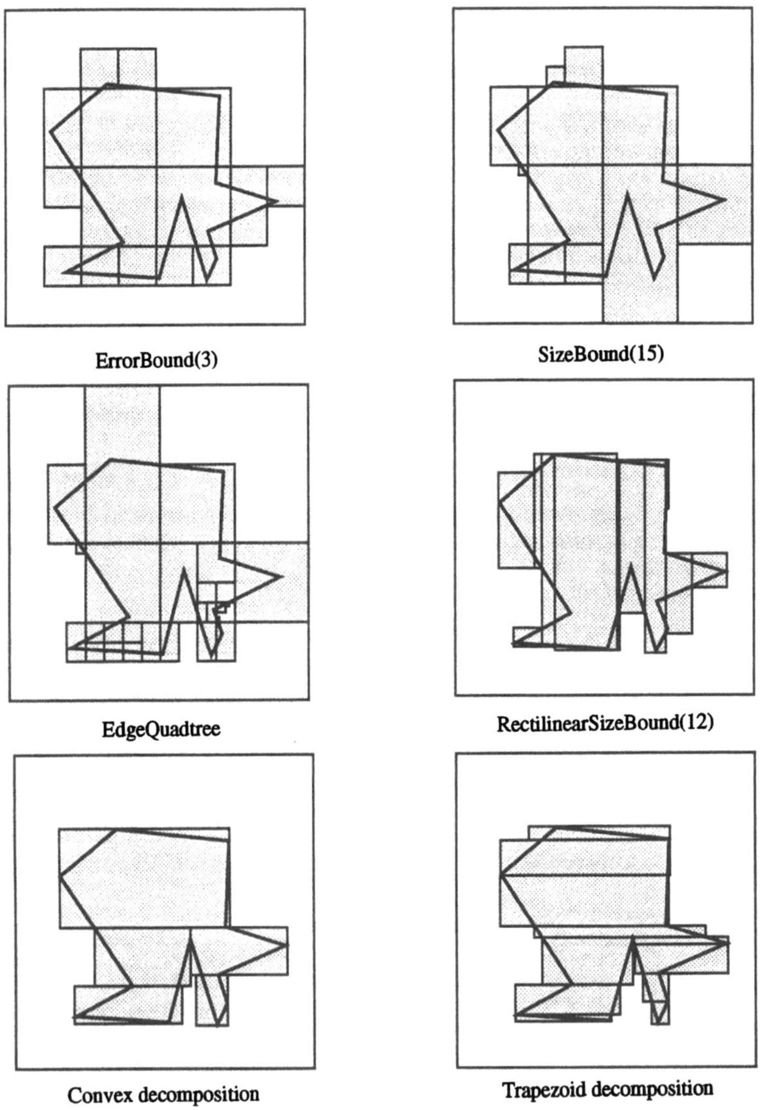


**Fig. 7** : Visualization of approximation techniques

### 3.5 Empirical Comparison of Container Approximations

In the last sections we introduced different techniques for container approximation. The decision which technique is the best choice is difficult, because each one has a specific behaviour with respect to different sets of data. What we need are comparison parameters allowing the selection of adequate techniques for given sets of data. The major parameters which influence the decision are the quality of approximation by containers and the number of components supplied by a technique. In figure 9 we present a comparison of these parameters for an artificial sampling of 'starshaped' polygons with respect to different approximation techniques. Figure 8 depicts three example polygons as well as a table containing more parameters of this sampling. In figure 9 the 'approximation index' indicates the ratio between the area of a container approximation and the area covered by the original object. Thus a value of 100% represents an exact approximation, whereas values more than 100% indicate a measure of the additionally covered area. The average number of components induced by a decomposition technique applied to a single SPH (i.e. redundancy) is shown in the column titled 'number of components'.

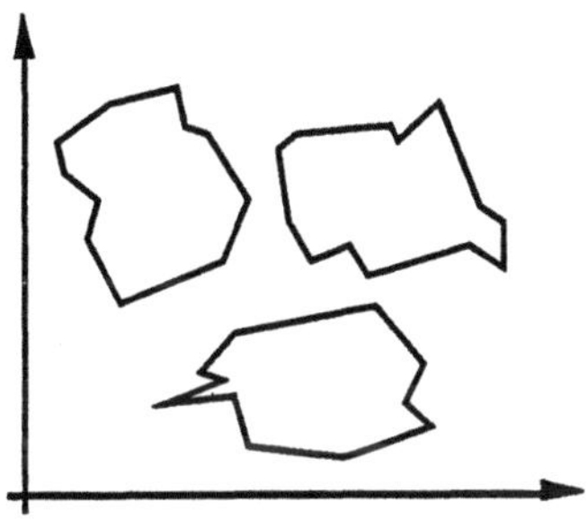

a) example polygons

| | |
|---|---|
| File | : starshaped polygons |
| # polygons | : 100 |
| Ø no. vertices | : 20 |
| Ø no. holes | : 0 |

b) table of parameters

**Fig. 8** : Polygon sampling

| Representation | Approximation Index | Ø Number of components |
|---|---|---|
| **Single container** | | |
| Minimum bounding box | 164 % | 1 |
| Minimum bounding sphere | 163 % | 1 |
| ConvexHull | 117 % | 1 |
| **Multiple container - grid oriented -** | | |
| ErrorBound(6) | 146 % | 19,9 |
| SizeBound(5) | 309 % | 4,3 |
| SizeBound(20) | 156 % | 13,5 |
| Edge quadtree | 168 % | 44,5 |
| **Multiple container - object oriented -** | | |
| RectilinearSizeBound(5) | 158 % | 2,6 |
| RectilinearSizeBound(20) | 115 % | 19,9 |
| Convex decomposition | 123 % | 10,0 |
| Trapezoid decomposition | 115 % | 19,3 |
| Triangle decomposition | 280 % | 18,3 |

**Fig. 9** : Approximation accuracy and number of components for different container approximations

### 3.6 Coupling SAMs with Decomposition Techniques - a Classification -

As indicated before, in order to handle the decomposition of a spatial object efficiently, it is necessary to minimize the total number of components to be considered by the refinement step of a query. Therefore, to achieve an efficient reduction of such components, the decomposition of spatial objects has to be coupled with spatial access methods to take advantage of the clustering provided by the SAM. In the following a combination of a SAM with a decomposition technique is called a representation of a spatial object. The efficiency of such a representation crucially depends on the decomposition technique as well as the adequate choice of the access method providing a special container type.

The most interesting combinations of the decomposition technique and access method are shown in figure 10. Each element of the matrix contains one or two (subclasses of) representations. One axis of the matrix corresponds to the type of container provided by the spatial access method whereas the other axis corresponds to the total amount of redundancy generated by a representation.

The combination of a grid decomposition into an at most constant number of elements (SizeBound(c)) with a ZB+-tree, as proposed by Orenstein [Ore 89], resides in the matrix element determined by container 'grid cell' and redundancy 'c'. As a generalization of the ZB+-tree we used the notion TG-method, to indicate that further spatial access methods may be used that are based on the transformation technique for grid cells (see figure 2).

| Redundancy | grid cell | rectangle | sphere | convex |
|---|---|---|---|---|
| d | Edge quadtree + TG-method<br>ErrorBound + TG-method | Structural Dec. + CR-method<br>Triangulation (SP) + TR/OR/CR-method | Triangulation (SP) + OS-method | Structural Dec. + CC-method |
| n | SizeBound(n) + TG-method | Structural Dec. + TR/OR-method<br>RectSizeBound(n) + TR/OR-method | Structural Dec. + OS-method | Structural Dec. + OC-method |
| c | SizeBound(c) + TG-method | RectSizeBound (c) + TR/OR-method | | |
| 1 | SizeBound(1) + TG-method | MBB + TR/OR-method | MBB + OS-method | ConvexHull + OC-method |

Container

**Fig. 10 :** Classification of access methods with respect to container types

As depicted in figure 2 each SAM is characterized by the simple spatial object type, which it was designed for. Therefore the simple spatial object type determines exactly the column of the matrix, where a SAM can be applied. However, decomposition techniques are not uniquely assigned to a row of the matrix. The structural decompositions for example generate n-redundancy. As the combination with a SAM using overlapping region or transformation technique is assigned to the row for n-redundancy, the coupling with a SAM using clipping is assigned to the row of d-redundancy, because clipping methods generate an additional and not predictable amount of redundancy. Therefore all representations using a SAM with clipping are identified by d-redundancy.

In the matrix element determined by rectangular containers and d-redundancy exists a double entry. In the case of the triangulation with Steiner Points the redundancy is induced by the decomposition technique and in the other case redundancy is induced by the access method, as described above. There are two empty elements in the matrix. This is due to the fact that there exists to the knowledge of the authors no decomposition technique which is restricted to c-redundancy and is tailored to sphere or convex container approximation.

Obviously, there is no representation mixing grid-based and non-grid based concepts. Coupling a grid decomposition with a non-grid access method, to our opinion yields no competitive representation, because some characteristics of grid cells (fixed size, shape and location), exploited in TG-methods, are lost by the combination with a non grid based access method. Contrarily, representations combining a structural decomposition and a TG-method suffer from the bad approximation through grid cells, which is documented by the SizeBound(1) decomposition. The alternative introducing additional redundancy for improving accuracy (e.g. SizeBound(c)) leads to a possible increase in the complexity of the components, which are then not simple any more.

Let us summarize the most interesting features of representations with respect to their use as a filter:

- the amount of redundancy generated by a representation is obvious from figure 10.
- the quality of the approximation of objects by the representations has to be determined experimentally; first results are presented in figure 9.
- the type of spatial object provided by a representation for the refinement step is determined by the decomposition technique used.

This last feature of representations is indicated in the matrix by shading. All representations providing more simple spatial objects than SPHs for the refinement step are shaded. At least n-redundancy has to be provided to yield more simple spatial objects, as indicated by the matrix. On the other hand, decomposition techniques generating at least n-redundancy not always provide simpler spatial objects as indicated by double entries in the two matrix elements determined by rectangular containers and n-redundancy as well as grid cell containers and d-redundancy.

# 4 The Architecture of a Spatial Query Processor

As shortly introduced in chapter 1 and 2, a spatial query processor is essential for efficient spatial query processing. Generally speaking, it performs queries by reducing large sets of spatial objects to a subset of those objects, known as filter step, and explicitely testing that subset of objects against the spatial query, known as refinement step, see chapter 2.4.

## 4.1 The Concept

The spatial query processor is in charge of selecting adequate object representations as described before. Different types of redundant and non-redundant object representations have been introduced and classified in the previous chapter. As there are different types of spatial queries as well as different types of spatial objects, there may not exist one single object representation which is for all types of queries the most efficient with respect to the filter and the refinement step. Therefore a spatial query processor will gain in using different types of representations of the same spatial object. One important ingredient is to use different types of spatial object representations describing different levels of abstractions, i.e. spatial approximations of that object. In the case of different levels of approximation of spatial objects the system is called a multi-level query processor. With this concept we provide a flexible filter mechanism which facilitates the qualitative improvement of the refinement step by the selection of appropriate representations. Thus we meet the two design goals for efficient query processing of complex spatial objects proposed in chapter 1.

It is the task of such a query processor to provide the management of different representations and, in particular, the coordination of access to these representations (see figure 11). In the case of a multi-level query processor different representations are queried and therefore the combination of the query results is required. A set of queries to representations as well as instructions for the combination of query results is called a query plan. In order to support different types of spatial queries, which are usually formulated in a query language in a non-algorithmic way, the interpretation of those queries then yields a query plan with respect to available representations. The construction of an optimal query plan is the task of a query optimizer which is beyond the scope of this paper. In the following, we concentrate our consideration on some combinations of representations, socalled occurrences of the query processor, and on selected queries, such as the point query and the region query, including corresponding query plans.

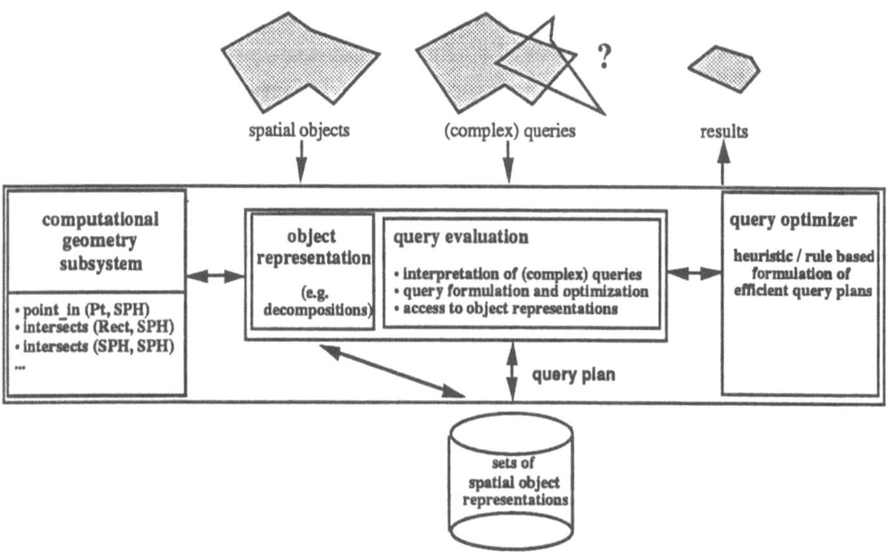


**Fig. 11** : Concept of a spatial query processor

## 4.2 Occurrences

One occurrence of a multilevel query processor is shown in figure 12. On the first level, a minimum bounding box representation combined with a spatial access method as outlined in section 2 is used to provide a set-oriented object selection. It reduces the set of objects to be examined further using the query conditions. The second level consists of a redundant and more exact approximation of one single object, e.g. an approximation using grid cells managed by a ZB+-tree. This representation allows for a simple object test algorithms, e.g. a spatial join, yielding a more exact query evaluation than the first level.

The filter step makes use of these two levels. The third level is the exact object representation, e.g. given by a triangulation of a spatial object, to be used in the refinement step. Query plans may, but do not need to use all three levels of object representation for query processing.

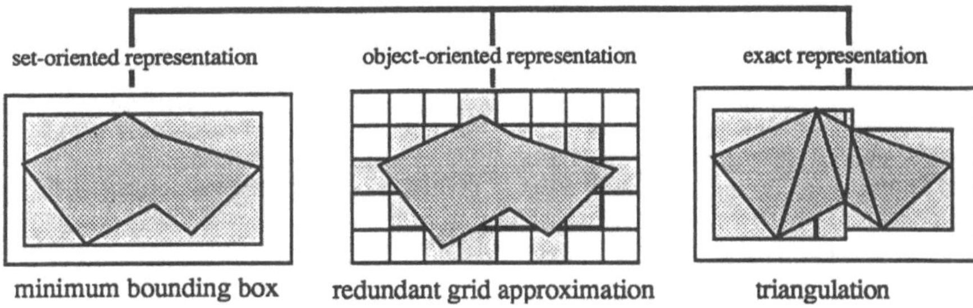


**Fig. 12** : Three-level object representation

Within the previous sections we presented a classification of different representations of spatial objects belonging to the type SPH, namely redundant and non-redundant, grid and vector approximations and decompositions of SPHs. Similar representations can be found for other types of spatial objects such as circles, spline-curves, etc. in the literature [Gue 88]. The representation components are assumed to be equipped with a suitable spatial access method for simple spatial objects. As most representations are based on some type of object approximation for the efficient use of access methods, a further step of object refinement using more or less complex algorithms from the field of computational geometry is necessary. The query processor selects alternative object representations which promise to be most efficient for a spatial query evaluation with respect to both, the filter step and the refinement step.

The most simple type of spatial query is the 'point query' (PQ). A PQ searches for spatial objects containing the given query point. Which is the best representation for performing that type of spatial query? As a point object is a special type of SPH, having an exact spatial location without extension, an object decomposition using redundancy seems to outperform a container representation due to two aspects. First, the object approximation in the case of an object decomposition will behave better with respect to the covered area compared to a single container representation. Therefore less objects will fulfill the filter condition. Second, the redundancy does not have much effect, as the spatial objects are decomposed, i.e. strictly partitioned into disjoint components. Therefore any point object is included by one or at most some of the components belonging to one object. For that reasons, decomposed representations yield less and generally more simple objects to be refined. First results achieved by implementing and testing this type of query under varying data distributions and object representations, confirm these reflections.

However, more detailed considerations are necessary for 'window queries' (WQ) or even 'region queries' (RQ). Queries of the type WQ search for all stored SPHs intersecting a given rectilinear query window, whereas in RQs the query window itself may be shaped as a SPH. Concerning these type of queries, with respect to redundancy in object representation, decomposition methods may lead to an exploding amount of components that must be tested against the query condition. Generally for RQ improvement, one has to decide how to manage the query region itself. Decomposing it as well as the objects may lead to a seriously rising amount of result components. Treating a RQ as WQ by using its minimum bounding box, necessitates additional complexity of computational geometry algorithms facing its more complex shape. Using more than one object approximation may tune the performance of those queries. This kind of tuning considering the type of spatial query as well as the choice of a suitable object representation is supported by the spatial query processor. We will show one special type of WQ evaluation in figure 13 using the complete three-level architecture from figure 12.

As outlined before, the first two steps of processing form the spatial filtering. Query evaluation starts making use of the spatial access method managing the set of spatial objects, each represented by its minimum bounding box. A major property of this access method is a spatial clustering of the MBBs and therefore it provides access to the SPHs with respect to spatial locality. This spatial locality is exploited efficiently extracting the container objects intersecting the query window. As an object is fully covered by its container, the set of extracted containers supplies a complete set of candidates which may meet the query condition. For a further examination of the set of candidates a second type of object approximation is used. Testing the query window and the elements of a grid approximation of each candidate for intersection reduces the set of candidates even more. Finally, an exact representation of those candidates, e.g. an object triangulation or even a direct object representation such as a list of points, is used for the refinement step. The intersection of the exact object representation with the query window yields the answers to the WQ.

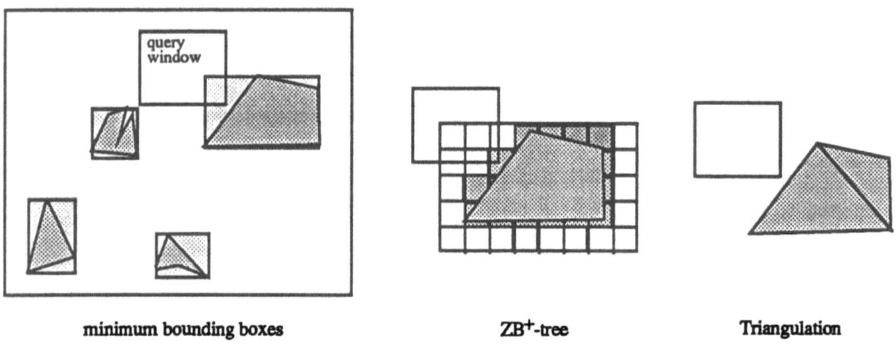


**Fig. 13** : Performing a window query in a three-level query processor

## 4.3 First Experimental Experience and Results

Because an extensive experimental comparison of our query processor is presently on the way but not yet available, we would like to point out the performance gain of our query processor, in particular the performance gain of structural decomposition, by analysing point queries in more detail.

Figure 14 depicts two different methods of processing a point query. *Method1*, the traditional approach, approximates the complex spatial objects by MBBs. The filter step is performed by using efficient SAMs for MBBs. Let us emphasize that we store pointers to the complex spatial object in the SAM whereas the complex spatial objects themselves, denoted by $O_1$, ...,$O_k$, are stored in a sequential file. *Method2* employs structural decomposition into simple components which are approximated by MBBs and stored in an efficient SAM for MBBs containing the components.

For an experimental comparison of query times we have to give a more detailed description of the experimental setup. We ran the performance comparisons on SUN 3/60 workstations under UNIX using Modula-2 implementations. Crucial for an efficient filter step using MBB approximation is an efficient and robust SAM. According to our comparisons [KSSS 89] and [BKSS 90] we have chosen the R*-tree because of its outstanding performance. Due to limited space we mention only the implementation detail that we have selected the page size for data pages and directory pages to be 1024 bytes which is at the lower end of realistic page sizes. Our experiments have shown that the choice of an adequate SAM is particularly important for storing the MBBs of the components following the decomposition. For example, the performance of the grid file degenerates in the test environment.

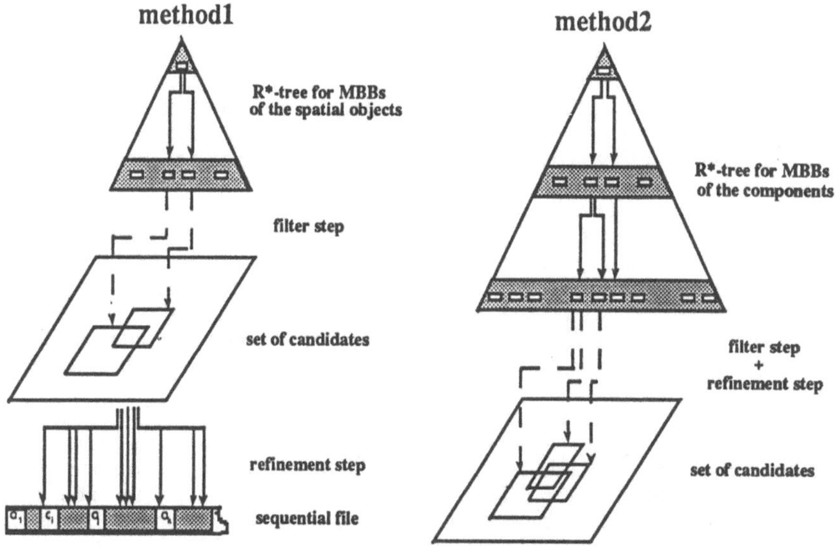


**Fig. 14** : Comparison of methods for point queries

Let us mention that for method2 we selected the trapezoid decomposition because this decomposition is very suitable for MBB approximation (see figure 9), exhibits n-redundancy and offers qualitative improvement of the refinement step.

In query processing we have measured two separate parameters:

- the number of disk accesses (usual parameter)
- CPU-time of the main memory operations
  a) CPU-time of the SAM performing the filter step
  b) CPU-time of the computational geometry algorithms
  (e.g. 'point in complex spatial object'-test vs. 'point in simple component'-test).

The following results for a comparison of method1 and method2 were obtained by using real cartography data. This cartography data represents the elevation lines in a 'rolling-hill-type' area in the Sauerland, Germany. The points are obtained as interpolation points of the elevation lines. In this data file 1883 simple polygons are stored. In the average one polygon consists of 45 vertices. The simple polygons overlap each other in such a way that for an arbitrary point query in the average 7.7 elevation lines are found. We thankfully acknowledge receiving this data from the Landesvermessungsamt NRW, Bonn, Germany.

In our comparison we ran 1000 uniformly distributed point queries against the real cartography data file. Each performance parameter is averaged with respect to one point query in table 1, which demonstrates the superiority of method2 over method1.

**Table 1** : Performance comparison of point queries

| | disk accesses for the filter step | found candidates | CPU-time of a point in object-test | CPU-SAM for the filter step | total disk accesses | total CPU-time |
|---|---|---|---|---|---|---|
| method1 | 2.0 | 10.4 | 14.3 ms | 4.2 ms | 12.4 | 153.0 ms |
| method2 | 4.8 | 9.3 | 0.14 ms | 10.1 ms | 4.8 | 11.4 ms |

In table 1 'found-candidates' denotes the number of candidates which result from the filter step. Due to the approximation with MBBs 'found-candidates' is higher than the degree of overlap. The gain of the structural decomposition by replacing complex computational geometry algorithms through simple ones is clearly demonstrated in the CPU-times for 'point in polygon'-test v.s. 'point in trapezoid'-test: 'point in trapezoid'-test is 100 times faster than 'point in polygon'-test. Let us mention that this ratio will be even worse for polygons with more than 45 vertices.

Now the reader might object that through decompostion the CPU-time for searching in the R*-tree will increase with the considerably higher number of components to be stored (on average 45 for one object). However it only increases from 4.2 ms to 10.1 ms as indicated in the column 'CPU-SAM' due to the logarithmic performance of the R*-tree. In the filter step of method1 2.0 disk accesses are carried out for searching the candidates, followed by 10.4 disk accesses to the actual objects, resulting in a total of 12.4 disk accesses per point query. With an average number of disk accesses of 4.8, method2 requires 2.6 times less disk accesses than method1.

The total CPU-time in ms for method1 is 'found candidates' * CPU-time for one 'point in polygon'-test + 'CPU-SAM' = 10.4 * 14.3 ms + 4.2 ms = 153.0 ms. The total CPU-time in ms for method2 is 'found candidates' * CPU-time for one 'point in trapezoid'-test + 'CPU-SAM' = 9.3 * 0.14 ms + 10.1 ms = 11.4 ms (see table 1). This calculation indicates that the superiority of our decomposition approach is even more clearly demonstrated for the CPU-time where our approach is 13.4 times faster than the traditional approach to answer a point query.

Let us emphasize that for a point query we do not even use all the possibilties and flexibilities of the query processor. This will be necessary in more complex queries. We are aware that no general statement may be made from this one comparison. However it indicates that it is worth studying our approach in more detail. A more extensive comparison considering more query types and more data files is presently on the way.

## 5 Conclusions

In this paper, we have proposed a multi-level, multi-representation query processor where the various levels correspond to various representations with different accuracies of approximating complex spatial objects. The efficiency of the spatial query processor was gained by two design paradigms: First, divide and conquer: decomposition of complex spatial objects into simple spatial objects (components), second: application of efficient and robust spatial access methods for simple spatial objects. The most powerful ingredient in our approach is the decomposition. First experimental results indicate that the query processor outperforms traditional query processing by an order of magnitude. A more extensive comparison which is presently on the way will demonstrate the general performance of the query processor. Additionally, such a performance comparison with many query types, operation types and many data files will be the fundamentals of an automatic physical database design tool for the query processor that chooses a special occurrence of the query processor, e.g. best decomposition, and then monitors the performance making changes as necessary. Furthermore, our query processor does not only exist on paper but also in an implementation. It is based on the implementation of a flexible and extensible index manager for complex spatial objects [KHHSS 91].

# References


[AA 83] Asano, Ta. & Te. Asano, 'Minimum Partition of Polygonal Regions into Trapezoids', in Proc. 24th IEEE Annual Symposium on Foundations of Computer Science, 233-241, 1983.

[BKSS 90] Beckmann, N., H.P. Kriegel, R. Schneider & B. Seeger, 'The R*-tree: An efficient and robust access method for points and rectangles', in Proc. 1990 ACM SIGMOD International Conference on Management of Data, 322-331, Atlantic City, USA, May 1990.

[Cla 89] Claassen, E., 'Spatial data structures in Geographic Information Systems' (in Dutch), Master's thesis, Department of Computer Science, University of Leiden, The Netherlands, 1989, an english short version can be found in [Oos 90].

[Fre 89] Freeston, M., 'A well-behaved structure for the storage of geometric objects', in Proc. "Symposium on the Design and Implementation of Large Spatial Databases", 287-300, Santa Barbara, USA, July 1989.

[GB 90] Günther, O. & J. Bilmes, 'Tree-based access methods for spatial databases: implementation and performance evaluation', in Proc. IEEE Trans. on Knowledge and Data Engineering, 1990.

[Gue 88] Günther, O., 'Efficient structures for geometric data management', Lecture Notes in Computer Science No. 337, Springer Verlag, Berlin, 1988.

[Gue 89] Günther, O., 'The design of the cell tree: an object-oriented index structure for geometric databases', in Proc. IEEE 5th Int. Conf. on Data Engineering, 598-605, Los Angeles, 1989.

[Gut 84] Guttman A., 'R-trees: a dynamic index structure for spatial searching', in Proc. ACM SIGMOD Int. Conf. on Management of Data, 47-57, June 1984.

[HSW 90] Hutflesz, A., H.W. Six & P. Widmayer, 'The R-file: An efficient access structure for proximity queries', in Proc. IEEE 6th Conf. on Data Engineering, 372-379, Los Angeles, USA, 1990.

[KHHSS 91] Kriegel, H.P., P. Heep, S. Heep, M. Schiwietz & R. Schneider, 'A Flexible and Extensible Index Manager for Spatial Database Systems', in Proc. 2nd Int. Conf. on Database and Expert Systems Applications (DEXA '91), Berlin, 1991.

[KHS 90] Kriegel, H.P., P. Heep & R. Schneider, 'Design and Implementation of a triangle decomposition for simple polygons with holes', Technical Report, University of Bremen, Germany, 1990.

[KSSS 89] Kriegel, H.P., M. Schiwietz, R. Schneider & B.Seeger, 'Performance Comparison of Point and Spatial Access Methods', in Proc. "Symposium on the Design and Implementation of Large Spatial Databases", 89-114, Santa Barbara, USA, July 1989.

[NHS 84] Nievergelt J., H. Hinterberger & K.C. Sevcik: 'The grid file: an adaptable, symmetric multikey file structure', ACM Trans. on Database Systems, Vol. 9, 1, 38-71, 1984.

[OM 86] Orenstein, J.A. & F.A. Manola, 'Spatial Data Modeling and Query Processing in PROBE', Technical Report CCA-86-05, Xerox Advanced Information Technology Devision, 1986.

[Ooi 88] Ooi, B.C., 'Efficient query processing in a geographic information system', PhD-thesis, Department of Computer Science, Monash University, Australia, 1988.

[Oos 90] Oosterom, P.J.M., 'Reactive Data Structures for Geographic Information Systems', PhD-thesis, Department of Computer Science at Leiden University, The Netherlands, 1990.

[Oren 89] Orenstein, J.A., 'Redundancy in Spatial Databases', in Proc. 1989 ACM SIGMOD International Conference on Management of Data, 294-305, Portland, USA, June 1989.

[Sam 90] Samet, H., 'The Design and Analysis of Spatial Data Structures', Addison-Wesley Publishing Company Inc., 1990.

[See 90] Seeger, B., 'Design and Implementation of multidimensional access methods' (in German), PhD-thesis, University of Bremen, Germany, 1990.

[SG 90] Smith, T.R. & P. Gao, 'Experimental Performance Evaluations on Spatial Access Methods', in Proc. 4th Int. Symp. on Spatial Data Handling, Zürich, Switzerland, 1990.

[SK 88] Seeger, B. & H.P. Kriegel, 'Techniques for design and implementation of efficient spatial access methods', in Proc. 14th Int. Conf. on Very Large Databases, 360-371, Los Angeles, USA, 1988.

[SK 90] Seeger, B. & H.P. Kriegel, 'The Buddy Tree: An Efficient and Robust Access Method for Spatial Databases', in Proc. 16th Int. Conf. on Very Large Data Bases, Brisbane, Australia, August 1990.

[SRF 87] Sellis, T., N. Roussopoulos & C. Faloutsos, 'The $R^+$-tree: a dynamic index for multi-dimensional objects', in Proc. 13th Int. Conf. on Very Large Databases, 507-518, Brighton, England, 1987.

[SW 88] Six, H.W. & P. Widmayer, 'Spatial searching in geometric databases', in Proc. IEEE 4th Int. Conf. on Data Engineering, 496-503, Los Angeles, USA, 1988.

[SV 89] Scholl, M. & A. Voisard, 'Thematic Map Modelling', in Proc. "Symposium on the Design and Implementation of Large Spatial Databases", 167-190, Santa Barbara, USA, July 1989.

[Tam 83] Tamminen, M., 'The extendible cell method for closest point problems', BIT 22, 27-41, 1982.

# A Multi-layered Approach to Query Processing in Geographic Information Systems

David W. Embley
George Nagy

A multi-layered approach to query processing in geographic information systems provides a bridge over the conceptually large gap between the top level, which is problem-oriented, and the bottom level, which handles the complexities of computational efficiency and robustness. The multiple layers allow conceptually smaller gaps to be bridged more easily and provide for a division of labor in solving the overall mapping problem. Principal issues addressed include languages and notation for describing each layer, faithful mappings to and from each layer, and time/space efficiency. To provide focus for such a broad topic, a unifying query, requiring multiple intersections and unions of polygons and area and perimeter-length computations, is traced through the system. Although many of the required transformations and algorithms have already been implemented at BYU and RPI, the overall system is only at the design stage.

## 1 Introduction

Following the lead provided by the development of other large information-handling systems, such as multi-tasking operating systems and the ISO Open Systems network model, we study a layered approach to bridging the (wide) conceptual gap between the high-level and low-level functions of a Geographic Information System (GIS). Each query is mapped through successive layers of the system, and the results of the data retrieval and computation are returned by inverse mappings. For specificity we focus on the representation of polygons and polygon intersections and unions at the various levels.

The principal issues we consider are:

(1) What are the appropriate formalisms for each level in the conceptual hierarchy of abstractions?

(2) How can high-level solutions be faithfully mapped into low-level solutions, and vice-versa?

(3) How can we produce the solution in reasonable time and space?

We propose an onion-layered approach with the following layers:

*User-Interface Model*: A human-computer interface that can accommodate problem-oriented paradigms based on application-specific user-models such as demography, wild-life management, and transportation.

*Conceptual Model*: An object-oriented model based on a domain analysis of geography that includes all the expected applications.

*Ideal-Representation Model*: Abstractions of the geographic (spatial) objects, relationships, operations, and constraints based on a continuous, infinite-precision view of the world.

*Computer-Representation Model*: Specific data structures and algorithms for manipulating geometric objects subject to finite-precision processor and memory limitations.

*Mass-Storage Model*: Preservation of original and computed data, possibly in compressed form, for archiving and transmission.

We illustrate the approach by tracing a single simplistic query requiring polygon intersections and unions through the various layers. We discuss design alternatives for each layer. In the process, we raise several problems and suggest some solutions.

## 2 Running Example

We assume that the United States Congress has just passed a law mandating that in any county where there are regions of (a) more than 20 square kilometers of alpine highlands, which are areas over 3000 meters in elevation, or (b) more than 35 square kilometers of coniferous forest, all the qualifying regions of alpine highlands and coniferous forest in that county become wilderness area.

Figure 1, for example, shows three counties in Utah along with regions of alpine highlands and coniferous forest. For our example, we suppose that the intersection of the large alpine highlands region with Summit County is greater than 20 $km^2$. We also suppose that the Summit County portion of the large coniferous forest region mostly in Summit County and the Duchesne County portion of the large coniferous forest region mostly in Duchesne County are each greater than 35 $km^2$. We assume that the other regions are each smaller than 10 $km^2$.

Under these assumptions, there are two wilderness areas satisfying the mandate. One is formed by the intersection of Summit County and the large coniferous forest region mostly in Summit County. The other is formed by a union of the large Summit County alpine highlands and the large Duchesne County coniferous forest. Observe that the Summit County forest wilderness area does *not* include the two small parts of the forest in Salt Lake County. Observe also that the wilderness area straddling the Summit County/Duchesne County line does *not* include the two small parts of the coniferous forest in Summit County. Only the qualifying regions in a county become wilderness area.

As a sample query we suppose that a user poses the following question to a GIS:

*My lumber yard serves customers in Summit and Duchesne Counties. To estimate the amount of lumber needed to fence the new wilderness areas in my local, I need to know the length of the perimeter of all the wilderness areas included in these two counties.*

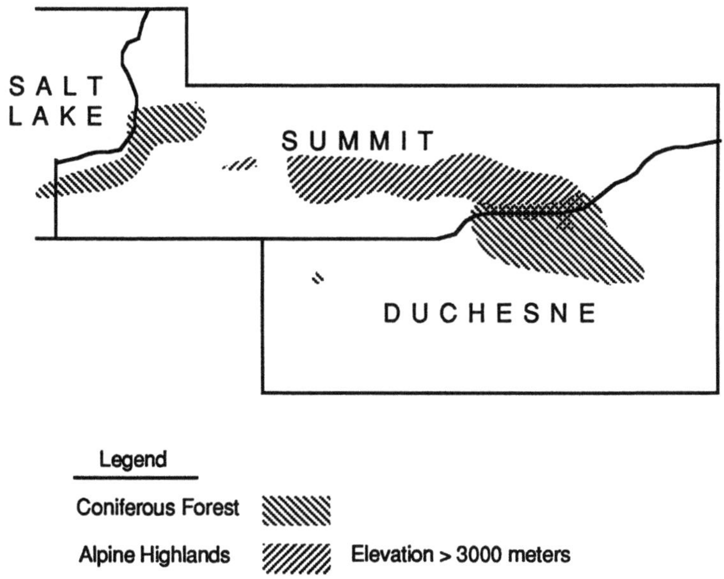


Fig. 1 Coniferous Forests and Alpine Highlands in Summit and Duchesne Counties.

The answer to this query requires us to find the wilderness area boundary. The boundary depends on the construction of the wilderness areas, which involves the identification of appropriate regions and some intersections and unions.

We now look more specifically at the problems we encounter from the perspective of each of the five layers in our model. Our focus is on the mappings between layers and on efficient and accurate computation. At the higher levels of the model, the main difficulty is the creation of a mapping. At the lower levels, the main difficulty is optimization and robustness. For all levels, we need a description of a domain and co-domain for our mappings.

# 3 Layers

## 3.1 User-Interface Model

At the user-interface level there are two main problems:

(1) Creation of an appropriately-expressive, user-friendly, interface language; and

(2) Proper translation of a user dialogue, communicated in the interface language, to a machine-understandable form.

Natural language may be an appropriate choice for a user language, but there are well-documented obstacles to mapping natural language to an underlying computational semantics. Progress is being made, however, and projects such as CYC [39] hold some promise that we may be able, eventually, to simply pose our query by giving the definition of the proposed wilderness area along with the user question as stated above.

When we back away from natural language, we find a wide diversity in the possible choices for a user interface. As we look to the future, this diversity is likely to increase. A reasonable approach for handling this diversity is through a User Interface Management System (UIMS) [33].

In a UIMS the application model is separate from the user interface. Hooks are established so that the two can communicate. The communication is direct and bidirectional. An action at the user-interface level causes a corresponding, complementary action at the application level and vice-versa. A prominent advantage of a UIMS is that if we change or modify our user interface, we do not need to change the application model: we only need to establish a connection to the hooks already present.

The top two layers of our model satisfy the requirements of a UIMS. The top layer is a user interface, and the next layer down is an application model. Since our application model is object-oriented, the hooks that establish the connection to a user interface can be built on the direct correspondence between objects in the real world and objects in the conceptual model. Whenever a user introduces a new real-world object into the system, the system creates a corresponding conceptual object; and whenever a user modifies or uses a real-world object, the system invokes an operation to change or access the corresponding conceptual object.

To illustrate the idea of a UIMS approach to mapping from our user-interface model to our conceptual model, we select a possible user interface and show how we can use it to answer our sample query. Any user-interface model that has similar features can replace the one we discuss here [11].

To show the complementary effects of user actions on our conceptual model, we first need to introduce our conceptual model. Since this is the topic of the next section, we defer until then the discussion of how user actions translate to system actions.

The user interface we present is based on ideas from NFQL (the Natural Forms Query Language) [19]. We assume an extension to NFQL that recognizes spatial types as well as lexical types. In NFQL, fields on a form correspond to object classes in a conceptual model that either exist or are derivable using operators defined for the object classes. The NFQL system can automatically derive the correspondence unless naming ambiguities or conceptual-model path ambiguities arise, in which case a user must help resolve the ambiguities.

For our query we first create the screen display in Figure 1. Starting with a map of the United States, we zoom into the counties shown. Next we specify the legend as shown including the constraint that alpine highlands are regions with an elevation greater than 3000 meters. Specification of the legend generates the regions to complete the display in Figure 1.

Continuing, we alter the legend by adding constraints to restrict the size of the regions and to split them by county. The legend in Figure 2 shows these changes, and the map shows the corresponding changes to the regions in the counties. We next create the table in Figure 2 and associate forest and highlands regions with wilderness-area blank slots as Figure 2 shows. Since the regions are disjoint, except for some of the boundaries, a union operation is

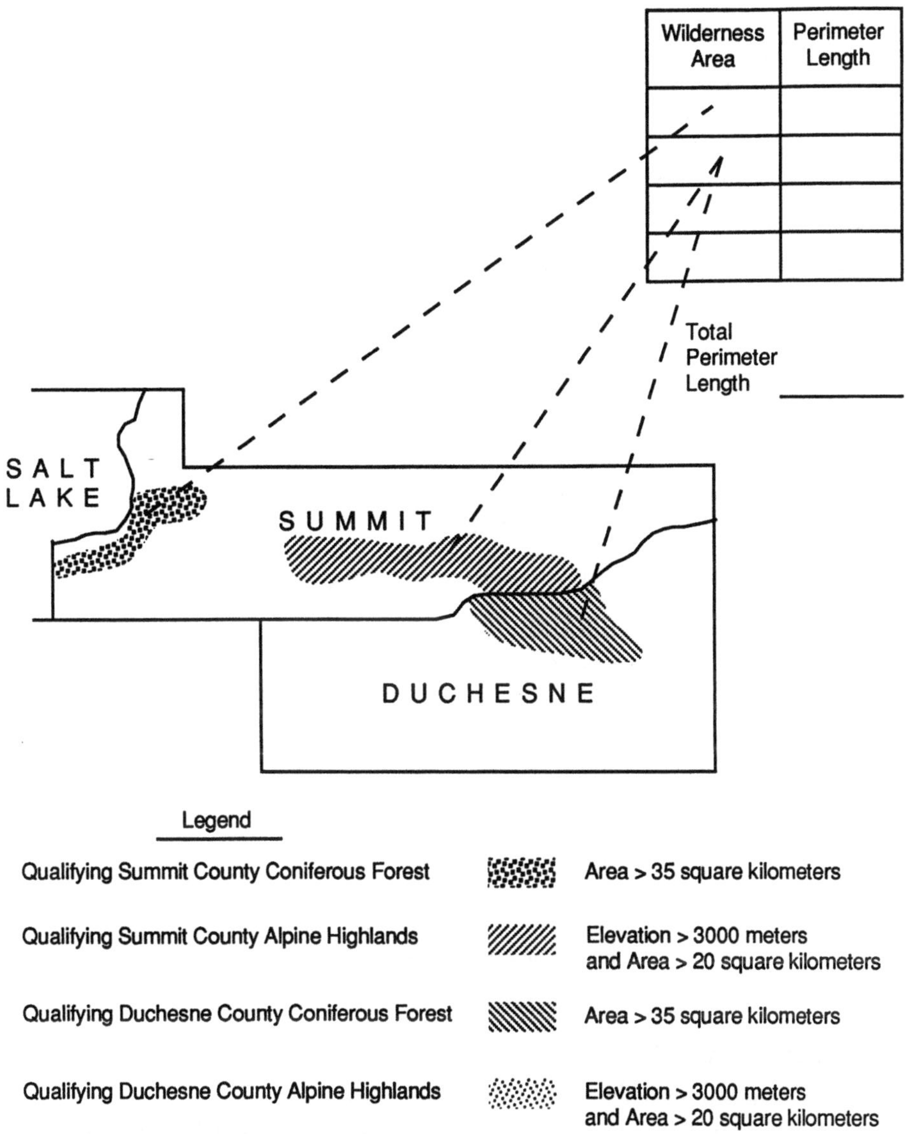


Fig. 2 Wilderness Area Perimeter Length Query.

assumed by the system and confirmed by the user. The table indicates that each wilderness area has a perimeter length. Since perimeter-length calculation is an available and applicable operation, the system chooses it to derive the values for the blank slots in the *Perimeter Length* column. We next create the single-entry blank for total perimeter length in Figure 2. The system chooses the summation operator for computing the total as a sum of the perimeter lengths. (For an explanation of how the system makes these choices, see [19].) When these calculations are done, the value in the *Total Perimeter Length* field is the answer sought by our lumber-yard owner.

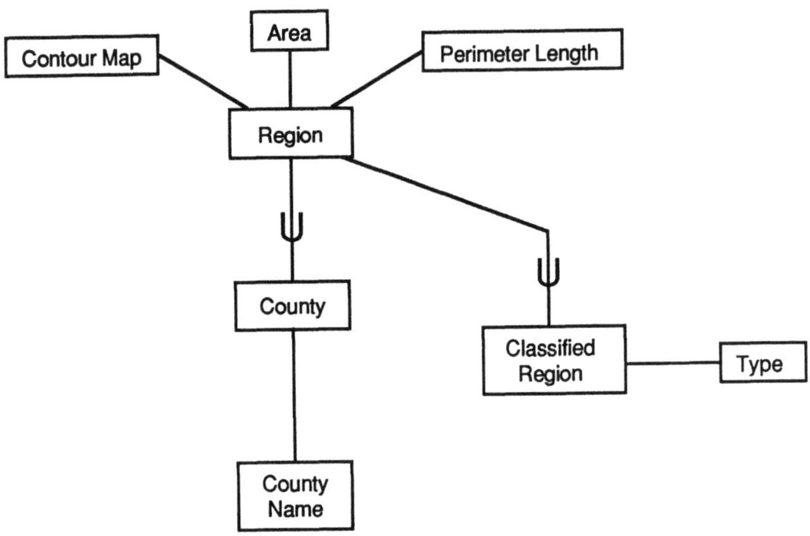


Fig. 3 Partial Conceptual Model.

### 3.2 Conceptual Model

At the conceptual level there are three main problems:

(1) Development of a conceptual model, rich enough to capture all essential concepts within the application domain;

(2) Specification of a direct correspondence to the user interface required by a UIMS; and

(3) Proper translation of actions, generated by the UIMS from a user dialogue, to a machine-understandable form.

Our conceptual model is the Object-oriented Systems Analysis (OSA) model [21,38], which extends analysis paradigms beyond programming concepts to naturally capture real-world semantics [54]. OSA has similarities with many of the popular semantic data models [35,46], but is much more extensive and also has several fundamental differences.

OSA allows a domain analyst to model the real world in an object-relationship model, an object-behavior model, and an object-communication model. The domain analyst provides a complete and detailed representation of some particular domain of interest [48,49]. A particular application would most likely draw on only a small part of a larger domain analysis.

For our sample application, we are interested only in the object-relationship model from OSA. Figure 3 shows the part of a (presumably much-larger) object-relationship model needed for our query. The rectangles represent object classes. Each object in the object class *Region*, for example, corresponds to a bounded region of the earth's surface. The correspondence is through a double projection, first to the horizontal datum and then by a cartographic projection to the plane. This object class may include political regions such as countries, states, and counties, and non-political regions such as forests, deserts, and lakes. For our application in this paper, we may assume that there is only one object in *Region*, namely the region representing the USA.

A region in *Region* has an area in square kilometers, a perimeter length in kilometers, and a contour map with elevations in hundred-meter increments. These related objects are connected by relationships in the relationship sets denoted by the lines in Figure 3 that connect the object classes *Region*, *Area*, *Perimeter Length*, and *Contour Map*.

The connections with subset symbols on them are *is-subset-of* relationship sets. As usual, the open side of the subset symbol is toward the superset and the closed side is toward the subset. An object in a subset object class is a subset of an object in a superset object class. In Figure 3, for example, each county in object class *County* is a subset of a region in object class *Region*, and thus, for our application, a county is a sub-region in the United States. Subsets inherit properties of their supersets. For example, each county has a perimeter length and area.

Having introduced our conceptual model, we are now able to explain how we map user actions to conceptual-model actions. The result will be the transformed conceptual model shown in Figure 4. We explain how we transform Figure 3 into Figure 4 step by step according to user actions. Since there is no space in this paper to show intermediate diagrams, we simply refer to the components in Figure 4 as we add them.

We begin with a screen containing the region representing the USA. One action the system expects is a zoom operation. With this operation we zoom into the region of Utah shown in Figure 1. Figure 4 shows the results of this zoom action as the constraint = *Zoom Scope* placed near the object class *Region*. Because of the subset constraints in Figure 3, this action also restricts the object classes *County* and *Classified Region* to those counties and classified regions within the zoom scope. At this point in the query development, the user screen shows only the three counties in Figure 1; the legend and regions are not yet there.

We next enter the legend. The system recognizes "Coniferous Forest" as a known type and imposes the restriction on the *Type* object class as Figure 4 shows. This restriction automatically propagates to the *Classified Region*

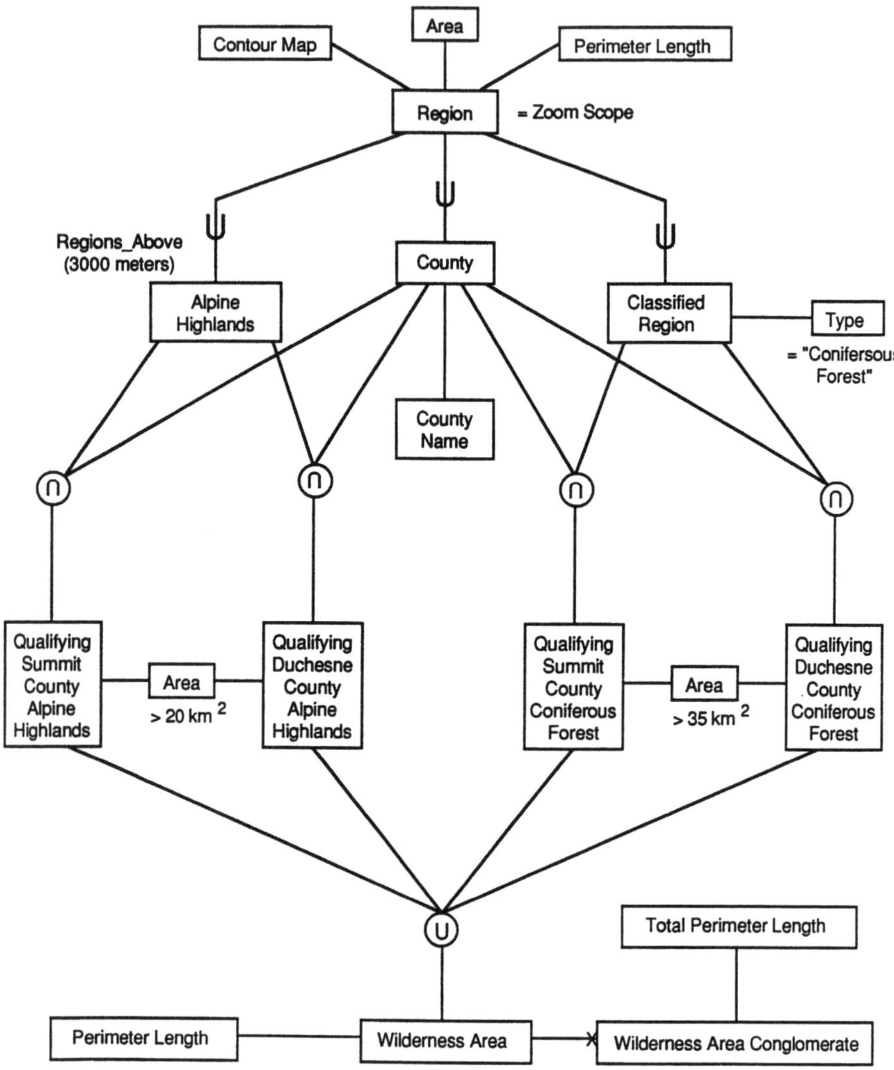


Fig. 4 Derived Conceptual Model for Sample Query.

object class, leaving only coniferous-forest regions in *Classified Region*. In the user interface, this action causes the screen to show the regions of coniferous forest. Although "Alpine Highlands" is not a recognized type, the user provides a way to recognize them by placing the constraint

$$Elevation > 3000\ meters$$

in the legend. These actions generate the new object class *Alpine Highlands* in Figure 4 and display the alpine highlands on the screen. Figure 1 shows the screen display at this point in the query development.

Altering the legend to be as shown in Figure 2 generates several new object classes and constraints. The system creates a new object class for each of the four types of qualifying regions. Objects in the *Qualifying Summit County Coniferous Forest* object class, for example, are the intersection of the Summit County and coniferous-forest regions restricted to have an area greater than 35 $km^2$ as Figure 4 shows. These actions also modify the screen so that it shows only the three regions in Summit and Duchesne Counties in Figure 2.

We next create the table with headers *Wilderness Area* and *Perimeter Length* and the single-entry blank for *Total Perimeter Length* as in Figure 2. Then, as Figure 2 shows, we draw lines to indicate that the wilderness area for the first slot is the Summit County coniferous forest and that the wilderness area for the second slot is composed of the alpine highlands in Summit County and the coniferous forest in Duchesne County. These actions add the four object classes *Wilderness Area, Perimeter Length, Total Perimeter Length,* and *Wilderness Area Conglomerate* and their relationship-set connections as in Figure 4.

Addition of the first three object classes is immediate, based on the creation of the table and single entry blank. The system adds the *Wilderness Area Conglomerate* object class when it determines that the total perimeter length is a sum of the perimeter lengths, which should be associated with a conglomerate of wilderness areas. The connecting relationship set between *Wilderness Area* and *Wilderness Area Conglomerate* is an *is-member-of* relationship set and denotes that the objects in *Wilderness Area Conglomerate* are sets of objects from the *Wilderness Area* object class. The union relationship set in Figure 4 results from determining that the dashed lines in Figure 2 denote unions of wilderness areas, potentially of all four kinds of regions listed in the legend.

Our next problem is the mapping from the conceptual level to the ideal-representation level. Our approach will be to make use of semantic model query languages, which construct query expressions based on the manipulation of high-level semantic data models [6]. Here again, before we can describe the mapping, we must first introduce our next-lower-level model.

### 3.3 Ideal-Representation Model

At the ideal-representation level there are two main problems:

(1) Packaging concepts in the conceptual model into ideal representations; and

(2) Mapping conceptual-model transformations to a computational algorithm over the ideal representations.

To create our ideal-representation model, we transform our OSA object-relationship model in Figure 3 into abstraction packages, called object modules. This involves semantic-model diagram normalization [15] and the generation of highly cohesive/loosely coupled object-modules [14, 16]. The result is a set of object modules.

As an example, Figure 5 shows part of the object module for the *Region* object class in our conceptual model. In the ideal-representation model, an object module has three parts: a *Domain*, a set of *Properties*, and a set of *Operations*. Each object in an object module is a member of the specified domain, satisfies the properties, and allows the operations to be applied. Thus, as Figure 5 shows, a region is a two-dimensional, compact set of points with unary operators for computing the perimeter length and area and binary operators for producing other regions by intersection, union, and relative complement.

Ideal Object Module: Region
Domain: Power-Set(Point)
Properties:
    Two-Dimensional
    Compact Set
Operations:
    Perimeter_Length: Region → Length (in kilometers)
    Area: Region → Area (in square-kilometers)
    Intersect: Region, Region → Region
    Union: Region, Region → Region
    Symmetric_Difference: Region, Region → Region
...

Fig. 5 Partial Ideal-Representation Object Module.

A two-dimensional, compact set has an infinite number of points that completely fill its space in a plane. Although the boundaries of compact sets could be represented by parametrized curves (such as splines), for simplicity we assume a piece-wise linear (i.e., Polygonal) representation. This representation is ideal, of course, because we can only store vertices in a computer that can be represented by finite-precision coordinates. Note that the compact-set property implies *regularization*. Dangling edges and isolated vertices, such as those formed by the intersection of two edge- or vertex-adjacent polygons, are eliminated [17], and perimeter-length and area functions are well-defined operations. Furthermore, since compact sets may have holes and islands and may be disjoint, so may our regions.

Given an appropriate group of interrelated object modules, algorithms using the operations in the object modules may be written. Using the *Region* object module and other object modules obtained from the conceptual model, the system can generate the algorithm we need to answer the lumber-yard owner's query. Figure 6 shows the resulting algorithm, generated from the transformation of the conceptual model in Figure 3 to the conceptual model in Figure 4.

1. Let Region = Intersect(Region, Zoom Scope)
2. Let County = Intersect(County, Zoom Scope)
3. Let Classified Region = Intersect(Classified Region, Zoom Scope)

4. Restrict(Type)
   where Type = "Coniferous Forest"
5. Restrict(Classified Region)
   where Type(Classified Region) = "Coniferous Forest"

6. Derive(Alpine Highlands)
   = Regions_Above(Contour Map(Region), 3000)

7. Derive(Qualifying Summit County Alpine Highlands)
   = Intersect(Alpine Highlands, County where County Name = "Summit")
8. Restrict(Qualifying Summit County Alpine Highlands)
   where Area(Qualifying Summit County Alpine Highlands) > 20
9. Derive(Qualifying Duchesne County Alpine Highlands)
   = Intersect(Alpine Highlands, County where County Name = "Duchesne")
10. Restrict(Qualifying Duchesne County Alpine Highlands)
    where Area(Qualifying Duchesne County Alpine Highlands) > 20
11. Derive(Qualifying Summit County Coniferous Forest)
    = Intersect(Classified Region, County where County Name = "Summit")
12. Restrict(Qualifying Summit County Coniferous Forest)
    where Area(Qualifying Summit county Coniferous Forest) > 35
13. Derive(Qualifying Duchesne County Coniferous Forest)
    = Intersect(Classified Region, County where County Name = "Duchesne")
14. Restrict(Qualifying Duchesne County Coniferous Forest)
    where Area(Qualifying Duchesne County Coniferous Forest) > 35

15. Derive(Wilderness Area)
    = {
        Qualifying Summit County Coniferous Forest$_1$,
        Union(Qualifying Summit County Alpine Highlands$_1$,
            Qualifying Duchesne County Coniferous Forest$_1$)
      }
16. Derive(Perimeter Length)
    = Perimeter_Length(Wilderness Area)
17. Derive(Wilderness Area Conglomerate)
    = Set_Of(Wilderness Area)
18. Derive(Total Perimeter Length)
    = Sum(Perimeter Length(Wilderness Area Conglomerate))

Fig. 6 Generated Algorithm for Lumber-Yard-Owner Query.

The technique used to generate this algorithm is similar to the technique used to generate query algorithms for semantic-model graphical query languages [4,5,6,18]. We associate with each object module and relationship set an expression that, when executed, yields the objects in the object module or relationship set. The initial expressions are usually degenerate, but as transformations take place generating derived object modules and relationship sets, the expressions represent the semantic derivation. When the query transformations are complete, the system combines the expressions to form an algorithm to solve the query.

In our example, zooming generates the first three statements in the algorithm. A *Let* statement alters each object in a set. Statement 1 in the algorithm modifies the USA region so that it contains only the region within the zoom scope. Statements 2 and 3 show the automatic propagation of the zoom scope to the *County* and *Classified Region* object classes.

The next two statements in Figure 6 show the restrictions caused by entering the first line in the legend of Figure 1. A *Restrict* statement restricts the objects in an object class to just those that satisfy the condition in its *where clause*. Thus Statement 4 limits the object class *Type* to the singleton set {"Coniferous Forest"}. Statement 5 reflects the automatic propagation we described earlier and restricts the set of classified regions to just those whose type is "Coniferous Forest".

Statement 6 in Figure 6 shows the generation of the *Alpine Highlands* object class in Figure 4, which happened as a result of entering the second line in the legend of Figure 1. We assume that the domain analyst has defined an operation *Regions_Above* for the *Contour Map* object class, which when invoked on a contour map for a given elevation produces a set of regions. Thus, the object class *Alpine Highlands* contains a set of regions within the zoom scope each of which is entirely above 3000 meters in elevation.

Statements 7 through 14 in Figure 6 derive the four "*Qualifying*" object classes. For example, the system derives the *Qualifying Summit County Alpine Highlands* as an intersection of the regions in the *Alpine Highlands* object class and the regions in the *County* object class restricted to "Summit" county. The results of this intersection are restricted to only those regions whose area is greater than 20 square kilometers.

The last group of statements in Figure 6 derives the wilderness-area and perimeter-length information. All the statements we have been using here, *Restrict*, *Let*, and *Derive*, apply to sets of objects in object modules. So does Statement 15, except that the *Derive* statement here makes use of a set constructor to create individual objects for the *Wilderness Area* object class. As explained earlier in our discussion of the user interface, the user identifies the individual regions that make up the wilderness-area objects. Internally, each object has a surrogate identifier. The system generates these and knows the association between these system-generated surrogate identifiers and the objects they reference. We assume that system-generated surrogate identifiers are formed by a concatenation of an object-class name and a subscript -- one for the first one generated, two for the second, and so forth. Thus, the user actions that identify the constituent parts of each wilderness shown by the dashed lines in Figure 2 are translated into the description of the set elements in Statement 15. *Qualifying Summit County Coniferous* $Forest_1$, for example, is the Summit County coniferous forest in Figure 2. Statement 16 derives a perimeter length for each wilderness area, and Statements 17 and 18 use operators assumed to be defined in appropriate object modules to derive the wilderness-area group of interest and the total perimeter length.

Having outlined how we derive an algorithm for object modules in the ideal-representation model, we must now translate the algorithm to the next lower level so that it can be executed. The translation is straightforward because we use the same algorithm. What is different is that we must now use object modules whose domain definition and operator implementations are computer-representable.

### 3.4 Computer-Representation Model

At the computer-representation level there are three main problems:

(1) Choice of data structure for representing each object;

(2) Choice of algorithm for executing the specified operations; and

(3) Ensuring that geometric degeneracies and finite-precision computation do not cause inconsistent results ("robustness").

In our computer-representation model, we provide multiple implementations for each object module. Especially in computational geometry, there are many different data structures and algorithms for each type of object [7,29,43,45,47]. These data structures and algorithms each have their strengths and weaknesses. We therefore provide not only a set of implementations for each object module, but also a characterization of their strengths and weaknesses. Our challenge is to capture these implementation properties in such a way that a system can automatically make appropriate choices about which implementations to use for efficient execution of a particular query. The strategy we use is similar to the strategy used for cost estimation for query optimization [36].

Without attempting to minimize the importance of the other types of data in a GIS (such as full-text, tabular lexical data, or raster images [17]), we concentrate here on vector data. We restrict our attention to objects in the object module *Region*, and to the binary boolean operation *Intersection*.

The partial computer-representation object module for *Region* in Figure 7 shows how we provide information about properties pertinent to the selection of appropriate implementations for a particular query. Each implementation gives a domain characterization, a list of properties for the domain including information about the domain's computer-storage characteristics, and a list of operations with their execution-time characteristics. We have not yet worked out a full syntax for recording all properties of interest.

Our main concerns are the computational resources required to perform an operation, and robustness. Computational resources typically include (primary storage) space and CPU time. We are interested in performance on typical queries, where sophisticated worst-case-optimal algorithms often yield poor performance. While disk access is also important, we assume that operations reference only primary storage, and defer secondary-storage considerations to the next section. In some instances, initialization (retrieval from mass storage), conversion (transforming objects in different computer-representations to a common representation), and preprocessing (setting up ancillary data structures) to speed up the response to a sequence of queries, may be significant factors.

Robustness attacks the problem of faithfully implementing ideal-representation operations as finite-precision, computer-representation operations [34]. Any query based on data within the specified domain should yield a consistent answer and, above all, should always return control to the layer

```
Object Module: Region
Implementation 1:
  Domain: Convex Polygons
  Properties:
    Space: O(N)
    ...
  Operations:
    Intersection: Region, Region → Region
      Time: O(N).
    Point_in_Polygon_Test: Point, Region → Boolean
  ...
Implementation 2:
  Domain: Simple Polygons
  Properties:
    Space: O(N)
    ...
  Operations:
    Intersection: Region, Region → Region
      Time: O(N + K)log N, where K is the number of intersections.
  ...
Implementation 3:
  Domain: Multiply-connected Polygons
  Properties:
    Space: O(N)
    ...
  Operations:
    Intersection: Region, Region → Region
      Time: O(N + K)log N, where K is the number of intersections.
  ...
Implementation 4:
  Domain: Smooth Partioned Polygons
  Properties:
    Space: O(N)
    ...
  Operations:
    Intersection: Region, Region → Region
      Time: O(N + K)log N, where K is the number of intersections.
  ...
Implementation 5:
  Domain: Zig-zag Partitioned Polygons
  Properties:
    Space: O(N)
    ...
  Operations:
    Intersection: Region, Region → Region
      Time: O(N + K)log N, where K is the number of intersections.
  ...
Implementation 6:
  Domain: Zig-zag Partitioned Polygons
  Properties:
    Space: O(N)
    ...
  Operations:
    Intersection_Area/Perimeter: Region, Region → Area, Perimeter_Length
      Time: O(N + K), where K is the number of intersections.
  ...
```

Fig. 7 Some Partial Implementations for a Computer-Representation Object Module.

above, rather than to the host operating system. While in earlier systems pathological cases (accidental multiple coincidences or collinearities) could be handled by operator intervention (for instance, by perturbing a data point), the adoption of computational-geometry algorithms for massive data manipulation requires accommodating such cases without interactive fixes. Another desirable aspect is consistency: the response should not depend, for instance, on the order in which objects are listed.

With respect to our lumber-yard owner's query, Figure 8 shows the various types of polygons of interest. Figures 8a, 8b, and 8c show isolated labeled polygons on an unlabeled "background." These correspond respectively to Implementations 1, 2, and 3 in Figure 7. They are typically overlays corresponding to distinct categories of land-usage. They may be *simply-connected*, or *multiply-connected* with islands within islands. Simply-connected polygons may be *convex* or *non-convex*.

Figures 8d and 8e show *partitioned polygons*, which consist of mutually exclusive, completely exhaustive coverage, such as county boundaries or thematic maps. These correspond to Implementations 4 and 5 in Figure 7.

Vertices in a polygon must clearly be of at least degree two. In partitioned polygon structures, however, some of the vertices (where more than two polygons meet), may be of higher degree. The ratio of the number of vertices of degree two to the number of vertices of higher degree is called the *zig-zag ratio*. In the western United States most counties have relatively smooth boundaries, while in the East they have a high zig-zag ratio. Strictly political boundaries generally have a lower ratio than natural boundaries. Two of the implementations differentiate between zig-zag polygons and smooth polygons, because the number of higher-degree vertices affects the choice of data structure.

For Implementations 1, 2, and 3 in Figure 7 our choice of internal data structure is a singly or doubly linked list of vertices-around-the-polygon. For isolated polygons, the boundary of the background is omitted. If the polygons are multiply-connected (Implementation 3), the inclusion hierarchy must also be preserved. We assume for our example that Alpine Highlands are of this type.

Implementation 4 is appropriate for county boundaries. For smooth partitioned polygons, we store the complete adjacency topology, which leads to simple, robust algorithms. The relations of interest are edge-vertex, vertex-edge, previous/next-edge-around-a-face, previous/next-edge-around-a-vertex, face-edge, edge-face, face-loop, and loop-face. Data structures based on some subset of these relations are the Winged-edge[3], Symmetric [53], and DCEL [47] structures. Although the storage increases only linearly with the number of polygons, the constant of proportionality is high.

Implementation 5 is appropriate for coniferous forests. For zig-zag polygons, the edge-pointers need be kept only for each *chain* linking higher-degree vertices. When the polygons result from raster images (the coniferous forest boundaries may have been obtained from winter satellite pictures), each edge may be represented by a chain-code [30, 31].

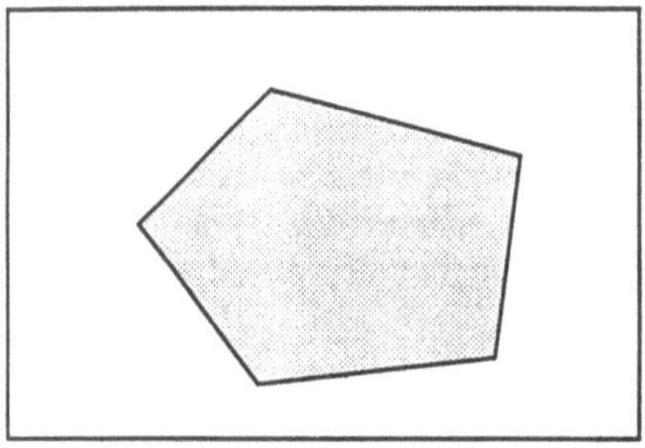

Figure 8a Simple Convex.

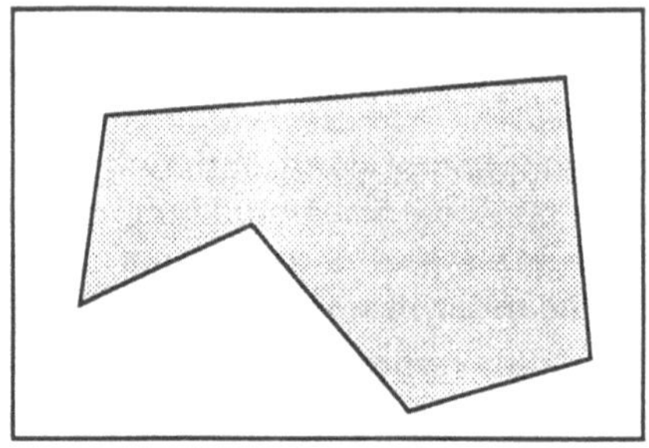

Figure 8b Simple Non-convex.

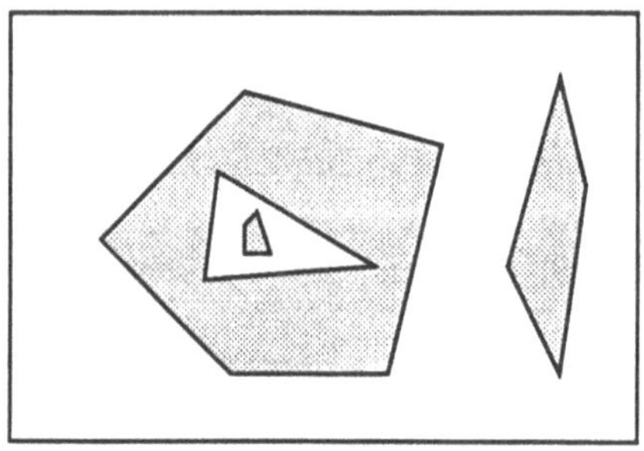

Figure 8c Multiply-connected.

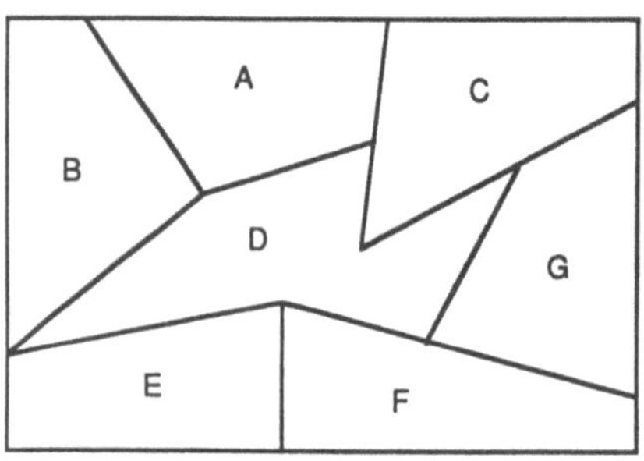


Figure 8d Smooth Partitioned.

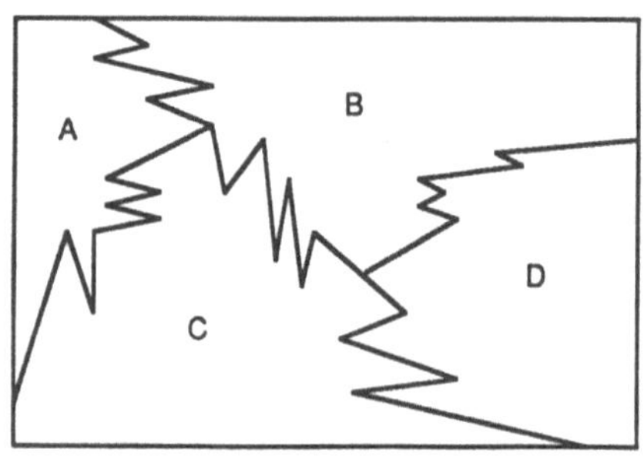


Figure 8e Zig-zag Partitioned.

Fig. 8 Some Polygon Types.

Implementation 6, an alternative to the complex data structure above, is the vertex-neighborhood data structure [24,27,44]. Here only the local topology is stored. With the appropriate algorithm, it can be used for computing mass properties (which obey the additive-set property) such as area and perimeter-length of the intersection polygons, without actually reconstructing the polygons. This implementation is eminently suitable for intersecting county boundaries with alpine highlands to determine if the area of their intersection is large enough to qualify as wilderness area.

The intersection of polygons has been of long-standing interest in computer graphics, automated cartography, geographic data processing, computer-aided design, and computational geometry. We therefore have a good selection for our operator implementations.

For Implementation 1, two convex polygons can be intersected in linear time if the vertices of each are given in order. The algorithms either subdivide the polygons into slabs or angular sectors, or advance around both polygons and separate the internal and external "sickles" formed by each pair of convex boundary fragments [47]. If most but not all the given polygons are convex, it may pay to divide the few that are not into convex subsets.

For Implementation 2, the intersection of two simple polygons can be transformed to the intersection of line segments, for which efficient sweep algorithms are available with $O((N+K)\log N)$ complexity, where $N$ is the total number of vertices, and $K < N^2$ is the number of intersections [47]. (The achievable lower bound is $O(N\log N+K)$ [42].) Once the intersections have been found, one can trace the boundary of the intersection polygon by alternating "left" and "right" turns. If there are no intersections, a point-in-polygon test is necessary to determine whether one polygon contains the other. Alternatively, half-plane intersection techniques with BSP (binary space partitioning) trees can be used [52]. The above algorithms are easily modified for Implementation 3, multiply-connected polygons.

For Implementations 4, 5, and 6, we can use Franklin's fast uniform grid technique [22,28]. The space is subdivided into rectangular cells so that intersection tests are applied only to pairs of line segments that traverse the same cell. The grid resolution depends (but not sensitively) on the average length of the segments. On cartographic examples with millions of segments this technique has performed extremely well. Potentially even greater speed-up can be attained with a hierarchical subdivision such as a quadtree, but the resulting storage and overhead is high, and the algorithms are far more complex [50,51]. When implemented on parallel processors, programs for line and polygon intersections based on the uniform grid and vertex neighborhood show near-optimal speed-up [23,37]. In general, parallel algorithms favor data structures with minimal topology.

Unfortunately, all the above algorithms are vulnerable to numerical imprecision, which creates inconsistency between the geometry (vertex locations) and the topology (incidence relationships). Such errors typically propagate and destroy the distributive, associative, and transitive properties of ideal operations.

Even when the computer representation of the vertices of both polygons is perfectly accurate, the intersection-point of two edges cannot be represented exactly with the same precision. This may cause inconsistencies in the database: for instance, one may find that edges *b* and *c* intersect, but edges *c* and *b* do not. In other examples a point near a boundary shifts from one side of the boundary to the other as the result of a finite-precision scale change or rotation. In general, with finite-precision computation it is not possible to determine on which side of a line a point lies.

Several families of techniques are available to increase robustness. Topological consistency can be achieved through non-redundant judgements (*one* result cannot be inconsistent). At a lower level, numerical inaccuracy is handled by traditional round-off analysis techniques. Some methods that have been applied to polygon intersections are the following.

*Rational and interval arithmetic.* The intersection of two line segments with integer-valued endpoints can be represented exactly as a rational number. However, the length of the rational numbers grows exponentially as the new polygons participate in further intersections. Rational-arithmetic geometry has been implemented in Prolog [25, 26]. Since we require only three consecutive intersections to answer our query, we could use this technique to avoid numerical problems.

*Location constraints.* We may approximate each vertex of the intersection polygon by the nearest grid point *inside* the intersection [40]. This would guarantee that every point in the generated Wilderness Area really belongs to it, but at the cost of excluding some qualifying slivers.

*Incidence constraints.* Sometimes there is information available in addition to the geometry. We may know that two vertices coincide, or two lines are collinear, even if their computer representation is slightly different. For instance, if a point is created by the intersection of two line segments, then regardless of how it is represented we should retain the information that it is collinear with the endpoints of the two segments, so that further results are consistent [20]. We show below that this method allows avoiding an egregious error in our example.

*Simplical geometry.* The topological integrity of simplices - which are triangles in 2-D - can be guaranteed under affine transformations [12, 13]. However, the overhead incurred in finding all the additional intersections is prohibitive for polygons with zig-zag edges.

*Simulation of simplicity.* Recently, it has been shown that many standard techniques for special cases, that were hitherto considered heuristic, can be formalized by considering the parameter space of the functions on which *sign tests* are performed to answer topological queries [10, 32]. Such special (degenerate) cases normally arise when the value of the function is zero to within the computer's accuracy. Simulation of simplicity has been applied to polygon intersections [41] and provides an alternative to complex "topological" data structures for dealing with degenerate cases.

We illustrate the vulnerability of our query using Figure 9, which shows a simplified version of the coniferous forests and alpine highlands in our two counties. The counties, forests, and highlands are shown as simple polygons.

The vertices of all given polygons are restricted to integer grid points. The vertices of the polygons resulting from intersecting the county polygons with the forest and highlands polygons are approximated by the *nearest* integer grid point. (The essential aspects of the example would not change if the coordinates were represented as floating-point numbers; floating-point representation also corresponds to a finite, discrete grid, albeit not a uniform one.)

In our example, the union of the intersection of Duchesne County with coniferous forest $F_1$, and of the intersection of Summit County with alpine highlands $H_1$, contains a gap (it is not singly connected), even though both the forest and the highlands straddle the boundary. The answer to the query will erroneously include twice the length of the fence along the county boundary! The error can be corrected, however, by applying incidence constraints to ensure that the four points $a$, $b$, $c$, and $d$ remain topologically collinear in spite of the vertex displacements caused by the finite-precision representation.

Clearly, automating the choice and combination of implementations will take considerable thought and effort. In our example, the system must determine that Implementation 3 of the alpine highlands and Implementation 4 of the county boundaries need to be transformed to Implementation 6 to find the area of intersection efficiently. The decision should be based on a count of the vertices and edges of the polygons that participate on the operation, the formulas for the time and space requirements of the alternative implementations, and the difficult-to-quantify robustness considerations. We are currently investigating database cost-estimation techniques along with database query optimization strategies for this purpose.

Leaving these difficult questions, we now consider the last layer and the necessary mapping.

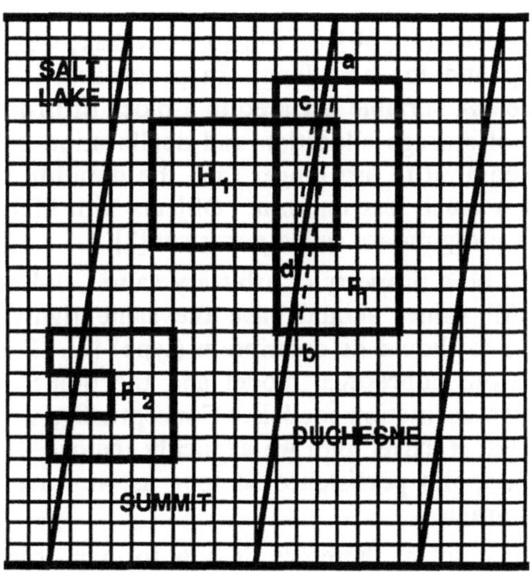


Fig. 9 Wilderness Area with Unwanted Gap.

### 3.5 Mass-Storage Model

At the mass-storage level there are three main problems:

(1) Analysis of the effects of the type of storage devices used;

(2) The interface between mass storage and working memory ("persistence"); and

(3) The application of data-independent compression and decompression schemes.

The nature of the mass storage devices available is seldom taken into consideration in the design of GIS's. However, they are responsible for a large part of both the equipment cost and the query processing time. We assume that the original data resides on read/only access devices such as optical jukeboxes, CD-ROM's, or magnetic tapes (the latter two media are also convenient for the distribution of data). The data from all such devices is usually staged through magnetic disk, therefore we will consider only the mapping from disk to the layer above.

We must also consider whether all the data resides at one location or, as is increasingly more likely, at distributed sites accessible through a network. Network access, in turn, may be the local-area network of a university, research institute, or government agency, or a wide-area network linking entire countries.

We have not yet worked out a suitable formalism for this level. For now, our model is a simple block-access device, where data can be stored only as one-dimensional, fixed-length word arrays. We assume that the operating system takes care of the directory structure, and that sequential files of arbitrary length may be opened for reading and writing. We may write objects to disk by storing them in files on disk, and we may read objects from disk by retrieving files.

Besides reading and writing objects to and from disk, we must also decide which objects to keep and for how long. In procedural languages, objects have a defined lifetime. A variable declared in a block or procedure will "persist" through the use of storage allocation and deallocation primitives. However, when data are required to last longer than the duration of program execution, the treatment is much less uniform. Objects with lifetime longer than a single execution of a program are called *persistent* [1].

Persistence should be a property of arbitrary values and not be limited to certain data types. All values should have the same rights to persistence. While a value persists, so should its description (type) [2]. Persistence is very much in the spirit of our conceptual model and the object modules in our ideal- and computer-representation models. Conceptually, when we place an object in an object class, it exists until we explicitly remove it. Thus, our layered model directly supports persistence so long as we have a way to store each object on mass storage.

A problem we must solve, therefore, is how to map objects to and from the data structures used during execution and the files in mass storage. We could naively make a copy of virtual memory on disk, but this approach is usually unsatisfactory. For some data structures, such as arrays, the mapping is straightforward. Singly-linked lists can also be readily transferred to and from disk by following the pointers and storing the value fields in sequential order in a file. The process also works in reverse, with pointers constructed between consecutive entries. Thus, for example, we can easily store our polygons with linked-list implementations on disk. Unfortunately, these simple ideas do not generalize to more complex data structures; therefore more sophisticated methods are needed [8, 9, 45].

Regardless of the encoding of the data objects, geographic data is LARGE. Therefore it must keep traversing the memory hierarchy - slow optical disk, fast local disk, extended memory, cache - whenever objects are modified. If the storage devices are not co-located, network transmission delays add to the data access time. It is therefore highly desirable to compress the data.

Some compression techniques for images are lossy, and the data is reconstructed only approximately. However, only exact compression is used for vector and lexical data. The most widely-used techniques, such as the CCITT fax algorithms, are based on run-length encoding. Many operating systems, including UNIX, routinely use compression algorithms for file storage. With the advent of special-purpose chips for compression and decompression, the trade-off is tilting more heavily towards storing and transmitting data in compressed form, even for interactive applications.

An outstanding problem with a potentially high payoff is the design of geometric algorithms that work on compressed data. Since the compression algorithms perform a predetermined mapping, in principle this is clearly feasible. The challenge is to design "compilers" that would transform an algorithm designed for uncompressed data into the algorithm for the compressed data.

## 4 Concluding Remarks

Users of geographic information systems rightly expect increased convenience and functionality based on rapid advances in processing and storage technologies. Their expectations can no longer be met with monolithic "programs" designed from a single perspective. We have attempted to demonstrate that a layered approach allows the application of expertise and tools from many dynamic areas such as user-interface management systems, semantic datamodeling, domain analysis, database systems, programming languages, computer graphics, and computational geometry.

Our own research benefits from this structured approach. The top layers are implemented at Brigham Young University, while the geometric algorithms are developed primarily at Rensselaer. Of course, the proof of the pudding must await assembling the various components into a single cohesive system.

For the upper layers, a prototype system for the Natural Forms Query Language including its graphic interface has been developed. Currently, NFQL works with a standard Entity-Relationship model. Thus, it does not have access to the richer semantic features available in our Object-oriented Systems Analysis Model. We are currently implementing OSA. A rudimentary prototype of a graphical query interface for manipulating semantic models to formulate database queries has been implemented.

For the computer-representation layer, geometric algorithms have been programmed for Triangulation; 2-D Convex Hull; Line-Segment, Polygon, Polyhedron, and 3-D Isothetic Cell-Complex Union/Intersection; Point-in-Polygon/Polyhedron Test; Polygon-to-Raster Conversion; Area and Perimeter Computation; Visible-Surface Determination; Watchtower Location; Uniform Grid; Vertex Neighborhoods; Shortest Visible Path; and Terrain-from-Horizons Reconstruction. Some of these algorithms have also been implemented on coarse-grained parallel machines. Robustness techniques, including Rational Arithmetic, Location Constraints, and Simulation-of-Simplicity, have been experimentally investigated.

We believe that among the many challenging problems posed by the inherent complexity of geographical information, one of the most exciting is the development of formal techniques for the synthesis of different strands of research in computer science and software engineering.

## Acknowledgements

We have benefited from many suggestions from our colleagues, W.R. Franklin (RPI), M. Ancona (University of Genova), S. Woodfield (BYU), B. Kurtz (Hewlett-Packard), and B. Czejdo (Loyola University, New Orleans) and the work of students S. Clyde, M. Kankanhalli, A. Maulik, M. Mukherjee, and C. Narayanaswami. We are also grateful for the support provided by National Science Foundation grants IRI-8704718, INT-8714578, and MCS-8219941, by the College of Physical and Mathematical Sciences at BYU, and by the New York State Science and Technology Council through the RPI Center for Advanced Technology.

## References

1. M. Ancona, University of Genova, Genova, Italy, 1990, (personal communication).
2. M.P. Atkinson and O.P. Buneman, "Types and persistence in database programming languages," *ACM Computing Surveys*, vol. 19, no. 2, pp. 105-190, June 1987.
3. B.G. Baumgart, "Winged-edge polyhedron representation," Technical Report STAN-CS-320, Computer Science Department, Stanford University, Stanford, California, 1974.
4. D.M. Campbell, D.W. Embley, and B. Czejdo, "Graphical query formulation for an entity-relationship model," *Data & Knowledge Engineering*, vol. 2, no. 2, pp. 89-121, June 1987.
5. B. Czejdo and D.W. Embley, "An approach to computation specification for an entity-relationship query language," *Proceedings of the Sixth International Conference on Entity-Relationship Approach*, pp. 307-321, New York, New York, November 1987.
6. B. Czejdo, R. Elmasri, M. Rusinkiewicz, and D.W. Embley, "A graphical data manipulation language for an extended entity-relationship model," *Computer*, vol. 23, no. 3, pp. 26-36, March 1990.
7. W.A. Davis and C.P. Gangadharaiah, "Data structures for spatial data processing," Technical Report 85-7, Computer Science Department, The University of Alberta, Edmonton, Canada, March 1985.
8. W.A. Davis and C.H. Hwang, "File organization schemes for geometric data," Technical Report 85-14, Computer Science Department, The University of Alberta, Edmonton, Canada, August 1985.

9. W.A. Davis and C.H. Hwang, "Organizing and indexing spatial data," *Proceedings of the Second International Symposium on Spatial Data Handling*, Seattle, Washington, 5-14 July 1986.

10. H. Edelsbrunner and E.P. Mucke, "Simulation of simplicity: a technique to cope with degenerate cases in geometric algorithms," *Proceedings of the ACM Symposium on Computational Geometry*, pp. 118-133, Champaign-Urbana, Illinois, June 1988.

11. M.J. Egenhofer and A.U. Frank, "Designing object-oriented query languages for GIS, human interface aspects," *Proceedings of the Third International Symposium on Spatial Data Handling*, pp. 79-96, Sydney, Australia.

12. M.J. Egenhofer, A.U. Frank, and J.P. Jackson, "A topological model for spatial databases," in *Preliminary Proceedings, NCGIA Symposium on Design and Implementation of Large Spatial Databases*, Santa Barbara, July 1989. (to be published in Lecture Notes in Computer Science, Springer-Verlag)

13. M.J. Egenhofer and J.R. Herring, "A mathematical framework for the definition of topological relationships," *Fourth International Symposium on Spatial Data Handling,*, pp. 803-813, Zurich, Switzerland, 23-27 July 1990.

14. D.W. Embley and S.N. Woodfield, "Assessing the quality of abstract data types written in Ada," *Proceedings of the Tenth International Conference on Software Engineering*, pp. 144-153, Raffles City, Singapore, April 1988.

15. D.W. Embley and T.W. Ling, "Synergistic database design with an extended entity-relationship model," *Proceedings of the Eighth International Conference on Entity-Relationship Approach*, pp. 118-135, Toronto, Canada, 18-20 October 1989.

16. D.W. Embley and S.N. Woodfield, "Assessing the quality of object modules," *Proceedings of the Object-Oriented Systems Symposium*, pp. D68-D117, Digital Consulting, Inc., Boston, Massachusetts, Los Angeles, California, and Chicago, Illinois, 1989.

17. D.W. Embley and G. Nagy, "Intersection operations in a multi-layered spatial data model," Technical Report BYU-CS-89-3, Department of Computer Science, Brigham Young University, Provo, Utah, April 1989.

18. D.W. Embley and G. Nagy, "On the integration of lexical and spatial data in a unified high-level model," *Proceedings of the International Symposium on Database Systems for Advanced Applications*, pp. 329-336, Seoul, Korea, April 1989.

19. D.W. Embley, "NFQL: the natural forms query language," *ACM Transactions on Database Systems*, vol. 14, no. 2, pp. 168-211, June 1989.

20. D.W. Embley, M. Mukherjee, and G. Nagy, "Making do with finite numerical precision in spatial data structures," *Fourth International Symposium on Spatial Data Handling,*, pp. 55-65, Zurich, Switzerland, 23-27 July 1990.

21. D.W. Embley, B.D. Kurtz, and S.N. Woodfield, *Object-Oriented Systems Analysis: A Model-Driven Approach,* Prentice Hall, Englewood Cliffs, New Jersey, 1992. (in press -- available after November 1991)

22. W.R. Franklin, "Efficient polyhedron intersection and union," *Proceedings of Graphics Interface '82*, pp. 73-80, May 1990.

23. W.R. Franklin and M. Kankanhalli, "Parallel object-space hidden surface removal," *Computer Graphics*, vol. 24, no. 4, pp. 87-94, August 1990.

24. W.R. Franklin, "Polygon properties calculated from vertex neighborhoods," *Proceedings of the Third Annual Symposium on Computational Geometry*, pp. 110-118, Waterloo, Canada, June 1987.

25. W.R. Franklin and P.Y.F. Wu, "A polygon overlay system in Prolog," *Autocarto 8: Proceedings of the Eighth International Symposium on Computer-Assisted Cartography*, pp. 97-106, Baltimore, Maryland, April 1987.

26. W.R. Franklin and P.Y.F. Wu, "A logic programming approach to cartographic map overlay," *Proceedings of the NATO '87 International Advanced Study Institute*, pp. 1-12, Ciocco, Italy, July 1987.

27. W.R. Franklin, "Localized polygon data structres," RPI Technical Report, April 1988.

28. W.R. Franklin, N. Chandrasekhar, M. Kankanhalli, M. Seshan, and V. Akman, "Efficiency of uniform grids for intersection detection on serial and parallel machines," in *New Trends in Computer Graphics*, Springer Verlag, Berlin, 1988.

29. W.R. Franklin, N. Chadrasekhar, M. Kankahalli, V. Akman, and P.F.Y. Wu, "Efficient geometric operations for CAD," in *Geometric Modelling for Product Engineering*, ed. K. Preiss, pp. 485-498, Elsevier Science Publishers, North Holland, 1990.

30. H. Freeman, "On the encoding of arbitrary geometric configurations," *IRE Transactions*, vol. EC-10, pp. 260-268, 1960.

31. H. Freeman and Computer processing of line drawing images, *ACM Computing Surveys*, vol. 6, no. 1, March 1974.

32. L. Guibas, D. Salesin, and J. Stolfi, "Epsilon-geometry: building robust algorithms for imprecise computations," *Proceedings of the ACM Symposium on Computational Geometry*, pp. 208-217, Saarbrücken, FRG, June 1989.

33. H.R. Hartson and D. Hix, "Human-computer interface development: concepts and systems," *ACM Computing Surveys*, vol. 21, no. 1, pp. 5-92, March 1989.

34. C.M. Hoffman,, "The problems of accuracy and robustness in geometric computation," *Computer*, pp. 31-41, March 1989.

35. R. Hull and R. King, "Semantic database modeling: survey, applications, and research issues," *ACM Computing Surveys*, vol. 19, no. 3, pp. 201-260, September 1987.

36. M. Jarke and J. Koch, "Query optimization in database systems," *ACM Computing Surveys*, vol. 16, no. 2, pp. 111-152, June 1984.

37. M. Kankanhalli, "Techniques for parallel geometric computations," Ph.D. Dissertation, RPI, 1990.

38. B.D. Kurtz, S.N. Woodfield, and D.W. Embley, "Object-oriented systems analysis and specification: a model-driven approach," *Proceedings of Compcon Spring 90*, pp. 328-332, San Francisco, California, 27 February - 1 March 1990.

39. D.B. Lenat, R.V. Guha, K. Pittman, D. Pratt, and M. Shepherd, "CYC: toward programming with common sense," *Communications of the ACM*, vol. 33, no. 8, pp. 30-49, August 1990.

40. S. Mehta, M. Mukherjee, and G. Nagy, "Constrained integer approximation to 2-D line intersections," *Proceedings of the Second Canadian Conference on Computational Geometry*, pp. 302-305, Ottawa, Canada, August 1990.

41. M. Mukherjee, "An efficient and simple polygon intersection algorithm," RPI-ECSE Technical Report 90-1122, Troy, New York, November 1990.

42. K. Mulmuley, "A fast planar partition algorithm," *Proceedings of the 29th Annual Symposium on Computer Science*, pp. 580-589, White Plains, New York, October 1988.

43. G. Nagy and S. Wagle, "Geographic data processing," *Computing Surveys*, vol. 11, no. 2, 1979.

44. C. Narayanaswami, "Parallel processing for geometric applications," Ph.D. Dissertation, RPI, 1990.

45. P. van Oosterom, "Reactive data structures for geographic information systems," Ph.D. Dissertation, University of Leiden, 1990.

46. J. Peckham and F. Maryanski, "Semantic data models," *ACM Computing Surveys*, vol. 20, no. 3, pp. 153-189, September 1988.

47. F.P. Preparata and M.I. Shamos, *Computational Geometry*, Springer Verlag, New York, New York, 1985.

48. R. Prieto-Diaz, "A software classification scheme," Ph.D. Dissertation, Department of Information and Computer Science, University of California - Irvine, Irvine, California, 1985.

49. R. Prieto-Diaz, "Domain analysis for reusability," *COMPSAC 87*, Tokyo, Japan, October 1987.

50. H. Samet, *The design and analysis of spatial data structures*, Addison Wesley, Reading, Massachusetts, 1990.

51. H. Samet, *Applications of spatial data structures*, Addison Wesley, Reading, Massachusetts, 1990.

52. W.C. Thibault and B.F. Naylor, "Set operations on polyhedra using binary space partition trees," *Computer Graphics*, vol. 21, no. 4, pp. 153-162, July 1987.

53. T.C. Woo and J.D. Wolter, "A constant expected time, linear storage data structure for representing three-dimensional objects," *IEEE SMC*, vol. SMC-14, no. 3, pp. 510-515, May/June 1984.

54. S.N. Woodfield, D.W. Embley, and B.D. Kurtz, "Extending analysis paradigms," *Proceedings of the Ninth International Phoenix Conference on Computers and Communications*, pp. 441-446, Scottsdale, Arizona, 21-23 March 1990.

# ESPRIT Basic Research Series

J. W. Lloyd (Ed.): **Computational Logic.** Symposium Proceedings, Brussels, November 1990. XI, 211 pages. 1990

E. Klein, F. Veltman (Eds.): **Natural Language and Speech.** Symposium Proceedings, Brussels, November 1991. VIII, 192 pages. 1991

G. Gambosi, M. Scholl, H.-W. Six (Eds.): **Geographic Database Management Systems.** Workshop Proceedings, Capri, May 1991. VIII, 320 pages. 1992

R. Kassing (Ed.): **Scanning Microscopy.** Symposium Proceedings, Wetzlar, October 1990. X, 207 pages. 1992

MIX
Papier aus verantwortungsvollen Quellen
Paper from responsible sources
FSC® C105338

If you have any concerns about our products,
you can contact us on
**ProductSafety@springernature.com**

In case Publisher is established outside the EU,
the EU authorized representative is:
**Springer Nature Customer Service Center GmbH**
**Europaplatz 3, 69115 Heidelberg, Germany**

Printed by Libri Plureos GmbH
in Hamburg, Germany